Cell Engineering

Cell Engineering

Volume 1

The titles published in this series are listed at the end of this volume

CELL ENGINEERING

Edited by

Mohamed Al-Rubeai

School of Chemical Engineering,
The University of Birmingham,
Edgbaston, Birmingham, U.K.

SPRINGER-SCIENCE+BUSINESS MEDIA, B.V.

Library of Congress Cataloging-in-Publication Data

ISBN 978-90-481-5254-4 ISBN 978-0-585-37971-5 (eBook)
DOI 10.1007/978-0-585-37971-5

Printed on acid-free paper

Preface

Cell engineering is developing rapidly as an exciting and enterprising field in biotechnology. Its emergence reflects the need to produce complex proteins for use as therapeutics to a very high degree of fidelity and has been facilitated by rapid progress in molecular biology, molecular genetics and bioengineering. As a consequence of product discovery and rapid development, the worldwide sales of biopharmaceuticals has significantly increased, over 40 therapeutic biologics have been approved with more than 400 human health care products in development, a significant number of these products are produced in animal cells. The estimated sales of biopharmaceuticals in the US alone will approach $16 billion by 2004. As the "gene therapy" and "tissue and organ engineering" revolution begins to spread and requires input from molecular and process biotechnology, the absence of publications that provide detailed coverage of the methodology for improving cellular properties has become increasingly apparent. Integrating the advances in molecular biology into bioprocesses presents a continuous challenge to scientists and engineers. This series is conceived to help us meet this challenge. It intends to examine and assess the feasibility of the new approaches for the modification of cellular function such as secretion, protein processing, glycosylation, proliferation and apoptosis as well as the systematic study of the metabolic genotype-phenotype relationship. It is hoped that this series will benefit both scientists working in this area and the biotechnology industries in their search for new and more effective products and processes. One target group is those biotechnologists interested in the introduction of specific modifications to biological functions, but many topics will also be of interest to other bioscientists and bioengineers.

The rapid expansion in our knowledge has led to an ever expanding literature. This presents us with a challenge to understand the fundamentals of basic research, their relation to various products and to consider the future potential for improving production and reducing cost. This volume which is based on presentations at the "European Workshop on Animal Cell Engineering" held in Costa Brava, Spain contains a collection of chapters relating to gene expression, protein synthesis and modification, cell proliferation, immortalisation and apoptosis by leading authorities in several different areas of basic research and the biopharmaceutical industry.

I applaud the fine efforts of the editors and the European Society for Animal Cell Technology (ESACT) that have produced this book, Mrs Lynn Draper for editorial assistance and Kluwer Academic Publishers for publishing it. I would also like to thank each contributor for their diligence and perseverance in preparing chapters which I am sure will be enjoyed and valued by readers.

Mohamed Al-Rubeai
(Managing Editor)

Cell Engineering Vol 1

With contributions by

P.P. Mueller
A.V.Carvalhal
J.L. Moreira
C. Geserick
K. Schroeder
M.J.T. Carrondo
H. Hauser
M. Himmelspach
U. Schlokat
B. Plaimauer
F.G. Falkner
F. Dorner
P. Pierandrei-Amaldi
B. Cardinali
A-C. Prats
H. Prats
B. Osborne
L. Paillard
G. Huez
V. Kruys
J-J. Toulme
S.L. McKenna
R.J. Carmody
T.G. Cotter
G.S. Jennings
M. Strauss
J.P. Magyar
H.M. Eppenberger
M. Fussengger
J.E. Bailey
A.J.Mastrangelo
M. Szyrach
O. Heidenreich

Contents

UNDERSTANDING THE TRANSLATION REGULATORY MECHANISMS TO IMPROVE THE EFFICIENCY AND THE SPECIFICITY OF PROTEIN PRODUCTION BY THE CELL FACTORY.

P. PIERANDREI-AMALDI & B. CARDINALI(1), Anne-Catherine & Hervé PRATS(2), Bev. OSBORNE & Luc PAILLARD(3), Georges HUEZ & Véronique KRUYS(4*), Jean-Jacques TOULME(5)

Istituto di Biologia Cellulare, CNR, Viale Marx 43 , 00137 Rome, Italy(1), INSERM U397, Endocrinologie et Communication Cellulaire, Institut Louis Bugnard, CHU Rangueil, bât L3, 31054 Toulouse (France)(2), CNRS UPR41/Université de Rennes 1,Faculté de Médecine,2, av. Pr. Léon Bernard, CS 34317, 35043 Rennes (France)(3), Université Libre de Bruxelles, Département de Biologie Moléculaire, Laboratoire de Chimie Biologique, rue des Chevaux, 67, 1640 Rhode-St-Genèse (Belgique)(4), Université Victor Segalen, INSERM U 386, 146, rue Léo Saignat, , 33076 Bordeaux cédex (France).(5)*

(4) Corresponding authors :*
tel.: 32 2 650 98 17/16 - fax : 32 2 650 98 39
e-mail : ghuez@dbm.ulb.ac.be and vkruys@dbm.ulb.ac.be

1. Introduction

Although transcription already constitutes a very diverse and flexible means for the eucaryotic cell to control gene expression, it is but the very first step in the long chain of events which ultimately leads to the product of gene expression: the protein.
Indeed, even if transcription is in itself a very complex process remarkably regulated by numerous factors whose presence or activation is governed by the differentiation status of the cell, it appears that the mechanisms involved in the posttranscriptional events which follow are equally or even more complex.

Posttranscriptional regulations provide to the cell a way to finely tune the genetic expression. In the eurkaryotic cell, the step at which such regulations can occur are numerous; splicing, transport of the mRNA to the cytoplasm, stability of the mRNA in the cytoplasm and last but not least translation of the mRNA into protein.
It is clear now that translation is a much more sophisticated process than it was anticipated before. One has discovered in the recent years that sequences which were somewhat curiously looked at in the past, the untranslated regions, are in fact at the heart of very elaborate mechanisms of control.

One knows now that a given messenger RNA will not be translated in the same way in different cellular environments precisely due to the structural features it carries in the untranslated or even translated regions of the molecule and which are recognised by the

factors associated with the translation process. It is now clear also that one messenger mRNA molecule can often be translated into several proteins whose functions can be extremely diverse. Some of these proteins may have a caryophilic character while the other will remain in the cytoplasm. Some of the mRNAs are so peculiar that the untranslated parts of the molecule can account for up to 90% of the RNA.
This means that translation is certainly one of the most elaborate mechanisms in all the chain of transfer of information from the DNA to the protein.

Obviously, the translation step as well as the mechanisms which govern the mRNA stability have to be taken seriously into account when the efficient expression of protein is mandatory. Moreover, playing with the translation mechanism offers an extraordinary and growing numbers of possibilities to express a gene with a high specificity. In the last few years, a number of progresses have been made in the understanding, not only of the translation process itself but also in the way it can be used to efficiently produce proteins of biotechnological interest. The following chapters describe some examples of the steps which have been made towards the understanding of the mechanisms which control the abundance of the ribosomes in the cell, the use of non-classical initiation mechanisms of translation, the stability of the mRNA, the influences of sequences located in the untranslated regions of the mRNA, and finally how specifically designed antisense oligonucleotides can be used to modulate translation.

2. Production control of the protein component of ribosomes

Optimisation of the translation process could benefit from the possibility of inducing an enhanced and balanced production of the numerous protein and RNA moieties involved. If the intricate regulatory processes underlying the cellular control of the translation apparatus production could be unravelled, one might find a way to optimise the protein synthesis capacity of a cell by adjusting one or few regulatory key elements. We will describe below the findings supporting the idea that all ribosomal protein genes, and many other genes involved in the production and function of the translation apparatus, are co-regulated at the translational level and share structural similarities.

2.1. Coordinated regulation of ribosomal protein genes at the translational level

The coordinated synthesis of ribosomal proteins (rp) in vertebrates is regulated at the translational level through a redistribution of the rp-mRNAs between active polysomes and untranslated messenger ribonucleoprotein-particles (mRNPs) in response to the cellular need for ribosomes (Pierandrei-Amaldi *et al.*, 1985; Pierandrei-Amaldi *et al.*, 1982). The *X. laevis* kidney cell line B 3.2 in which ribosomal protein synthesis decreases rapidly after a downshift to serum free medium and recovers quickly after serum addition (Loreni and Amaldi, 1992) was used to study some aspects of this translational regulation. During these growth rate changes, a shift of rp-mRNA distribution between polysomes and mRNPs was observed. Similar changes have also been observed in various mammalian systems following growth rate variations: mouse fibroblast after serum starvation, mouse lymphosarcoma cells after dexamethasone treatment and mouse myoblasts during differentiation (for a review see (Amaldi and Pierandrei-Amaldi, 1997; Meyuhas *et al.*, 1996). The rp-mRNA relocation between the two compartments is very fast, specific and occurs simultaneously for all the rp-mRNAs (Loreni and Amaldi, 1992; Pierandrei-Amaldi *et al.*, 1982). The rapidity and the reversibility of the regulation allow a rapid adjustment in response to extra cellular stimuli and suggest a modification at the post-transcriptional level, since the synthesis of new transcripts would require a longer time (Agrawal and Bowman, 1987; Aloni *et al.*, 1992).

2.2. *Cis* regulatory elements in the r-protein mRNAs translation

The specific and simultaneous response of the numerous rp-mRNAs at the translational level implies some common features in the mRNAs that respond at the same time to the same stimuli. In fact it was observed that the 5'UTR of the numerous rp-mRNAs of different vertebrates so far analyzed share similarity in the 5'UTR: they are all short and start with a typical Tract of Oligo Pyrimidine at the 5' end (5' TOP). This conservation suggested an important role in the coordinated translational regulation of this class of, mRNAs. This notion has been proved by introducing, in vertebrate cells, chimeric constructs containing a reporter gene preceded by the 5'UTR of an rp-mRNA. The reporter mRNA responded to the translational regulation as an rp-mRNA does (Mariottini and Amaldi, 1990). In addition, site directed mutagenesis showed that the 5' oligopyrimidine sequence, as well as a downstream sequence, are necessary to maintain the control (Levy *et al.*, 1991). In this way two regions of the rp-mRNA 5'UTR were identified, they are responsible for regulating r-protein synthesis and considered to be the *cis*-acting elements controlling the rp-mRNA at the level of translation.

2.3. Some proteins interact with the regulatory sequences of rp-mRNAs

It is possible that common factors, acting in *trans* on similar regulatory sequences present in the different members of the rp-mRNA class, could concert a coordinated regulation. Indeed, it was found that some proteins interacted *in vitro* with the pyrimidine tract and with the downstream region of the 5'UTR of mammals and amphibia (Cardinali *et al.*, 1993; Kaspar *et al.*, 1992) . The analysis was carried out by band shift and UV cross-linking experiments and the specificity was proven by competition (Cardinali *et al.*, 1993; Pellizzoni *et al.*, 1997). The protein that binds the pyrimidine tract was identified as La (Pellizzoni *et al.*, 1996), a 56 kD RNA-binding protein previously reported to be implicated in RNA polymerase III transcription termination and in translation initiation of poliovirus and immunodeficiency virus type 1 RNAs. The protein that binds the downstream region was identified as the Cellular

Nucleic Acid Protein, CNBP (Pellizzoni *et al.*, 1997), a 21kD cytoplasmic protein containing seven zinc finger motives, whose function is still unknown. Mutations in the pyrimidine tract and deletions of the downstream region of the 5'UTR impair the binding of La and CNBP *in vitro* (Pellizzoni *et al.*, 1996; Pellizzoni *et al.*, 1997). The same mutations result in the loss of translational repression *in vivo* (Avni *et al.*, 1994; Levy *et al.*, 1991), consistently with the role that the binding proteins may play in the translational control of rp-mRNAs. While it is known that La and CNBP are also able to bind other RNAs, their binding to the rp-mRNA 5'UTRs is peculiar. In fact the binding to these RNAs, but not to other RNAs, requires the assistance of a factor. The assisting factor appears to be common to both proteins and to show features pertaining to a ribonucleoprotein of the RoRNP family (Pellizzoni *et al.*, 1998; van Venrooij *et al.*, 1993).

2.4. La and CNBP interact with different structural conformations of rp-mRNA 5'UTR

La and CNBP cannot bind simultaneously to the same RNA molecule, their binding being mutually exclusive. Moreover, CNBP, which interacts with its target sequence of the 5'UTR as a 21 kD protein, binds to the whole 5'UTR as a homodimer of 42 kD. This might be related to a conformation assumed by the 5'UTR in the presence of the protein. Binding competition experiments between La and CNBP in RNA excess showed that La has a higher affinity for the 5'UTR than CNBP and that their alternative binding depends on a different affinity for the assisting factor (Pellizzoni *et al.*, 1998). This alternative binding of La and CNBP can affect the conformation and presumably the translatability of the mRNA. One of the possible conformations implies base pairing of the typical pyrimidine tract, located at the 5' end, and the region preceding the AUG. Base pairing disruption by mutations in the pyrimidine tract results in the loss of La binding, but also of CNBP dimer formation suggesting a change in RNA conformation unable to maintain the CNBP dimeric form (Pellizzoni *et al.*, 1998). These observations agree with *in vivo* data showing that the integrity of the pyrimidine tract is a pre-requisite for keeping the translational control of the rp-mRNAs (Avni *et al.*, 1994). These *in vitro* and *in vivo* results can be integrated in a model (Figure 1) in which the 5'UTR is in equilibrium between open and closed conformations that can behave differently in terms of recognition by the translational machinery. A closed CNBP-binding form would result in translation inhibition, while an open La-bound form would instead allow translation. The reciprocal concentration and/or activity of La, CNBP and assisting factor in living cells could account for the transition between the two conformations and consequently from an inefficient to an efficient translational state.

2.5. Do La and CNBP have a role in translation efficiency of rp-mRNAs?

Following the indications outlined by the experiments described above, functional experiments were performed to see if and how the proteins affect the translation of the rp-mRNAs. Stably transfected cell lines were constructed that overexpress wild type La or mutant forms that might exert a dominant negative effect. In serum starved cells, the overexpression of normal La increases the efficiency of translation of rp-mRNAs. On the contrary, the expression of a mutant form inhibits the translation of these mRNAs (unpublished results). This suggests that indeed La is involved in rp-mRNA translational regulation and that, in this regulation, it has a positive effect, supporting the above mentioned model. It should be emphasised, though, that this model for growth dependent translational regulation of the rp-mRNA could be more complex, with several effectors responding to numerous stimuli. In fact other important effectors have been described and others are expected to be found in the future.

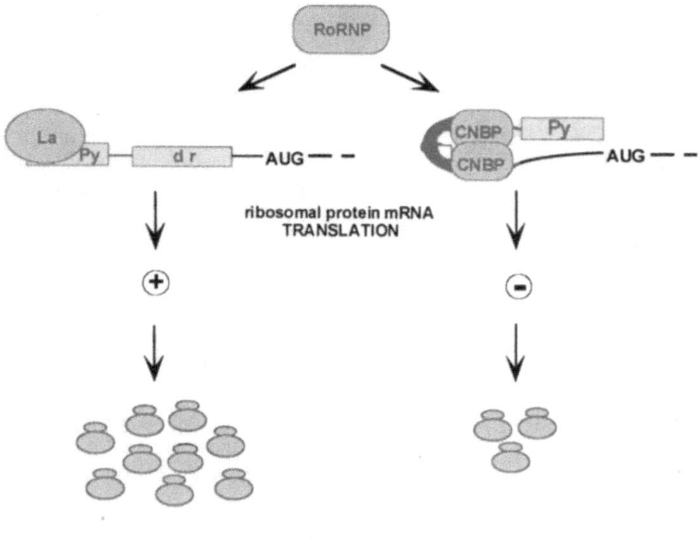

Figure 1. Modulation of ribosome production by regulating the synthesis of ribosomal proteins. Model of the interaction of La and CNBP and the rp-mRNA 5'UTR. The alternative binding of La and CNBP proteins, assisted by the factor RoRNP, should stabilise one conformation and lead to opposite effects on the translation of the rp-mRNAs and consequently on the amount of ribosomes. Py=pyrimidine tract, dr=downstream region.

2.6. Unified control of the translational apparatus components operated by growth signals

Besides the rp-mRNAs, several mRNAs encoding proteins related to the synthesis or function of the translational apparatus are also endowed with two features: that of carrying the typical 5'UTR, containing the 5'TOP sequence, and that of being co-regulated at the translational level with rp-mRNA in a growth-dependent manner (Amaldi and Pierandrei-Amaldi, 1997; Meyuhas *et al.*, 1996). Moreover eEF-1a, the only one of these mRNAs that has been analysed in La transfection experiments, has also been found to be positively affected by La, similarly to rp-mRNAs (unpublished results). These co-regulated genes, that include those coding for all r-proteins and those for several other proteins implicated in translation, are now recognised as belonging to a class of genes called TOP genes (Amaldi and Pierandrei-Amaldi, 1997). Interestingly, in recent years, it has been shown that the translation of TOP mRNAs is selectively regulated by the phosphorylation state of r-protein S6 (Jefferies *et al.*, 1997; Jefferies *et al.*, 1994; Terada *et al.*, 1994), being thus connected, through the FRAP/p70^{S6k} signal transduction pathway, to the incoming signals controlling cell growth and proliferation. In this view, by specifically recognising regulatory regions of the 5'UTR of TOP mRNAs, La and CNBP could have a mediating role between these signals and the synthesis of the components of the translational machinery. The identification of activators or inhibitors that operate at a nodal step of the complex production of this machinery, opens the possibility to usefully modify the protein synthetic properties of cells, as for instance to engineer cell lines that maintains constitutively a high synthetic activity, regardless of growth conditions.

3. IRESes, boxes to selectively activate protein expression in biotechnology.

3.1. IRESes, translational regulation boxes in natural mRNAs.

Until the end of the nineteen eighties, the unchallenged model was that eukaryotic mRNAs could only be translated according to the so-called scanning mechanism (Kozak, 1978). The hypothesis was that ribosome recruitment could only occur upon recognition of the capped 5' mRNA end by the initiation factor eIF-4F, followed by ribosome scanning along the RNA molecule, until it encountered an initiation codon. However, such a mechanism could not explain the efficient translation of picornavirus mRNAs, which are uncapped and contain 800 to 1000 nt-long 5' untranslated regions with numerous AUG start codons upstream of the authentic start codon. Furthermore such viruses produce a protease that cleaves an eIF-4F subunit, the initiation factor 4G, thereby generating a shut-off of the cellular cap-dependent translation while the synthesis of viral proteins remains unaffected. This suggested that cap-dependent scanning was not a general rule. This dogma was in fact simultanously discredited in two laboratories when it was found that poliovirus and encephalomyocarditis virus mRNAs were translated in a cap-independent manner, by an internal ribosome entry mechanism involving an mRNA element called ribosome landing pad or internal ribosome entry site, respectively (Jang et al., 1988; Pelletier and Sonenberg, 1988). Strikingly, internal initiation of translation of these viral mRNA was shown to involve cellular proteins. The first ones found to be involved in IRES activity were a p52, and a p57, identified later as the La autoantigen and the splicing factor pyrimidine tract binding protein (PTB), for poliovirus and EMCV respectively (Jang et al., 1988; Meerovitch et al., 1989). This involvement of cellular proteins in the regulation of picornavirus activity suggested that certain cellular mRNAs could be translated by an IRES-dependent mechanism. The scanning mechanism rule was also questioned in the case of cellular mRNAs following the discovery of an IRES in the Bip mRNA, which codes for an immunoglobulin chaperon of the heat shock protein family (Macejak and Sarnow, 1991). This asked an interesting question: why should a cellular capped mRNA be translated by a cap-independent mechanism? One hypothesis was that such a mechanism could allow the translation of specific cellular mRNAs in conditions when cap-dependent translation was blocked, as would occur under stress conditions and during the M phase of the cell cycle (Jackson, 1991). This suggested that IRESes could be found in mRNAs coding for proteins involved in a stress response (like Bip) or in the control of cell growth.

Several IRESes were identified in cellular mRNAs over the following years, starting with fibroblast growth factor 2, insulin-like growth factor 2 and translation initiation factor 4G mRNAs as regards mammalian cells (Gan and Rhoads, 1996; Teerink et al., 1995; Vagner et al., 1995), while another IRES was identified in the Drosophila Antennapedia mRNA, which encodes a transcription factor (Oh et al., 1992). More recently, other cellular IRESes have also been found in the mRNAs of Drosophila Ultrabithorax, and of mammalian platelet derived growth factor (PDGF/*c-sis*), proto-oncogene c-*myc* and vascular endothelial growth factor (VEGF), thereby confirming that IRES-dependent translation concerns mRNAs coding for regulatory proteins (Akiri et al., 1998; Bernstein et al., 1997; Huez et al., 1998; Miller et al., 1998; Nanbru et al., 1997; Stein et al., 1998; Stoneley et al., 1998; Ye et al., 1997). These mRNAs have long 5' UTRs that are inefficiently translated by the classical scanning mechanism. The presence of an IRES, possibly regulated by cellular trans-acting factors, allows precise activation of their expression in response to specific stimuli. FGF-2 and PDGF expression, for example, is translationally activated in response to stress and to cell differentiating agents, respectively (Bernstein et al., 1997; Vagner et al., 1996). IRES activity also seems to be constitutive in cancerous cells (Galy et al., 1999). The binding of cellular factors to these IRESes has been demonstrated: p60 or p110 binds to FGF-2 IRES in response to stress or upon cell transformation, whereas p43-45 identified as hnRNPC binds to PDGF/*c-sis* IRES upon K562 cell induced differentiation (Elroy-Stein, personal communication OR Sella et al 1999; Vagner et al., 1996). However the role played by such proteins in IRES function has remained unelucidated up to now.

At the same time, numerous new IRESes were found not only in various picornaviruses (including human rhinovirus, hepatitis A...), but also in flavivirus (hepatitis C virus), pestivirus and in murine, avian and human retroviruses (Berlioz and Darlix, 1995; Berlioz et al., 1995; Jackson and Kaminski, 1995; Vagner et al., 1995)(Waysbort et al, unpublished results). Mutagenesis and comparison of the different picornavirus IRESes provided a better insight into the internal ribosome entry mechanism. IRES involves the RNA secondary and/or tertiary structure rather than the primary sequence (Jackson and Kaminski, 1995) . This structural element allows ribosome recruitment at an AUG triplet located in 3' of the structure. At the present time picornavirus IRESes are divided into three groups: in cardiovirus (EMCV) the ribosome recruiting codon is used as a start codon; in entero/rhinovirus (poliovirus), ribosome entry is followed by scanning and initiation occurs at an AUG downstream, whereas in aphtovirus initiation occurs both at the 3' end of the IRES and at an AUG downstream. A fourth class of IRES is found with hepatitis C and pestivirus IRESes which require a portion of the coding sequence in order to function. These latter IRESes are able to recruit ribosomes in the absence of specific trans-acting factors, and in this way resemble prokaryotic translation initiation sites (Pestova et al., 1998). In contrast, picornavirus IRESes generally involve trans-acting proteins: PTB requirement for internal initiation of translation of cardiovirus mRNAs is conditional rather than absolute, while this protein is necessary but not sufficient for efficient internal initiation of translation of human rhinovirus-2 (HRV) mRNA (Hunt and Jackson, 1999; Kaminski and Jackson, 1998). HRV IRES activity also requires a protein of the cold shock family, unr (Hunt et al., 1999).

3.2. Biotechnological use of IRESes.

The approach generally used to demonstrate IRES existence is the bicistronic mRNA strategy (Pelletier and Sonenberg, 1988). Eukaryotic mRNAs are monocistronic and the cap-dependent mechanism predicts that a bicistronic mRNA will express only its first cistron. The expression of a downstream cistron will be conditioned by the presence of an IRES upstream.

It is often necessary to express two or more genes from expression vectors, for instance the gene of interest and a selection gene. The use of two independent transcription units in a given vector often creates a source of interference between the two promoter activities and can lead to the extinction of one of them. The only way to be sure that two proteins are synthesized together is to express them from the same mRNA (Fig 2). The discovery of IRES provided a solution to this problem and now made it possible to synthesize two proteins from a bicistronic mRNA.

The first bicistronic vector which allowed successful gene transfer used EMCV IRES (Ghattas et al., 1991; Kaufman et al., 1991; Moss et al., 1990). Bicistronic vectors were successfully used not only to constitutively express genes in stable cell lines, but also to express antisense RNAs directed against a cellular gene, such as FGF-2 (Maret et al., 1995). They have also been used in vivo, in transgenic mice and in gene therapy protocols. Bicistronic retrovirus vectors and less frequently, adenovirus-associated virus vectors have been assayed in gene therapy programs. One of several interesting studies involved the co-expression of human adenosine deaminase and multidrug resistance gene (MDR1) in a retroviral vector using EMCV IRES (Zhou et al., 1998). Functional ADA was shown to be co-expressed in proportion to the MDR1 level. In an antitumoral suicide approach, the co-expression of MDR1 and herpes simplex virus thymidine kinase allowed selective killing of MDR1-transduced human tumors transplanted in nude mice (Sugimoto et al., 1997). Comparison of the bicistronic retroviral and adeno-associated viral vectors designed to express human clotting factor IX showed that both kinds of vectors produced biologically active factor IX; however lower transduction efficiency and poorer expression was generally observed with AAV vectors versus retroviral vectors (Chen et al., 1997).

One of the advantage of IRES-containing vectors is the possibility of producing two subunits of a heterodimeric complex (Fig. 2). In particular, interleukin12 (IL-12) is a heterodimeric cytokine which has been shown to cause tumor regression and to stimulate long-term antitumor immunity in treated animals (Brunda et al., 1993; Scott, 1993). Strategies involving regulated splicing or introducing exogenous promoters raised substantial concern about a discordant production of the two chains from transfected cells. This is a problem because an excess of one subunit can inhibit the formation of the active heterodimer. It was overcome by using polycistronic vectors: in an antitumoral immunotherapy strategy, tricistronic retroviral vectors were constructed that expressed active IL-12 (the two subunits in an adequate ratio) together with the *Neo* gene conferring G418 resistance (Zitvogel et al., 1994). Such vectors, expressing one of the IL-12 subunits from the first cistron, while the other IL-12 subunit and neomycin resistance were expressed from the second and third cistrons, respectively, both under the control of an EMCV IRES, were able to effectively eradicate murine tumors with IL-12 gene therapy (Tahara et al., 1995).

While IRESes provide the crucial ability to express several proteins in adequate stoechiometry in gene therapy programs based on viral vectors, they have also been used to design polycistronic vectors in industrial biotechnological programs aimed to optimize the production of recombinant proteins. Several laboratories in the European Commission Biotechnology animal cell technology program have used bicistronic and tricistronic vectors to co-express proteins of interest and a resistance gene. Tricistronic vectors have been particularly successful in a program of controlled proliferation: co-expression of the cyclin-dependent kinase inhibitor p27 (enabling CHO cell-cycle arrest in the G1 phase) and of the antiapoptotic gene bcl-x_l resulted in a 30-fold stimulation of the production of secreted alkaline phosphatase, the model heterologous protein expressed from the third cistron (Fussenegger et al., 1998). Stable growth regulation involving the addition of b-estradiol to the medium was also obtained in BHK-21 cells by tightly linking, on a bicistronic construct, the expression of a drug resistance gene to the growth regulator IRF-1 fused to estrogen receptor (Müller et al., 1998). IRES have also been useful in an o-glycosylation protein engineering program for the production of receptor blockers in which tricistronic constructs with neomycin resistance in the third cistron were used to obtain the coordinate expression of two glycosyltransferases (A.

Dinter, personal communication). Thus it is becoming increasingly apparent that IRESes will have a wide utilization in protein production as well as in gene therapy programs.

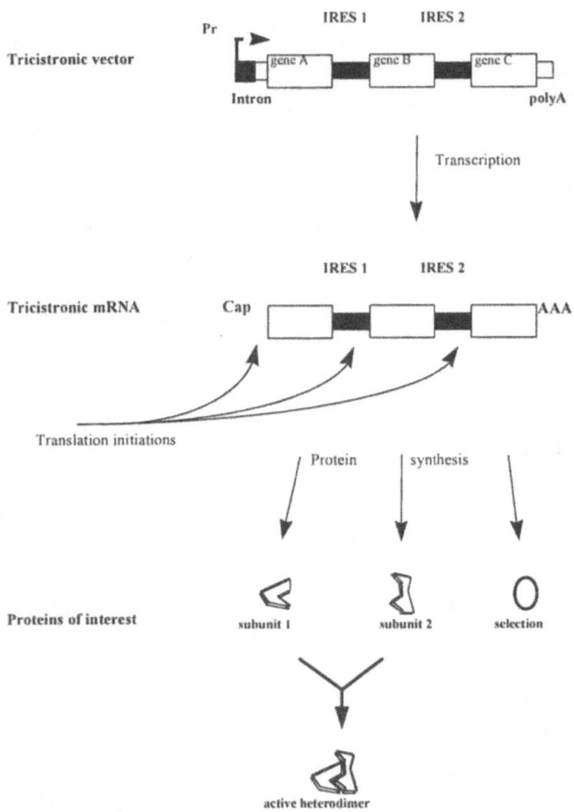

Figure 2: Design of tricistronic vectors for protein production in industry and gene therapy.

Tricistronic vectors can be constructed from classical viral or non viral expression vectors, by introducing between the promoter (Pr) and the polyadenylation site (poly A) the cDNA sequences coding for the genes to express, separated by IRESes. As regards the non viral vectors, the presence of an intron will allow a better maturation and expression of the polycistronic mRNA. Transcription of the tricistronic mRNA is followed by translation of the three cistrons: translation initiation occurs in a cap-dependent manner for gene A and is controlled by IRES 1 and 2 for gene B and C, respectively. If genes A and B codes for two subunits of an heterodimeric protein, their expression in a correct stoechiometry from the tricistronic mRNA will result in expression of an active heterodimer, while expression of gene C allows the selection of producing cells.

3.3. Polycistronic vectors with cellular and retroviral IRESes.

All the bicistronic and tricistronic vectors used up to now have been based either on EMCV or poliovirus IRESes. While these picornavirus IRESes allow efficient expression of the proteins of interest in the conditions used, we have little information about their tissue specific efficiency. One transient transfection study reports that EMCV IRES was efficient in six cultured cell lines of different origins whereas poliovirus IRES was efficient in HeLa, FRhK4 and HepG2 cells but extremely inefficient in SK-NBE, BHK21 and Neuro-2A cells (Borman et al., 1997). This reveals that IRES efficiency can vary in a tissue specific manner, probably in relation to the availability of regulatory trans-acting factors and that this feature could condition the choice of IRES to be used in a given program. Furthermore in tricistronic vectors, and despite the fact that the vectors used up to now have always contained two identical IRESes, the use of two different IRESes may ensure a better stability of expression. One disadvantage of picornavirus IRESes is their size of about 450 nt which can be a problem in viral vectors when the genes of interest are large.

All these points have led to the emergence of vector design programs based on cellular or retroviral IRESes. These IRESes offer an interesting diversity, with different sizes and specificities of expression. One recent study has compared the efficiencies of different cellular and retroviral IRESes with EMCV IRES, and shows that whereas EMCV IRES remained the most active in transiently transfected HeLa, 3T3, COS, SK-Hep-1 and CHO, several cellular retroviral IRESes were more active in other cell types such as Jurkat cells (Créancier et al, unpublished results). Furthermore vectors with EMCV IRES seem to be less stably expressed in permanently transfected cells than vectors with cellular or retroviral IRESes (Créancier et al, unpublished results). Both their tissue-specificity and apparent better stability make the use of cellular and retroviral IRESes of real interest. One important purpose is to obtain correct stoechiometry, for example in the industrial production of two enzymes in a given biosynthesis chain, or of two subunits of heterodimeric enzyme.

We could predict that to design a vector in permanently transfected cells, the use of a moderately efficient IRES to control expression of the resistance gene would be a good choice to force expression of the vector whithin the cell. However, the IRESes controlling the genes of interest must be selected according to the requirements of each specific program. The best choice will be based on tissue-specificity data for IRES in cell culture and in vivo (Borman et al., 1997)(Créancier et al, unpublished). In the case of a non toxic protein, the IRES which is most efficient in the target cell type should be used, remembering that the term of "efficiency" needs to be considered with caution as transient efficiency does not necessarily mean permanent efficiency.

If two proteins have to be expressed with equal stoechiometry, an IRES giving an expression level corresponding to that of the first cistron must be chosen. In non viral vectors the first cistron is very abundantly expressed, whereas this is not the case in retroviral vectors in which the first cistron is preceded by several hundred nucleotides containing the packaging sequence, which partially coincides with that of MuLV IRES (Berlioz and Darlix, 1995; Vagner et al., 1995). Retrovirus vectors which have been constructed using retroviral IRESes to control expression of the second cistron have, in fact, resulted in very high titers of retrovirus vector production, despite the moderate efficiency of these IRESes in vitro and in transiently transfected cells (Berlioz et al., 1995; Lopez-Lastra et al., 1997; Torrent et al., 1996).

The availability of a broad range of IRESes also allows the creation of tri- or multicistronic vectors without repeating the same IRES in a given vector, this providing better vector stability and preventing possible competition of IRESes for the trans-acting factors that may exist in limiting amounts.

3.4. A close perspective on the horizon: inducible IRESes.

One major problem in protein expression by gene transfer concerns the putative toxicity of various proteins when overexpressed, or their ability to prevent cell growth. This has made it necessary to use inducible systems, and such a point is under investigation with regards to IRES-controlled expression. On the one hand, inducible IRESes could be derived from natural IRESes such as that of FGF-2 mRNA, which seems to be naturally activated by stress or cell transformation (Vagner et al., 1996). Alternatively, artificial IRESes could be designed, based on the hypothesis that the C-terminal two thirds of the initiation factor 4G should suffice to achieve basal ribosome recruitment to an mRNA and, consequently, that tethering such a polypeptide to a site 5' from an open reading frame should mediate the internal initiation of translation (M. Hentze, personal communication). Strategies will also need to be developed to completely block the genes of interest in uninduced conditions. The design of new aptamers directed against IRESes could be one of the directions to follow (Tinévez et al., 1998; Toulmé and Giegé, 1998).

4. Controlled mRNA deadenylation as a means to regulate protein production in cell factories.

We describe here the way in which changes of a regulatory modification, namely changes in mRNA adenylation status, can affect mRNA translation and stability and why mastering the control of this process may be important for the production of proteins in cells.

4.1. Background.

It has been known for a long time that polyadenylated (polyA$^+$) mRNAs, that is mRNAs with a stretch of adenosine at their 3' end, are translated more efficiently and are more stable than their deadenylated (polyA$^-$) counterparts. One of the first experimental demonstrations of this was obtained by the injection of in vitro deadenylated rabbit globin mRNAs into *Xenopus laevis* oocytes (Huez et al., 1974). This was substantiated by experiments using mRNA transfection into various cells (Gallie, 1991) and cell free extracts (reviewed in Ross, 1993). Subsequently, it was shown that the degradation of short lived mRNAs in somatic cells is correlated, in general, with their rapid and sequence specific deadenylation (reviewed in Greenberg and Belasco, 1993).

The data accumulated over the last ten years, initially in yeast (Sachs and Davis, 1989; Decker and Parker, 1993) and then extended to mammalian cells (Couttet et al., 1997), has established a molecular basis for the dual effect of the poly(A) tail on mRNA translation and stability (see Wickens et al., 1997 for a recent review). In eukaryotic cells all mRNAs, except those encoding histones, are polyadenylated. In most cells this 3' extension, which is added to pre-mRNAs in the nucleus, is complexed with the highly conserved 70 kDa Poly(A) Binding Protein (PABP). PABP, in addition to binding to the poly(A) tail, can associate with the proteins that form a complex on the 5' terminal structure of a mRNA or cap (Tarun and Sachs, 1996; Le et al., 1997; Craig et al., 1998). The cap-dependent translation of mRNAs is initiated by the binding of the 43S ribosomal particle to the cap-binding complex. It has been suggested that the association of PABP with this complex may stabilise the ensemble with a concomittant increase in the translational efficiency (LaGrandeur and Parker, 1996; Sachs and Buratowski, 1997). This of course implies that the 5' and 3' termini of a mRNA are in close proximity and, indeed, circular mRNAs have been visualised by electron microscopy (Christensen et al., 1987). In vitro the formation of circular mRNAs is dependent on the presence of the cap-binding complex and PABP (Wells et al., 1998).

The interaction of the cap-binding complex with PABP may also protect the mRNA against degradation. Indeed, the removal of PABPs from a mammalian degradation competent cell extract renders unstable the otherwise highly stable globin mRNA (Bernstein et al., 1989). It was also observed that deadenylation preceeds the degradation of short lived mRNAs in mammalian cells (Wilson and Treisman, 1988, Laird-Offringa et al., 1990; Peppel and Baglioni, 1991). More recently it was demonstrated that in both yeast and mammalian cells deadenylation preceeds decapping which renders the mRNA available for attack by a 5'-3' exonuclease (Decker and Parker, 1993; Couttet et al., 1997). In addition, in mutant yeast strains lacking PABP decapping occurs without deadenylation (Caponigro and Parker, 1995) suggesting that PABP acts as an inhibitor of decapping and thereby stabilises the mRNA. There is also evidence that mRNAs can be degraded by 3'-5' exonucleases and that the poly(A)-PABP complex also protects a mRNA against the attack of this exonuclease (Ford et al., 1997). The degradation of a mRNA can also be initiated by endonucleolytic sequence specific cleavage but, in at least two cases, this initial cleavage is independent of the adenylation status of the mRNA (Binder et al., 1989; Binder et al., 1994).

The adenylation status of a mRNA is determined by the presence or the absence of specific cis-elements. These are most commonly situated in the 3' untranslated region (3'UTR). Two types of such cis-elements have been identified in mammalian somatic cells: those that permit maintenance of a long poly(A) tail and subsequently stabilise the mRNA and those whose presence leads to a rapid deadenylation and degradation. The high stability of globin mRNA in cells of the erythrocyte lineage is due to a C-rich element situated just 3' to the stop codon (Wang et al., 1995). Readthrough of the stop codon, as occurs in Constant Spring (CS) thalassemia, causes destabilisation of the mRNA presumably by hindering the binding of proteins to this C-rich element. In transgenic mice the lesser stability of the globin mRNA is correlated with a shorter poly(A) tail when compared to that of the normal human globin mRNA also expressed in transgenic mice (Morales et al., 1997). This element may therefore stabilise mRNAs by protecting them against deadenylation. The high instability of many mRNAs that encode protooncogenes or cytokines is due to the presence of an AU-rich element (ARE) in the 3'UTR of these mRNAs (reviewed in Greenberg and Belasco, 1993). The presence of this element is required for the transitory or low level of expression of these proteins. Mutations that render these cis-elements non-functional can cause an over expression of the encoded protein which, in turn, can lead to cell transformation (see Schiavi et al., 1992 for a review). The initial step in ARE-dependent degradation is a shortening of the poly(A) tail (Wilson and Treisman, 1988, Laird-Offringa et al., 1990; Peppel and Baglioni, 1991). This presumably leads either to decapping and attack by a 5'-3' exonuclease or directly to attack by a 3'-5' exonuclease.

Chen and Shyu (1995) have defined two classes of AREs, those containing an AUUUA motif and the non-AUUUA class.A number of factors have been identified that bind to the AREs of the AUUUA class (reviewed in Jarzembowski and Malter, 1997). However to date there is no definitive experimental evidence to show that any of these factors are required for the rapid deadenylation or degradation of these mRNAs. In fact it was recently shown that the overexpression of HuR, an RNA-binding protein of the Elav family that binds to the AUUUA-motif, actually stabilises RNAs containing this element without affecting the rate at which the mRNAs are deadenylated (Fan and Steitz, 1998 ; Peng et al., 1998).

4.2. *Xenopus laevis* - a model system to study post-transcriptional regulation.

The difficulty encountered in identifying cis-elements and trans-factors that control mRNA deadenylation and stability is, in part, due to the tight temporal and functional coupling between these two events in somatic cells. In mammalian cells it is virtually impossible to observe the deadenylated intermediates. *Xenopus laevis* oocytes and embryos present a particularly useful experimental model for studying post-transcriptional control mechanisms. In the oocytes and embryos of this amphibian the deadenylation and the degradation of mRNAs, although being functionally coupled

(Audic et al., 1997) are temporally dissociated (Duval et al., 1990). In addition, transcription is essentially silent during the period that encompasses the completion of meiosis (oocyte maturation), fertilisation and the first twelve mitotic cell divisions (Newport and Kirschner, 1982). In the maturing oocyte and early embryo a number of processes coexist that produce sequence specific changes in the polyadenylation of maternal mRNAs that have accumulated during oogenesis. These changes, which occur in the cytoplasm, are tightly correlated with changes in the translational status of the mRNA (reviewed in Osborne and Richter, 1997). A further advantage of *Xenopus laevis* oocytes and embryos is that extracts can be made of undiluted cytoplasm. These extracts, which were first developped for the study of cell cycle regulatory processes (Murray et al., 1989), also reproduce many aspects of the post-transcriptional changes to mRNAs (Legagneux et al., 1995). The different uses of these extracts are depicted in Figure 3.

Cytoplasmic polyadenylation is correlated with the polysomal recruitement of mRNAs and requires the presence of a cytoplasmic polyadenylation element (CPE) and the AAUAAA nuclear polyadenylation signal in the 3' UTR of the mRNA (McGrew et al., 1989; Fox et al., 1989). Deadenylation is correlated with translational arrest and can be of two types, either sequence specific or a default process. The default deadenylation is active on all mRNAs devoid of a functional CPE (Fox and Wickens, 1990; Varnum and Wormington, 1990). The *Xenopus laevis* deadenylase or Poly(A) specific Ribonuclease (PARN) responsible for default deadenylation has been identified by homology to a poly(A)-nuclease purified from calf thymus and the cloned human cDNA (Körner et al., 1998). Whether the mammalian or *Xenopus laevis* proteins are also involved in sequence specific deadenylation is not known. The human and *Xenopus laevis* PARNs appear to have sequence homologues in diverse organisms, but surprisingly, there is no yeast gene with obvious similarity to these PARNs (Körner et al., 1998). The sequence specific deadenylation of a number of maternal mRNAs after fertilisation (Paris et al., 1988; Bouvet et al., 1994) is dependent on the presence of a cis-element named an Embryo Deadenylation ElemeNt (EDEN) (Paillard et al., 1998). As the name implies, the EDEN targets mRNAs for deadenylation in the embryo and not in the oocyte. It also has the property of overriding a CPE in the embryo (Bouvet et al., 1994; Paillard et al., 1998). Therefore, a mRNA that contains a CPE and an EDEN in the 3' UTR is polyadenylated during oocyte maturation (due to the CPE) and then deadenylated after fertilisation (due to the EDEN).

As mentioned above deadenylation and degradation in the *Xenopus laevis* embryo are functionally coupled but temporally dissociated. The mRNAs that are deadenylated after fertilisation are not degraded until several hours latter. During this intermediate period the deadenylated maternal mRNAs are quantitatively conserved (Paris et al., 1988; Duval et al., 1990). The degradation of these mRNAs requires the translation of at least one other mRNA (Duval et al., 1990; Bouvet et al., 1991) which implies that one or more components of the degradation machinery is absent or present in very limiting amounts in early embryos. This missing activity could be the decapping enzyme as uncapped RNAs, injected into oocytes or embryos, are immediately degraded (Gillian-Daniel et al., 1998; and Osborne, unpublished data). The functional connection between deadenylation and degradation was demonstated by injecting synthetic mRNAs into two cell embryos. Only those that were injected as poly(A^-) molecules, or which were deadenylated after injection, were degraded (Audic et al., 1997). In addition, the presence of a CPE in the 3'UTR of the injected mRNA, which caused its in vivo adenylation, protected the mRNA against degradation (Audic et al., 1997).

The factors that bind to these elements (CPE and EDEN) have been isolated and in all cases the mammalian homologues have been cloned. Two types of CPE have been characterised, a UUUUUAU- type motif (McGrew et al. 1989; Fox et al., 1989) and a poly(U) motif (Simon et al., 1993). We have previously named these UA-rich CPEs and U-rich CPEs respectiverly (Osborne and Richter, 1997). In *Xenopus laevis* oocytes the UA-CPE binds a 64 kDa protein (CPEB) (Hake and Richter, 1994). Immunodepletion of CPEB from a polyadenylation competent extract abolished CPE-driven polyadenylation. This activity was restored by the addition of the recombinant protein. However, CPEB appears to be degraded during oocyte maturation (Hake and Richter, 1994) and its amount is much reduced in embryos where UA-CPE driven polyadenylation is still robust (Audic et al., 1998). In addition, in an in vitro test Bilger et al. (1994) showed that UA-CPE driven polyadenylation can occur in the absence of CPEB; only the nuclear polyadenylation factor, CPSF, and a poly(A) polymerase (PAP) are required. Concerning the U-rich CPE, the protein elrA is able to bind to this element (Simon et al., 1996) but at present there is no direct experimental demonstation of its role. Interestingly, this protein is highly related to the human protein HuR whose overexpression stabilises mRNAs with AUUUA-containing AREs (Fan and Steitz, 1998 ; Peng et al., 1998).

The protein that associates with an EDEN (EDEN-BP) has recently been characterised (Paillard et al., 1998) and, from a biotechnology point of view, may be the most interesting. As mentioned previously a cell free extract of *Xenopus laevis* embryos has been developped that allows the EDEN-dependent deadenylation process to be studied in vitro. The immunodepletion of EDEN-BP from these extracts abolished the EDEN-dependent deadenylation but did not affect default deadenylation (Paillard et al., 1998). The EDEN-dependent deadenylation can be restored by the additon of recombinant

EDEN-BP (Paillard, unpublished results). The sequence of the *Xenopus laevis* protein is highly (88%) identical to that of a human protein CUG-BP/hNab50 (Timchenko et al., 1996) and this human protein may be a functional homologue of EDEN-BP.

EDEN-BP is part of a large complex (probably >106 Da) that presumably contains the catalytic activity and several as yet unidentified cofactors (Paillard et al., 1996). Whether the catalytic activity in this complex is provided by the default PARN has yet to be established. The activation of EDEN-BP dependent deadenylation at fertilisation is not associated with a change in the amount of EDEN-BP. This suggests that EDEN-dependent deadenylation could be controlled by the post-translational modification of at least one of the components of the active complex.

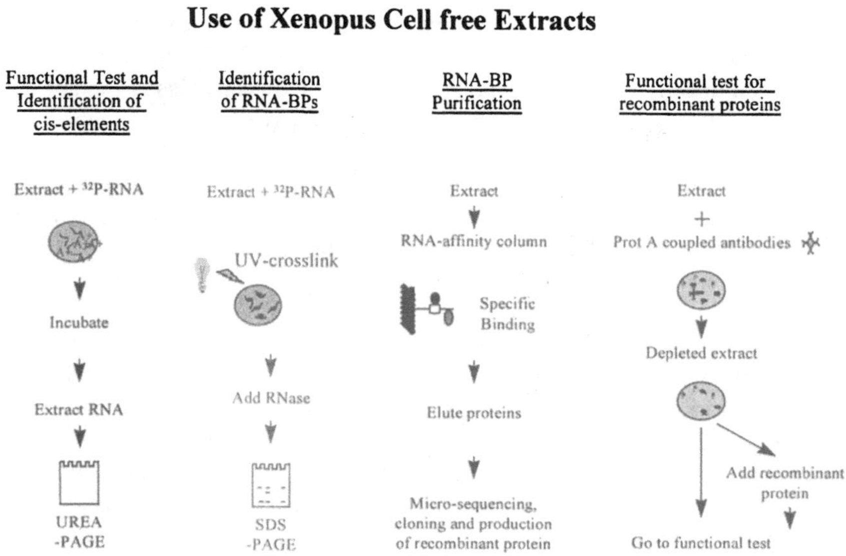

Figure 3
The different ways in which Xenopus embryo extracts can be used to study the cis-elements and trans-factors implicated in the contol of mRNA adenylation/stability are depicted schematically. In the functional test either wild-type or mutant ^{32}P-labelled RNA are added to the extract. The changes in adenylation status are determined by electrophoresis in a denaturing urea-polyacylamide gel. RNA-binding proteins (RNA-BPs) are detected by uv-crosslinking of the endogeneous or recombinant proteins to ^{32}P-labelled RNAs. After identification, these proteins can be purified using a RNA-affinity chromatography where the RNA ligand contains the specific binding site. The bound proteins can be eluted with either a salt gradient or with RNase. Specificity can be controlled by using a column where the ligand does not posses a binding site for the desired protein. The function of the cloned protein(s) can be ascertained by first immunodepleting the endogeneous protein in a functional extract and then rescuing the activity by the addition of recombinant protein.

4.3. Applications in biotechnology.

The conditional regulation of mRNA stability is a way of significantly enhancing the control on the expression of exogeneous proteins in cells. This would be particularly interesting when the exogeneous protein is detrimental to the growth or the viability of the cells and can only be expressed in pulses. Such a pulsed expression can be readily achieved by the combined action of a conditional promotor and a "conditional" instability element. It is evidently necessary to ensure that modifications in the activity of the deadenylation / degradation machinery targetted by the "conditional" instability element do not directly affect the general metabolism of the cell. This is a major interest of the EDEN motif. In addition, synthetic EDENs have been developped, whose presence in a RNA leads to a more efficient deadenylation than that observed with the "best" natural EDEN so far characterised, and elements that enhance the action of an EDEN have been indentified (Audic et al., 1998). Therefore, by engineering the instability element it will be possible to render the deadenylation, and hence the degradation, of the exogeneous mRNA more efficient than that of the endogeneous mRNAs.

Another advantage of EDEN-motifs is that one EDEN-associated trans-factor, clearly required for deadenylation (EDEN-BP), has been characterised (Paillard et al., 1998). The expression of a dominant negative mutant of EDEN-BP would block the function of the endogeneous protein and the EDEN-dependent deadenylation. Such mutants could be produced by engineering changes into the domains of EDEN-BP that interact with cofactors or by point mutations that disrupt eventual regulatory post-translational modifications.

By using a combination of these approaches we predict that it will be possible to construct a conditional mRNA stability package or kit that would contain, for exemple, the following components: (1) a cis-element cassette composed of an optimum EDEN motif with the appropriate enhancer element(s) and (2) a plasmid allowing the expression of a mutant EDEN-BP driven by a conditional promoter. EDEN-BP and the human sequence homologue (CUG-BP/hNab50) are expressed almost ubiquitously (Paillard et al, unpublished results). Hence there should not be any important restriction on the cell type with which this instability package can be used.

There are undoubtedly a number of refinements that could be introduced as progress is made in the characterisation of the cofactors that participate in this deadenylation/degradation pathway and the way in which they interact. For instance, due to the modular nature of RNA-binding proteins it may be possible to engineer the sequence of EDEN-BP so that it very specifically recognises a cis-element distinct from the natural EDEN motifs.

5. Identification of cis and trans-acting elements involved in mRNA translational regulation: innovative means to control protein production in biotechnology.

A growing number of translational regulatory mechanisms involving sequences located in the 3' untranslated region (UTR) of mRNAs are being discovered. The aim of this chapter is not to give an exhaustive list of these regulatory processes but to give from two different examples a perspective on how translational control of gene expression by specific sequence elements located in mRNA 3'UTR can be exploited for protein production by "cell factories".

5.1. Translational regulation of 15-lipoxygenase mRNA during erythroid differentiation

Erythroid 15-lipoxygenase (LOX) is an important enzyme during erythroid cell differentiation as it can attack intact phospholipids and appears to participate in the breackdown of internal mitochondrial membranes during the last stages of reticulocyte maturation (Rapoport and Schewe, 1986). LOX mRNA is abundantly transcribed from early stages of erythropoiesis, however LOX protein is only synthezised during the final steps of reticulocyte maturation indicating an uncoupling between LOX mRNA accumulation and translation during the early stages of erythropoiesis (Thiele et al., 1981). A Differentiation Control Element (DICE) composed of a CU-rich repetitive sequence and located in the 3' UTR of LOX mRNA was demonstrated to mediate translational silencing (Ostareck-Lederer et al., 1994). The specific proteins binding to the 3' UTR differentiation control element (DICE) of LOX mRNA was purified by DICE affinity chromatography. The proteins were sequenced and identified as hnRNPs K and E1. The proteins were expressed in E. coli. The purified recombinant proteins were found to function as specific translational silencers in vitro. Importantly, hnRNP K and/or E1 expression vectors specifically silenced the translation of DICE-containing reporter mRNAs in transiently transfected HeLa cells. Further analysis of the regulatory mechanism revealed that the binding of hnRNPs K and E1 to DICE causes the specific inhibition of 80S ribosome assembly (Ostareck et al., 1997).

5.2. AU-rich elements (AREs) as translation regulatory elements

AU-rich elements (AREs) were first discovered in the 3'UTR of several cytokine and oncoprotein genes. These sequences which are composed of several copies of the AUUUA pentamer control gene expression at the post-transcriptional level. AREs can be grouped into two categories (see also chapter 3.1). Class I AREs contain one to three copies of the AUUUA pentamer in a U-rich context. Prototypes of these class I AREs are found in the 3'UTR of c-myc and c-fos mRNAs and confer mRNA instability. Class II AREs are defined by the presence of multiple clustered pentamers and are preferentially found in the 3'UTR of cytokine mRNAs. Depending on the system considered, these AREs mediate mRNA instability and/or translational blockade (Kruys and Huez, 1994). The mechanism by which AREs mediate instability is described in chapter 3.1. Translational regulation by AREs has been so far best characterized in the case of Tumor Necrosis Factor-α (TNF-α) mRNA. TNF-α is a cytokine predominantly produced by macrophages. The most powerful inducers of TNF-α production by macrophages are the lipopolysaccharides (LPS) which are membrane components released by Gram negative bacteria in the course of infection. It is now well established that the induction of TNF-α production upon stimulation of macrophages by LPS results from both an enhancement of TNF-α gene transcription and a translational derepression of the mRNA. In unstimulated macrophages, TNF-α mRNA translation is blocked.

Upon stimulation with LPS, this repression is overcome and TNF-α mRNA becomes efficiently translated. The key element involved in this regulation is the ARE located in the 3'UTR of TNF-α mRNA (Han et al., 1990). In order to identify proteins specifically binding to TNF-α ARE, RNA probes corresponding to various parts of TNF-α mRNA 3' untranslated region. were synthezised. These labeled RNAs were incubated with cytoplasmic extracts of unstimulated or LPS-stimulated RAW 264.7 macrophages. The RNA/protein complexes formed were analyzed by Electrophoretic Mobility Shift Assay (EMSA). TNF-α ARE can form two complexes (1 and 2) of different electrophoretic mobilities. Whereas complex 1 is formed with cytoplasmic extracts obtained from macrophages independently of their activation state, complex 2 is detected only with cytoplasmic extracts obtained from LPS-stimulated macrophages. Interestingly, formation of complex 2 is triggered by the Sendaï virus, another potent activator of TNF-α mRNA translation in RAW 264.7 macrophages. In contrast, complex 2 is not detected with cytoplasmic extracts obtained from B and T cell lines which are unable to produce TNF in response to LPS. Protein tyrosine phosphorylation is required for LPS induced TNF-α mRNA translation. Remarkably, the protein tyrosine phosphorylation inhibitor herbimycin A abolishes LPS-induced complex 2 formation. Complex 2 is already detectable after 1/2 hour of LPS treatment and is triggered by a minimal LPS dose of 10 pg/ml (Gueydan et al., 1996). Further studies demonstrated that complex 1 requires a large fragment of the ARE containing clustered AUUUA pentamers. The complex 2 binds to a minimal UUAUUUAUU nonamer and is composed of a 55-kDa protein (Lewis et al., 1998). By a cloning strategy based on the differential screening of a macrophage cDNA expression library with TNF-α mRNA 3'UTR riboprobes containing or not the ARE, we isolated the cDNA encoding the short 40-kDa isoform of the RNA-binding protein TIAR. Moreover, the addition of anti-TIAR antibody in the EMSA markedly alters the electrophoretic mobility of complex 1 indicating that TIAR is involved in the formation of this complex. TIAR is a RNA-binding protein which was previously shown to bind U or A homopolymers. TNF-α ARE constitutes thus the first natural target sequence of this protein (Gueydan et al., in press).
The identification of the 55-kDa protein of complex 2 along with the further characterization of TIAR function will provide further insight into the mechanism TNF-α mRNA translational control.

5.3. Biotechnological relevance of translational control for protein production

Specific translational control by sequences located in untranslated regions provides a new approach to regulate protein production by cell factories. This is particularly valuable when the recombinant protein is deleterious for cell growth and viability and can thus be expressed only conditionally. The combination of transcriptional and translational controls would ensure a very inducible expression system in which no recombinant protein would be allowed to be synthesized constitutively but would be massively produced in induction conditions. Indeed, although many conditional promoters are available, none of them ensures a zero level of transcripition in uninduced conditions. Therefore, an additional downstream level of control would impose a stricter supresssion of expression in absence of induction. The identification of cis/trans pairs of specific 3' UTR translational regulators such as DICE/hnRNP K or E1 and TNF ARE/ TIAR (or other trans-acting factors to be identified) would allow the design of expression vectors containing the cis-acting element which in modifed cell factories expressing the trans-acting factor would be translationally regulated.

6. Synthetic oligonucleotides as artificial modulators of translation.

Over the last 15 years several strategies have been developped, using the genetic information itself to artificially modulate the expression of a given gene (Hélène and Toulmé, 1990 ; Toulmé, 1996). Synthetic oligonucleotides are now routinely used to selectively control the synthesis of a protein of interest. This very generally results in the reduction of the expression of the target gene (let's say gene « a »). However if the oligonucleotide interferes with a repression mechanism, this might secondarily lead to the overexpression of a gene « b » under the control of the artificially down-regulated gene « a ». For instance, the translation of many mRNAs such as those encoding cytokines is repressed likely through the binding of proteins to A-U rich elements located in the 3' untranslated region of the RNA (Kruys et al, 1989 ; Han et al, 1990 ; Lewis et al, 1998). Synthetic oligonucleotides able to outcompete the protein-RNA complex would lead to the increased translation of this mRNA. Therefore synthetic oligonucleotides constitute powerful tools to manipulate gene expression.

6.1. Chosing a strategy.

The antisense strategy rests upon the use of a synthetic sequence (antisense) complementary to a selected site on the RNA (sense) of interest (Zamecnik et al, 1978 ; Stein and Cheng, 1993). The formation of the sense-antisense duplex will prevent the « reading » of the genetic information, either splicing or translation, depending on whether pre-mRNA or mRNA is targeted. A related strategy makes use of catalytic antisense sequences: the portion of the oligonucleotide complementary to the target drives an RNA enzyme - a ribozyme - which cleaves the RNA sequence once the complex is formed (Uhlenbeck, 1987 ; Kashani-Sabet and Scanlon, 1995). In the triplex approach, a double-stranded target is recognised by a third strand through the formation of base triplets (Le Doan et al, 1987 ; Moser and Dervan, 1987). This strategy restricted to the targetting of $(purine)_n$-$(pyrimidine)_n$ sequences has been essentially used to control transcription although triple helices made of both RNA and DNA strands have been described (Roberts and Crothers, 1992). Indeed, a decanucleotide 5'G_2AG_7, targeted to a hairpin structure whose stem contained exclusively purines on one strand, was shown to prevent translation of the downstream reporter gene in cell-free assay (Le Tinévez, Chabas and Toulmé, unpublished). Whether this selective effect is due to triple helix formation is not demonstrated yet. It has also been suggested to use oligonucleotides to specifically trap nucleic acid binding proteins. In this so-called sense approach the oligomer, either single- or double-stranded acts as a decoy and depletes the pool of the targeted protein thus preventing it to bind to its natural site and subsequently to play its biological role (Clusel et al, 1993). Up to now this strategy received limited attention probably due to the fact that nucleic acid binding proteins such as transcription factors frequently have pleiotropic effects. Combinatorial methods have considerably expanded the number and kind of targets for oligonucleotides. *In vitro* selection of RNA or DNA sequences in randomly synthesized libraries allows the identification of a ligand, termed aptamer, against any kind of molecules (Ellington and Conrad, 1995 ; Gold et al, 1995). This includes species involved in the control of gene expression: proteins, nucleic acid structure, ... It constitutes a very promissing strategy and offers an alternative to rational approaches (antisense, sense, ribozymes). We will briefly describe the salient characteristics of antisense oligonucleotides and aptamers.

6.2. Antisense oligoncleotides.

Since the first suggestion by Belikova and co-workers in 1967 (Belikova et al, 1967) that oligonucleotides could be used to target a complementary nucleic acid, thousands of papers have been devoted to the study of antisense oligomers. Two months ago the first therapeutic oligonucleotide was approved by FDA for the treatment of retinitis induced by cytomegalovirus infection in immunodeficient patients (Crooke, 1998). The use of unmodified DNA or RNA in living cells and organisms is limited due to their sensitivity to nucleases (Hélène and Toulmé, 1990). Chemists prepared hundreds of analogues which show increased lifetime compared to DNA (Milligan et al, 1993). The most widely used are phosphorothioate (PS) oligodeoxynucleotides in which one non-bridging oxygen of the phosphodiester linkage is substituted by a sulfur atom (Stein and Cheng, 1993). The anti-CMV oligonucleotide is actually a PS derivative, underlining the non-toxicity of these compounds. Several recently developped nuclease-resistant molecules (N3'→5' phosphoramidates, peptide-nucleic acids, 2'-methoxyethoxy) show in addition an increased affinity for RNA compared to unmodified nucleic acids. However, none of these high affinity antisense oligomers are able to prevent the elongation of the polypeptide chain. Any oligonucleotide targeted more than about 15 nucleotides downstream of the initiator AUG is displaced by the elongating ribosome (Gee et al, 1998). But these oligomers are efficient at inhibiting the initiation step of translation *i.e.* when the target sequence is located upstream of the initiator AUG (Larrouy et al, 1995 ; Boiziau et al, 1991). This holds true both for cap-dependent and -independent processes. Indeed the most efficient antisense oligonucleotides against the hepatitis C virus RNA are complementary to the Internal Ribosomal Entry Site, close to the initiation codon (Vidalin et al, 1996 ; Lima et al, 1997). In this case, the inhibition likely results from the competition between the antisense sequence and the initiation complex. A second mechanism accounts for antisense effects: an antisense oligodeoxynucleotide bound to an RNA sequence generates a substrate for ribonucleases H (RNases H), ubiquitous enzymes which cleave RNA in RNA-DNA duplexes (Crouch and Toulmé, 1998). In this case, of course, the location of the target sequence with respect to the initiator AUG is not important. PS oligos are the only modified analogues able to elicit RNase H activity. Hence the design of « sandwich » or « chimeric » oligomers in which a central part triggers RNase H whereas the two flanks are made with another chemistry (Larrouy et al, 1992 ; Giles and Tiddl, 1992). Such molecules combine different properties and might provide more efficient antisense sequences than uniformly modified ones.

6.3. Aptamers : a way for designing oligomers able to bind to an RNA structure.

Another limitation to the use of antisense oligonucleotides resides in the availability of the target sequence. Although RNA is made of a single chain, a large part of the molecule is not single-stranded due to secondary and tertiary structures. The competition between intramolecular (structure) and intermolecular (sense-antisense complex) base pairing will weaken antisense effects (Verspieren et la, 1990). Several possibilities are being explored to circumvent this problem (Toulmé et al, 1996). Very generally there is no need to precisely target a position of the RNA of interest. Therefore multiple attempts against different regions of the target RNA will reveal a sensitive site. However, as many RNA structures play a key role in the regulation of gene expression, it is worth to design oligomers able to bind to such structures in order to interfere with the regulatory processes. On one hand one can take advantage of weak parts in RNA structures such as bulges or loops. On the other hand one can increase the affinity of the oligonucleotide allowing it to invade RNA structures. This was achieved by using « selective binding complementary » nucleic acid bases (Kutyavin et al, 1996); oligomers containing 2-thio,T and 2 amino,A instead of conventional T and A bases were shown to disrupt an RNA structure and to inhibit *in vitro* translation more efficiently than unmodified molecules (Compagno et al., submitted).

As mentioned above the triplex strategy can also be used against an appropriate double-stranded region, for instance the stem of a hairpin. One may also think using an *in vitro* selection procedure (Ellington and Conrad, 1995 ; Gold et al, 1995) to identify an oligonucleotide able to selectively interact with a folded RNA motif. In this case no hypothesis is made about the interactions which will be involved in the complex. Neither does one need to know the actual structure of the target. In this SELEX strategy a library comprising up to 10^{14}-10^{15} different sequences (the candidates) is screened for the property of interest (let say the affinity for an RNA structure). In a first step the libray is mixed with the target in limiting concentration. The candidate/target complexes are isolated, for instance by affinity chromatography. Each candidate is made of a random region inserted between two fixed sequences. These latter regions are used to PCR-amplify the isolated candidates which are submitted to a second round of selection. Each selection-amplification cycle enriches the starting pool with respect to the desired property. After several rounds (generally 8 to 15) the candidates are cloned and sequenced. The selected aptamers are then synthesized and studied individually. Such a procedure allowed the identification of aptamers able to recognize nucleic acid hairpins (Mishra and Toulmé, 1994 ; Mishra et al, 1996 ; Boiziau et al, 1997). In the case of an imperfect RNA hairpin (the TAR RNA motif of HIV-1) the aptamers selected in both DNA and RNA libraries, displayed themselves a stem-loop structure. Interestingly one class of aptamers (in each library) was characterized by a consensus sequence located in the apical loop, which was complementary to the loop region of the target RNA hairpin. Both RNA and DNA aptamers could potentially form 6 base pairs with the target RNA but the base pairs were not the same (Toulmé et al, 1997). Moreover, the sequence at the top of the aptamer stem, next to the loop appears to play a key role for the formation of loop-loop complexes (Boiziau et al. submitted). In addition RNA aptamers which did not display a sequence complementary to the target RNA hairpin are able to bind with a high affinity to this hairpin, suggesting the involvement of non-canonical interactions (Ducongé and Toulmé, unpublished). The use of *in vitro* selection is not restricted to the identification of ligands for RNA structures. Numerous examples are described in several recent reviews.

Therefore, synthetic oligonucleotides offer multiple ways to generate artificial regulators of gene expression. Different strategies lead, through the high affinity binding of oligomers to various key target molecules, to the specific inhibition of the gene of interest. Antisense, triplex-forming oligonucleotides, ribozymes and aptamers constitute promising tools in biotechnology and potential therapeutic agents.

7. Conclusion

Our better understanding of the translation mechanism, of the role of structural features which are carried by every mRNA molecule and which govern either its capacity to be translated or its stability in the cytoplasm offers an astonishing and often unsuspected number of possibilities for the biotechnologist. He has at his disposal a large number of elements which can be used to design new vectors that will allow an improved translation efficiency and specificity. This also makes possible the design of new adapted cells expressing the regulatory factors which interact with the identified structural features of the mRNA. This will prove in the years to come to be of major interest specially in the field of gene therapy where the highest specificity is required in the expression of a therapeutical protein.

One can assume that in a near future, significantly improved gene therapy vectors will be designed by combining the use of specific promoters with the introduction of sequences which allow the translation of the mRNA to be modulated and/or its stability to be controlled.

Figure 4 gives a summary of the different tools which are already available for the construction of such vectors and adapted cells.
No doubt that we are only at the beginning of a novel area in biotechnology in which, one can assume that translation and mRNA stability will play a pivotal role.

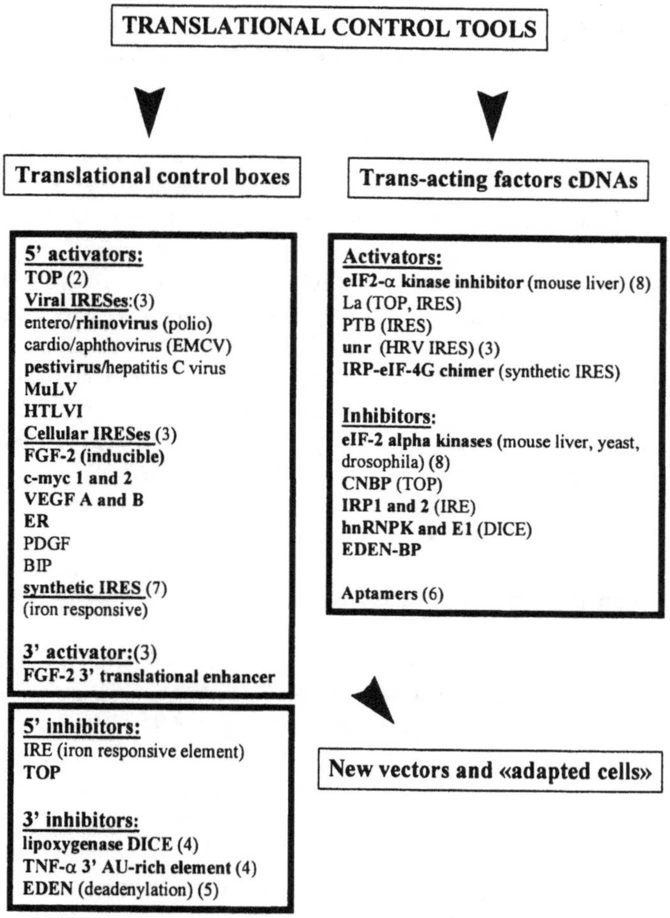

Figure 4 : Translational control tools already available to build new vectors and/or adapted cells. Numbers refer to chapters in this review except (7) Jackson R. and Hentze M., personal communication and (8) Santoyo J.et al (1997)

References

Agrawal, A. G. and Bowman, L. H. (1987). Transcriptional and translational regulation of ribosomal protein formation during mouse myoblast differentiation. *J. Biol. Chem.* **262**, 4868-4875.

Akiri, G., Nahari, D., Finkelstein, Y., Le, S. Y., Elroy-Stein, O., and Levi, B. Z. (1998). Regulation of vascular endothelial growth factor (VEGF) expression is mediated by internal initiation of translation and alternative initiation of transcription. *Oncogene* **17**, 227-36.

Aloni, R., Peleg, D. and Meyuhas, O. (1992). Selective translational control and nonspecific posttranscriptional regulation of ribosomal protein gene expression during development and regeneration of rat liver. *Mol. Cell. Biol.* **12**, 2203-2212.

Amaldi, F. and Pierandrei-Amaldi, P. (1997). TOP genes: a translationally controlled class of genes including those coding for ribosomal proteins. In *Cytoplasmic fate of eukaryotic mRNA* (Jeanteur, P., ed.), Vol. **18**, pp. 1-17. Springer-Verlag, Heidelberg.

Audic, Y., F. Omilli and H.B. Osborne. (1997) Postfertilization deadenylation of mRNAs in Xenopus laevis embryos is sufficient to cause their degradation at the blastula stage. *Mol. Cell. Biol.* **17**, 209-218.

Audic, Y., F. Omilli and H.B. Osborne. (1998) Embryo Deadenylation Element-Dependent Deadenylation is enhanced by a cis element containing AUU repeats. *Mol. Cell. Biol.* **18**: in press.

Avni, D., Shama, S., Loreni, F. and Meyuhas, O. (1994). Vertebrate mRNAs with a 5'-terminal pyrimidine tract are candidates for translational repression in quiescent cells: Characterization of the translational cis-regulatory element. *Mol. Cell. Biol.* **14**, 3822-3833.

Belikova, A. M., Zarytova, V. F. and Grineva, N. I. (1967) Synthesis of ribonucleosides and diribonucleoside phosphates containing 2-chloroethylamine and nitrogen mustard residues. *Tetrahedron Lett.* **37**, 3557-3562.

Berlioz, C., and Darlix, J. L. (1995). An internal ribosomal entry mechanism promotes translation of murine leukemia virus gag polyprotein precursors. *J Virol* **69**, 2214-22.

Berlioz, C., Torrent, C., and Darlix, J. L. (1995). An internal ribosomal entry signal in the rat VL30 region of the Harvey murine sarcoma virus leader and its use in dicistronic retroviral vectors. *J Virol.* **69**, 6400-7.

Bernstein, J., Sella, O., Le, S. Y., and Elroy-Stein, O. (1997). PDGF2/c-sis mRNA leader contains a differentiation-linked internal ribosomal entry site (D-IRES). *J Biol Chem.* **272**, 9356-62.

Bernstein, P., S.W. Peltz, and J. Ross. (1989) The Poly(A)-Poly(A)- Binding Protein Complex Is a Major Determinant on mRNA Stability In Vitro. *Mol. Cell. Biol.* **9**, 659-670.

Bilger, A., C.A. Fox, E. Wahle, and M. Wickens. (1994) Nuclear polyadenylation factors recognize cytoplasmic polyadenylation elements. *Genes Dev.* **8**, 1106-1116.

Binder, R., S.P.L. Hwang, R. Ratnasabapathy, and D.L. Williams. (1989) Degradation of Apolipoprotein II mRNA Occurs via Endonucleolytic Cleavage at 5'-AAU-3'/5'-UAA-3' Elements in Single-stranded Loop Domains of the 3'-Non-coding Region. *J. Biol. Chem.* **264**, 16910-16918.

Binder, R., J.A.Horowitz, J.P. Basillion, D.M. Koeller, R.D. Klausner and J.B.Harford. (1994) Evidence that the pathway of transferrin receptor mRNA degradation involves an endonucleolytic cleavage within the 3'UTR and does not involve poly(A) tail shortening. *EMBO J.* **13**, 1969-1980.

Boiziau, C., Dausse, E., Mishra, R., Ducongé, F. and Toulmé, J. J. (1997) Identification of aptamers against the DNA template for the HIV-1 TAR element. *Antisense Nucleic Acid Drug Dev* **7**, 369-380.

Boiziau, C., Kurfurst, R., Cazenave, C., Roig, V., Thuong, N. T. and Toulmé, J. J. (1991) Inhibition of translation initiation by antisense oligonucleotides via an RNase-H independent mechanism. *Nucleic Acids Res.* **19**, 1113-1119.

Borman, A. M., Le Mercier, P., Girard, M., and Kean, K. M. (1997). Comparison of picornaviral IRES-driven internal initiation of translation in cultured cells of different origins. *Nucleic Acids Res* **25**, 925-32.

Bouvet, P., F. Omilli, Y. Arlot-Bonnemains, V. Legagneux, C. Roghi, T. Bassez, and H.B. Osborne. (1994) The deadenylation conferred by the 3' untranslated region of a developmentally controlled mRNA in Xenopus embryos is switched to polyadenylation by deletion of a short sequence element. *Mol. Cell. Biol.* **14**, 1893-1900.

Bouvet, P., J. Paris, M. Philippe, and H.B. Osborne. (1991) Degradation of a Developmentally Regulated mRNA in Xenopus Embryos Is Controlled by the 3' Region and Requires the Translation of Another Maternal mRNA. *Mol. Cell. Biol.* **11**, 3115- 3124.

Brunda, M. J., Luistro, L., Warrier, R. R., Wright, R. B., Hubbard, B. R., Murphy, M., Wolf, S. F., and Gately, M. K. (1993). Antitumor and antimetastatic activity of interleukin 12 against murine tumors. *J Exp Med* **178**, 1223-30.

Cardinali, B., Di Cristina, M. and Pierandrei-Amaldi, P. (1993). Interaction of proteins with the mRNA for ribosomal protein L1 in *Xenopus*: structural characterization of in vivo complexes and identification of proteins that bind in vitro to its 5'UTR. *Nucl. Acids Res.* **21**, 2301-2308.

Caponigro, G. and R. Parker. (1995) Multiple functions for the poly(A)-binding protein in mRNA decapping and deadenylation in yeast. *Genes Dev.* **9**, 2421-2432.

Chen, L., Perlick, H., and Morgan, R. A. (1997). Comparison of retroviral and adeno-associated viral vectors designed to express human clotting factor IX. *Hum Gene Ther* **8**, 125-35.

Chen, C.Y.A., and A.B. Shyu. (1995) AU-rich elements: characterization and importance in mRNA degradation. *Trends Biochem. Sci.* **20**, 465-470.

Clusel, C., Ugarte, E., Enjolras, N., Vasseur, M. and Blumenfeld, M. (1993) Ex vivo regulation of specific gene expression by nanomolar concentration of double-stranded dumbbell oligonucleotides. *Nucleic Acids Res.* **21**, 3405-3411.

Christensen, A.K., L.E. Kahn, and C.M. Bourne. (1987) Circular polysomes predominate on the rough endoplasmic reticulum of somatotropes and mammotropes in the rat anterior pituitary. *Amer. J. Anat.* **178,** 1-10.

Couttet, P., M. Fromont-Racine, D. Steel, R. Pictet, and T. Grange. (1997) Messenger RNA deadenylation preceeds decapping in mammalian cells. *Proc. Natl. Acad. Sci. U. S. A.* **94,** 5628-5633.

Craig, A.W.B., A. Haghighat, A.T.K. Yu, and N. Sonenberg. (1998) Interaction of polyadenylate-binding protein with the eIF4g homologue PAIP enhances translation. *Nature* **392,** 520-523

Crooke, S. T. (1998) Vitravene^{TM}-Another piece in the mosaic. *Antisense Nucleic Acid Drug Dev.* **8,** 4,vii-viii.

Crouch, R. J. and Toulmé, J. J. (1998) "Ribonucleases H." Les Editions INSERM. Paris.

Decker, C.J., and R. Parker. (1993) A turnover pathway for both stable and unstable mRNAs in yeast : evidence for a requirement for deadenylation. *Genes Dev.* 7, 1632-1643.

Duval, C., P. Bouvet, F. Omilli, C. Roghi, C. Dorel, R. LeGuellec, J. Paris, and H.B. Osborne. (1990) Stability of maternal mRNA in Xenopus embryos: Role of transcription and translation. *Mol. Cell. Biol.* **10,** 4123-4129.

Ellington, A. D. and Conrad, R. (1995) Aptamers as potential nucleic acid pharmaceuticals. *Biochim. Biophys. Acta* **1049,** 99-125.

Elroy-Stein, O. (1998). The D-IRES of c-sis binds to hnRNPC which undergoes differentiation-induced hyper-phosphorylation (*Cold Spring Harbor, New York*)

Fan, X.C., and J.A. Steitz. (1998) Overexpression of HuR, a nuclear- cytoplasmic shuttling protein, increases the in vivo stability of ARE-containing mRNAs. *EMBO J.* **17,** 3448-3460.

Fox, C.A., and M. Wickens. (1990) Poly(A) removal during oocyte maturation : a default reaction selectively prevented by specific sequences in the 3' UTR of certain maternal mRNAs. *Genes Dev.* **4,** 2287-2298.

Fox, C.A., M.D. Sheets, and M. Wickens. (1989) Poly(A) addition during maturation of frog oocytes: Distinct nuclear and cytoplasmic activities and regulation by the sequence UUUUUAU. *Genes Dev.* **3,** 2151-2162

Ford, L.P., P.S. Bagga, and J. Wilusz. (1997) The poly(A) tail inhibits the assembly of a 3'- to -5' exonuclease in an in vitro RNA stability system. *Mol. Cell. Biol.* **17,** 398-406.

Fussenegger, M., Schlatter, S., Datwyler, D., Mazur, X., and Bailey, J. E. (1998). Controlled proliferation by multigene metabolic engineering enhances the productivity of Chinese hamster ovary cells [see comments]. *Nat Biotechnol* **16,** 468-72

Gallie, *D.R. (1991) The cap and poly(A) tail function synergistically to regulate mRNA translational efficiency. Genes Dev.* **5,** 2108-2116.

Gillian-Daniel, D.L., N.K. Gray, J. Astrom, A. Barkoff and M. Wickens. (1998) Modifications of the 5' Cap of mRNAs during Xenopus oocyte maturation; Independence from changes in Poly(A) length and impact on translation. *Mol. Cell. Biol.* **18,** 6152-6163.

Galy, B., Maret, A., Prats, A. C., and Prats, H. (1999). Cell transformation results in the loss of the density dependent translational regulation of FGF-2 isoforms expression. *Cancer Res.* **(in press).**

Gan, W., and Rhoads, R. E. (1996). Internal initiation of translation directed by the 5' untranslated region of the mRNA for eIF4G, a factor involved in the picornavirus-induced switch from cap-dependent to internal initiation. *J. Biol. Chem.* **271**, 623-626.

Gee, J. E., Robbins, I., van der Laan, A. C., van Boom, J. H., Colombier, C., Leng, M., Raible, A. M., Nelson, J. S. and Lebleu, B. (1998) Assessment of high-affinity hybridization, RNase H cleavage, and covalent linkage in translation arrest by antisense oligonucleotides. *Antisense Nucleic Acid Drug Dev.* **8**, 103-11.

Ghattas, I. R., Sanes, J. R., and Majors, J. E. (1991). The encephalomyocarditis virus internal ribosome entry site allows efficient coexpression of two genes from a recombinant provirus in cultured cells and in embryos. *Mol Cell Biol* **11**, 5848-59.

Giles, R. V. and Tidd, D. M. (1992) Enhanced RNase-H activity with methylphosphonodiester/ phosphodiester chimeric antisense oligodeoxynucleotides. *Anti-Cancer Drug Des.* 7, 37-48.

Gold, L., Polisky, B., Uhlenbeck, O. and Yarus, M. (1995) Diversity of oligonucleotide functions. *Ann. Rev. Biochem.* **64**, 763-797.

Greenberg M.E. and Belasco, J.G. (1993) Control of the Decay of Lbaile Protooncogenes and Cytokine mRNAs. In Control of Messenger RNA Stability, J. Belasco and G. Brawerman (Eds.) *Academic Press*, San Diego, USA. pp199-218

Gueydan C., L. Houzet, A. Marchant, A. Sels, G. Huez and V. Kruys (1996) Engagement of TNF mRNA by an endotoxin-inducible cytoplasmic protein. *Molecular Medicine* **2,** 479-488.

Gueydan C., L. Droogmans, P. Chalon, G. Huez, D. Caput, and V. Kruys. (in press) Identification of TIAR as a protein binding to TNF-α Mrna au-rich regulatory sequence. *J. Biol. Chem.*

Hake, L.E., and J.D. Richter. (1994) CPEB is a specificity factor that mediates cytoplasmic polyadenylation during Xenopus oocyte maturation. *Cell* **79,** 617-627.

Han, J., Brown, T. and Beutler, B. (1990) Endotoxin-responsive sequences control cachectin/tumor necrosis factor biosynthesis at the translational leve. *J. Exp. Med.* **171,** 465-475.

Hélène, C. and Toulmé, J. J. (1990) Specific regulation of gene expression by antisense, sense and antigene nucleic acids. *Biochim. Biophys. Acta* **1049**, 99-125.

Huez. G., G. Marbaix, E. Hubert, M. Leclercq, U. Nudel, H. Soreq, R. Salomon, B. Lebleu, M. Revel, and U.Z. Littauer, (1974). Role of the polyadenylate segment in the translation of globin messenger mRNA in Xenopus oocytes. *Proc. Natl. Acad. Sci. USA* **71,** 3143-3146.

Huez, I., Creancier, L., Audigier, S., Gensac, M. C., Prats, A. C., and Prats, H. (1998). Two independent internal ribosome entry sites are involved in translation initiation of vascular endothelial growth factor mRNA [In Process Citation]. *Mol Cell Biol* **18**, 6178-90.

Hunt, S. L., Hsuan, J. J., Totty, N., and Jackson, R. J. (1999). Unr, a cellular cytoplasmic RNA-binding protein with five cold shock domains, is required for internal initiation of translation of human rhinovirus RNA. *Genes & Development (in press).*

Hunt, S. L., and Jackson, R. J. (1999). Polypyrimidine-tract binding protein (PTB) is necessary, but not sufficient, for efficient internal initiation of translation of human rhinovirus-2 RNA. *RNA (in press).*

Jackson, R. J. (1991). mRNA translation. Initiation without an end. *Nature* **353**, 14-5.

Jackson, R. J., and Kaminski, A. (1995). Internal initiation of translation in eukaryotes: the picornavirus paradigm and beyond. *RNA* **1,** 985-1000.

Jang, S. K., Krausslich, H. G., Nicklin, M. J., Duke, G. M., Palmenberg, A. C., and Wimmer, E. (1988). A segment of the 5' nontranslated region of encephalomyocarditis virus RNA directs internal entry of ribosomes during in vitro translation. *J Virol* **62,** 2636-43.

Jarzembowski, J.A. and J.S. Malter. (1997) Cytoplasmic Fate of Eukaryotic mRNA: Identification and Characterization of AU-Binding Proteins. In Cytoplasmic Fate of Messenger RNA, P. Jeanteur (Ed), *Progress in Molecular and Subcellular Biology* Vol. **18**. pp141-172.

Jefferies, H. B., Fumagalli, S., Dennis, P. B., Reinhard, C., Pearson, R. B. and Thomas, G. (1997). Rapamycin suppresses 5'TOP mRNA translation through inhibition of p70s6k. *Embo J* **16**(12), 3693-704.

Jefferies, H. B., Reinhard, C., Kozma, S. C. and Thomas, G. (1994). Rapamycin selectively represses translation of the 'polypyrimidine tract' mRNA family. *Proc. Natl. Acad. Sci. USA* **91**, 4441-4445.

Kaminski, A., and Jackson, R. J. (1998). The polypyrimidine tract binding protein (PTB) requirement for internal initiation of translation of cardiovirus RNAs is conditional rather than absolute. *RNA* **4**, 626-638.

Kashani-Sabet, M. and Scanlon, K. J. (1995) Application of ribozymes to cancer gene therapy. *Cancer Gene Therapy* **2**, 213-223.

Kaspar, R. L., Kakegava, T., Cranston, H., Morris, D. R. and White, M. W. (1992). A regulatory cis element and a specific binding factor involved in the mitogenic control of murine ribosomal protein L32 translation. *J. Biol. Chem.* **267**, 508-514.

Kaufman, R. J., Davies, M. V., Wasley, L. C., and Michnick, D. (1991). Improved vectors for stable expression of foreign genes in mammalian cells by use of the untranslated leader sequence from EMC virus. Nucleic *Acids Res* **19,** 4485-90.

Körner, C.G., M. Wormington, M. Muckenthaler, S. Schneider, E. Dehlin and E. Wahle. (1998) The deadenylating nuclease (DAN) in involved in poly(A) tail removal during the meiotic maturation of Xenopus oocytes. *EMBO J.* **17,** 5427-5437.

Kozak, M. (1978). How do eucaryotic ribosomes select initiation regions in messenger RNA? *Cell* **15**, 1109-23.

Kruys V. and G. Huez. (1994) Translational control of cytokine expression by 3' UA-rich sequences. *Biochimie* **76**, 862-866.

Kruys, V., Marinx, O., Shaw, G., Deschamps, J. and Huez, G. (1989) Translational blockade imposed by cytokine-derived "UA-rich" sequences. *Science* **245**, 852-855.

Kutyavin, I. V., Rhinehart, R. L., Lukhtanov, E. A., Gorn, V. V., Meyer, R. B. and Gamper, H. B. (1996) Oligonucleotides containing 2-aminoadenine and 2-thiothymine act as selectively binding complementary agents. *Biochemistry* **35**, 11170-11176.

LaGrandeur, T.E., and R. Parker. (1996) mRNA decapping activities and their biological roles. *Biochimie* **78**, 1049-1055.

Laird-Offringa, I.A., C.L. De Wit, P. Elfferich, and A.J. Van der Eb. (1990) Poly(A) tail shortening is the translation-dependent step in c-myc mRNA degradation. *Mol. Cell. Biol.* **10,** 6132-6140.

Larrouy, B., Boiziau, C., Sproat, B. and Toulmé, J. J. (1995) High affinity or selectivity: a dilemma to design antisense oligodeoxynucleotides acting in RNaseH-containing media. *Nucleic Acids Res.* **23**, 3434-3440.

Larrouy, B., Blonski, C., Boiziau, C., Stuer, M., Moreau, S., Shire, D. and Toulmé, J. J. (1992) RNase-H mediated inhibition of translation by antisense oligodeoxynucleotides : a way to improve specificity. *Gene* **121**, 189-194.

Le, H., R.L. Tanguay, M.L. Balasta, C.C. Wei, Browning, K.S., A.M. Metz, D.J. Goss, and D.R. Gallie. (1997) Translation initiation factors eIF-iso4G and eIF-4B interact with the poly(A)-binding protein and increase its RNA binding activity. . *J. Biol. Chem.* **272,** 16247-16255.

Le Doan, T., Perrouault, L., Praseuth, D., Habhoub, N., Decout, J.-L., Thuong, N. T., Lhomme, J. and Hélène, C. (1987) Sequence-specific recognition, photocrosslinking and cleavage of the DNA double helix by an oligo-alpha-thymidylate covalently linked to an azidoproflavine derivative. *Nucleic Acids Res.* **19**, 7749-7760.

Legagneux, V., F. Omilli, and H.B. Osborne. (1995) Substrate- specific regulation of RNA deadenylation in Xenopus embryo and activated egg extracts. *RNA* **1,** 1001-1008.

Levy, S., Avni, D., Hariharan, N., Perry, R. P. and Meyuhas, O. (1991). Oligopyrimidine tract at the 5' end of mammalian ribosomal protein mRNAs is required for their translational control. *Proc. Natl. Acad. Sci. USA* **88,** 3319-3323.

Lewis T., C. Gueydan, G. Huez, J.J. Toulmé, and V. Kruys.(1998). Mapping of minimal AU-rich sequence required for LPS-induced binding of a 55 kD protein on TNF-α mRNA. T. *J. Biol. Chem.* **273**, 13781-13786.

Lima, W. F., BrownDriver, V., Fox, M., Hanecak, R. and Bruice, T. W. (1997) Combinatorial screening and rational optimization for hybridization to folded hepatitis C virus RNA of oligonucleotides with biological antisense activity. *J Biol Chem* **272**, 626-638.

Lopez-Lastra, M., Gabus, C., and Darlix, J. L. (1997). Characterization of an internal ribosomal entry segment within the 5' leader of avian reticuloendotheliosis virus type A RNA and development of novel MLV-REV-based retroviral vectors. *Hum Gene Ther* **8**, 1855-65.

Loreni, F. and Amaldi, F. (1992). Translational regulation of ribosomal protein synthesis in *Xenopus* cultured cells: mRNA relocation between polysomes and RNP during nutritional shifts. *Eur. J. Biochem.* **205**, 1027-1032.

Macejak, D. G., and Sarnow, P. (1991). Internal initiation of translation mediated by the 5' leader of a cellular mRNA [see comments]. *Nature* **353,** 90-4.

Maret, A., Galy, B., Arnaud, E., Bayard, F., and Prats, H. (1995). Inhibition of fibroblast growth factor 2 expression by antisense RNA induced a loss of the transformed phenotype in a human hepatoma cell line. *Cancer Res* **55,** 5075-9.

Mariottini, P. and Amaldi, F. (1990). The 5' untranslated region of mRNA for ribosomal protein S19 is involved in its translational regulation during *Xenopus* development. *Mol. Cell. Biol.* **10,** 816-822.

McGrew, L.L., E. Dworkin-Rastl, M.B. Dworkin, and J.D. Richter. (1989) Poly(A) elongation during Xenopus oocyte maturation is required for translational recruitment and is mediated by a short sequence element. *Genes Dv* **3(6),** 803-815.

Meerovitch, K., Pelletier, J., and Sonenberg, N. (1989). A cellular protein that binds to the 5'-noncoding region of poliovirus RNA: implications for internal translation initiation. *Genes Dev* **3,** 1026-34.

Meyuhas, O., Avni, D. and Shama, S. (1996). Translational control of ribosomal protein mRNAs in eukaryotes. In *Translational control* (Hershey, J. W. B., Mathews, M. B. & Sonenberg, N., eds.), pp. 363-388. Cold Spring Harbor *Laboratory Press, Cold Spring Harbor, N.Y.*

Miller, D. L., Dibbens, J. A., Damert, A., Risau, W., Vadas, M. A., and Goodall, G. J. (1998). The vascular endothelial growth factor mRNA contains an internal ribosome entry site [In Process Citation]. *FEBS Lett* **434,** 417-20.

Milligan, J. F., Matteucci, M. D. and Martin, J. C. (1993) Current concepts in antisense drug design. *J. Med. Chem.* **36,** 1923-1937.

Mishra, R. K. and Toulmé, J. J. (1994) In vitro selection of antisense oligonucleotides targeted to a hairpin structure. *C. R. Acad. Sci. Paris* **317,** 977-982.

Mishra, R. K., Le Tinévez, R. and Toulmé, J. J. (1996) Targeting nucleic acid secondary structures by antisense oligonucleotides designed through in vitro selection. *Proc. Natl. Acad. Sci. USA* **93,** 10679-10684.

Moser, H. E. and Dervan, P. B. (1987) Sequence-specific cleavage of double helical DNA by triple helix formation. *Science* **238,** 645-650.

Moss, B., Elroy-Stein, O., Mizukami, T., Alexander, W. A., and Fuerst, T. R. (1990). Product review. New mammalian expression vectors. *Nature* **348,** 91-92.

Müller, P. P., Kirchhoff, S., and Hauser, H. (1998). Sustained expression in proliferation controlled BHK-21 cells, in New developments and new applications in animal cell technology, (O.-W. Merten, Perrin, P, and and B. Griffiths, eds., eds.: Kluwer Academic Publishers).

Nanbru, C., Lafon, I., Audigier, S., Gensac, M. C., Vagner, S., Huez, G., and Prats, A. C. (1997). Alternative translation of the proto-oncogene c-myc by an internal ribosome entry site. *J Biol Chem* **272,** 32061-6.

Newport, J., and M. Kirschner. (1982) A Major Developmental Transition in Early Xenopus Embryos: II. Control of the Onset of Transcription. *Cell* **30,** 689-696.

Oh, S. K., Scott, M. P., and Sarnow, P. (1992). Homeotic gene Antennapedia mRNA contains 5'-noncoding sequences that confer translational initiation by internal ribosome binding. *Genes Dev* **6,** 1643-53.

Osborne, H.B. and J.D. Richter. (1997) Translational Control by Polyadenylation During Early Development. In Cytoplasmic Fate of Messenger RNA, P. Jeanteur *(Ed), Progress in Molecular and Subcellular Biology* Vol. **18**. pp173-198.

Ostareck-Lederer A. Ostareck, D.H., Standard, N., and Thiele, B.J.. (1994) Translation of 15-lipoxygenase mRNA is inhibited by a protein that binds to a repeated sequence in the 3' untranslated region. *EMBO J.* **13**, 1476-1481.

Ostareck D., Ostareck-Lederer, A., Wilm, M., Thiele, B.J., Mann, M., and Hentze, M. mRNA silencing in erythroid differentiation hnRNP K and hnRNP E1 regulate 15-lipoxygenase translation from the 3' end. . (1997) *Cell* **89**, 597-606.

Paillard, L., F. Omilli, V. Legagneux, T. Bassez, D. Maniey, and H.B. Osborne. (1998). EDEN and EDEN-BP, a cis element and an associated factor that mediates sequence-specific mRNA deadenylation in Xenopus embryos. *EMBO J.* **17,** 278-287.

Paillard, L., V. Legagneux, and H.B. Osborne. (1996) Poly(A) metabolism in Xenopus laevis embryos: Substrate-specific and default poly(A) nuclease activities are mediated by two distinct complexes. *Biochimie* **78,** 399-407.

Paris, J., H.B. Osborne, A. Couturier, R. LeGuellec, and M. Philippe. (1988) Changes in the polyadenylation of specific stable RNAs during the early development of Xenopus laevis. *Gene* **72,** 169- 176.

Pelletier, J., and Sonenberg, N. (1988). Internal initiation of translation of eukaryotic mRNA directed by a sequence derived from poliovirus RNA. *Nature* **334,** 320-5.

Pellizzoni, L., Cardinali, B., Lin-Marq, N., Mercanti, D. and Pierandrei-Amaldi, P. (1996). A *Xenopus laevis* homologue of the La autoantigen binds the pyrimidine tract of the 5'UTR of ribosomal protein mRNAs *in vitro*: implication of a protein factor in complex formation. *J. Mol. Biol.* **259**, 904-915.

Pellizzoni, L., Lotti, F., Maras, B. and Pierandrei-Amaldi, P. (1997). Cellular nucleic acid binding protein binds a conserved region of the 5'UTR of Xenopus laevis ribosomal protein mRNAs. *J. Mol. Biol.* **267**, 264-275.

Pellizzoni, L., Lotti, F., Rutjes, S. A. and Pierandrei-Amaldi, P. (1998). Involvement of the Xenopus laevis Ro60 autoantigen in the alternative interaction of La and CNBP proteins with the 5'UTR of L4 ribosomal protein mRNA. *J. Mol. Biol.* **281**, 593-608.

Peng, S.S., C.Y. Chen, N. Xu, and A.B. Shyu. (1998) RNA stabilization by the AU-rich element binding protein, HuR, an ELAV protein. *EMBO J* **17,** 3461-70

Peppel, K., and C. Baglioni. (1991) Deadenylation and turnover of interferon-ß mRNA. *J. Biol. Chem.* **266,** 6663-6666.

Pestova, T. V., Shatsky, I. N., Fletcher, S. P., Jackson, R. J., and Hellen, C. U. (1998). A prokaryotic-like mode of cytoplasmic eukaryotic ribosome binding to the initiation codon during internal translation initiation of hepatitis C and classical swine fever virus RNAs. *Genes Dev* **12,** 67-83.

Pierandrei-Amaldi, P., Beccari, E., Bozzoni, I. and Amaldi, F. (1985). Ribosomal protein production in normal and anucleolate *Xenopus* embryos: regulation at the posttranscriptional and translational levels. *Cell* **42,** 317-323.

Pierandrei-Amaldi, P., Campioni, N., Beccari, E. and Bozzoni, I. (1982). Expression of ribosomal protein genes in *Xenopus laevis* development. *Cell* **30**, 163-171.

Rapoport S.M., and Schewe T. (1986) The maturational breakdown of mitochondria in reticulocytes. *Biochem. Biophys. Acta* **864,** 471-495.

Roberts, R. W. and Crothers, D. M. (1992) Stability and properties of double and triple helices - dramatic effects of RNA or DNA backbone composition. *Science* **258,** 1463-1466.

Ross, J. (1993) mRNA decay in cell-free systems. In: Brawermann G. and Belasco, J. (Eds) , Control of Messenger RNA stability. *Academic Press, New York, NY.* pp 417-448.

Sachs, A.B., and R.W. Davis. (1989) The Poly(A) Binding Protein Is Required for Poly(A) Shortening and 60S Ribosomal Subunit- Dependent Translation Initiation. *Cell* **58,** 857-867.

Sachs, A.B., and S. Buratowski. (1997) Common themes in translational and transcriptional regulation. *Trends Biochem. Sci.* **22,** 189-192.

Santoyo, J., Alcalde, J., Mendez, R., Pulido, D. and de Haro, C. (1997) Cloning and characterization of a cDNA encoding a protein synthesis initiation factor-2α (eIF2α) kinase from *Drosophila melanogaster*. Homology to yeast GCN2 protein kinase. *J. Biol. Chem* **272**, 12544-12550.

Schiavi, S.C., J.G. Belasco, and M.E. Greenberg. (1992) Regulation of proto-oncogene mRNA stability. *Biochim. Biophys. Acta* **1114,** 95- 106.

Scott, P. (1993). IL-12: initiation cytokine for cell-mediated immunity. *Science* **260,** 496-497.

Simon, R., J.P. Tassan, and J.D. Richter. (1993) Translational control by poly(A) elongation during Xenopus development : differential repression and enhancement by a novel cytoplasmic polyadenylation element. *Genes Dev.* **6,** 2580-2591.

Simon, R., L. Wu, and J.D. Richter. (1996) Cytoplasmic polyadenylation of activin receptor mRNA and the control of pattern formation in Xenopus development. *Dev. Biol.* **179,** 239-250.

Stein, I., Itin, A., Einat, P., Skaliter, R., Grossman, Z., and Keshet, E. (1998). Translation of vascular endothelial growth factor mRNA by internal ribosome entry: implications for translation under hypoxia. *Mol Cell Biol* **18,** 3112-9.

Stein, C. A. and Cheng, Y. C. (1993) Antisense oligonucleotides as therapeutic agents - is the bullet really magical. *Sciencc* **261**, 1004-1012.

Stoneley, M., Paulin, F. E., Le Quesne, J. P., Chappell, S. A., and Willis, A. E. (1998). C-Myc 5' untranslated region contains an internal ribosome entry segment. *Oncogene* **16,** 423-8.

Sugimoto, Y., Sato, S., Tsukahara, S., Suzuki, M., Okochi, E., Gottesman, M. M., Pastan, I., and Tsuruo, T. (1997). Coexpression of a multidrug resistance gene (MDR1) and herpes simplex virus thymidine kinase gene in a bicistronic retroviral vector Ha-MDR- IRES-TK allows selective killing of MDR1-transduced human tumors transplanted in nude mice. *Cancer Gene Ther* **4**, 51-8.

Tahara, H., Zitvogel, L., Storkus, W. J., Zeh, H. J., 3rd, McKinney, T. G., Schreiber, R. D., Gubler, U., Robbins, P. D., and Lotze, M. T. (1995). Effective eradication of established murine tumors with IL-12 gene therapy using a polycistronic retroviral vector. *J Immunol* **154**, 6466-74.

Tarun, S.Z. and A.B. Sachs. (1996) Association of the yeast poly(A) tail binding protein with translation initiation factor eIF-4G. *EMBO J.* **15**, 7168-7177.

Teerink, H., Voorma, H. O., and Thomas, A. A. (1995). The human insulin-like growth factor II leader 1 contains an internal ribosomal entry site. *Biochim Biophys Acta* **1264**, 403-8.

Terada, N., Patel, H. R., Takase, K., Kohno, K., Nairn, A. C. and Gelfand, E. W. (1994). Rapamycin selectively inhibits translation of mRNAs encoding elongation factors and ribosomal proteins. *Proc. Natl. Acad. Sci. USA* **91** 11477-11481.

Thiele B.J., Andree, H., Höhne, M., and Rapoport, S.M.. (1981) Regulation of the synthesis of lipoxygenase in erythroid cells. *Acta Biol. Med. Germanica* **40**, 597-602.

Timchenko, L.T., J.W. Miller, N.A. Timchenko, D.R. DeVore, K.V. Datar, L. Lin, R. Roberts, C.T. Caskey and M.S. Swanson. (1996) Indentification of a (CUG)n triplet RNA-binding protein and its expression in myotonic dystrophy. *Nucl. Acids Res.* **24,** 4407-4414.

Tinévez, R. L., Mishra, R. K., and Toulmé, J. J. (1998). Selective inhibition of cell-free translation by oligonucleotides targeted to a mRNA structure. *Nucleic Acids Res . 26*, 2273-2278.

Torrent, C., Berlioz, C., and Darlix, J. L. (1996). Stable MLV-VL30 dicistronic retroviral vectors with a VL30 or MoMLV sequence promoting both packaging of genomic RNA and expression of the 3' cistron. *Hum Gene Ther* 7, 603-12.

Toulmé, J. J., and Giegé, R. (1998). Les aptamères : des ligands et des catalyseurs oligonucléotidiques obtenus par sélection in vitro. *Médecine Sciences* **14**, 155-166.

Toulmé, J. J. (1996) In Weiss, B. (ed.), Antisense oligodeoxynucleotides and antisense RNA as novel pharmacological and therapeutic agents. *CRC Press*, pp. 1-16.

Toulmé, J. J. (1996) "Historical aspects of antisense oligonucleotides."in Antisense oligodeoxynucleotides and antisense RNA as novel pharmacological and therapeutic agents. Weiss, B. (ed.), pp sous presse, CRC Press;

Toulmé, J. J. (1996) "Historical aspects of antisense oligonucleotides."in Antisense *Biotechnology Annual Review* **1**, 185-215.

Toulmé, J. J., Le Tinévez, R. and Brossalina, E. (1996) Targeting RNA structures by antisense oligonucleotides. *Biochimie* **78**, 663-673.

Toulmé, J. J., Le Tinévez, R., Boiziau, C. and Dausse, E. (1997) Rational and combinatorial strategies for designing oligonucleotides targeted to RNA structures. *Nucleic Acids Symp. Ser.* 39-41.

Toulmé, J. J. and Tidd, D. (1998) In Crouch, R. J., and Toulmé, J. J. (eds.), Ribonucleases *H. John Libbey*, Paris, pp. 225-250.

Uhlenbeck, O. C. (1987) A small catalytic oligoribonucleotide. *Nature* **328**, 596-600.

Vagner, S., Gensac, M. C., Maret, A., Bayard, F., Amalric, F., Prats, H., and Prats, A. C. (1995). Alternative translation of human fibroblast growth factor 2 mRNA occurs by internal entry of ribosomes. *Mol. Cell. Biol.* **15**, 35-44.

Vagner, S., Touriol, C., Galy, B., Audigier, S., Gensac, M. C., Amalric, F., Bayard, F., Prats, H., and Prats, A. C. (1996). Translation of CUG- but not AUG-initiated forms of human fibroblast growth factor 2 is activated in transformed and stressed cells. *J Cell Biol* **135**, 1391-402.

Vagner, S., Waysbort, A., Marenda, M., Gensac, M. C., Amalric, F., and Prats, A. C. (1995). Alternative translation initiation of the Moloney murine leukemia virus mRNA controlled by internal ribosome entry involving the p57/PTB splicing factor. *J Biol Chem* ***270***, 20376-83.

van Venrooij, W. J., Slobbe, R. L. and Pruijn, G. J. M. (1993). Structure and function of La and Ro RNPs. *Mol. Biol. Rep.* **18**, 113-119.

Varnum, S.M., and W.M. Wormington. (1990) Deadenylation of maternal mRNAs during Xenopus oocyte maturation does not require specific cis-sequences : a default mechanism for translational control. *Genes Dev.* **4,** 2278-2286.

Verspieren, P., Loreau, N., Thuong, N. T., Shire, D. and Toulmé, J. J. (1990) Effect of RNA secondary structure and modified bases on the in vitro inhibition of trypanosomatid protein synthesis by antisense oligodeoxynucleotides. *Nucleic Acids Res.* **18**, 4711-4717.

Vidalin, O., Major, M. E., Rayner, B., Imbach, J. L., Trepo, C. and Inchauspé, G. (1996) In vitro inhibition of hepatitis c virus gene expression by chemically modified antisense oligodeoxynucleotides. *Antimicrob Agents Chemother* **40**, 2337-2344.

Wang, X., M. Kiledjian, I.M. Weiss, and S.A. Liebhaber. (1995) Detection and characterization of a 3'untranslated region ribonucleoprotein complex associated with human (-globin mRNA stability. *Mol. Cell. Biol.* **15,** 1769-1777.

Wells, S.E., P.E. Hilner, R.D. Vale, and A.B. Sachs. (1998) Circularization of mRNA by Eukaryotic Translation Initiation Factors. Molecular. *Cell* **2,** 135-140.

Wickens, M., P. Anderson, and R.J. Jackson. (1997) Life and death in the cytoplasm: messages from the 3' end. Current opinion in *Genetics and Development.* **7,** 220-232.

Wilson, T., and R. Treisman. (1988) Removal of poly(A) and consequent degradation of c-fos mRNA facilitated by 3' AU-rich sequences. *Nature (London)* **336,** 396-399.

Ye, X., Fong, P., Iizuka, N., Choate, D., and Cavener, D. R. (1997). Ultrabithorax and Antennapedia 5' untranslated regions promote developmentally regulated internal translation initiation. *Mol. Cell. Biol.* **17,** 1714-1721.

Zamecnik, P. C. and Stephenson, M. L. (1978) Inhibition of Rous sarcoma virus replication and cell transformation by a specific oligodeoxynucleotide. *Proc. Natl. Acad. Sci. USA* **75**, 280-284.

Zhou, Y., Aran, J., Gottesman, M. M., and Pastan, I. (1998). Co-expression of human adenosine deaminase and multidrug resistance using a bicistronic retroviral vector. *Hum Gene Ther* **9**, 287-93.

Zitvogel, L., Tahara, H., Cai, Q., Storkus, W. J., Muller, G., Wolf, S. F., Gately, M., Robbins, P. D., and Lotze, M. T. (1994). Construction and characterization of retroviral vectors expressing biologically active human interleukin-12. *Hum Gene Ther* **5**, 1493-506.

Acknowledgements :

The work of J.J. Toulmé was supported by INSERM, the ANRS, the Conseil Régional d'Aquitaine and the Fondation pour la Recherche Médicale.
J.J. T. and A.C. & H. P. gratefully acknowledge their collaborators for sharing unpublished results.

The work of P. Pierandrei-Amaldi was supported by P.S. CNR « Post-translational control of gene expression ».

The work of H.B. Osborne & L. Paillard was supported by grants from the « Association pour la Recherche Contre le Cancer » and the « Ministère de la Recherche et Technology (ACC SV4).

The work of P. Pierandrei-Amaldi & B. Cardinal, A-C. & H. Prats, B. Osborne & L. Paillard, G. Huez & V. Kruys and J.J. Toulmé was supported by EC Biotech. Program Grant BIO4-CT95-0045.

MODULATION OF GENE EXPRESSION BY RIBOZYMES

MARA SZYRACH AND OLAF HEIDENREICH*

Department of Molecular Biology, Institute for Cell Biology, Eberhard-Karls-Universität Tübingen, Auf der Morgenstelle 15, 72076 Tübingen, Germany;
**Corresponding author, fax 49-7071-295459,*
e-mail: olaf.heidenreich@uni-tuebingen.de

1. Introduction

A reliable and specific modulation of gene expression in a cell would greatly facilitate the study of unknown gene functions, to combat pathogenic gene products or to control processes in biotechnology. One exciting option to achieve this goal is the application of antisense molecules such as antisense RNAs, antisense oligonucleotides or catalytic RNAs named ribozymes. All of these molecules bind to their target RNAs via complementary sequences, thereby forming base-paired duplices. Thus, in theory they permit a sequence-specific inhibition of gene expression. It is therefore not surprising that antisense approaches are increasingly used to understand the function of a given gene product, or to interfere with virus replication or cancerogenesis. Significant progress has been made in the design of antisense molecules, their delivery to cells, and in the understanding of their mode of action. However, compared to *in vitro* conditions, inside the cell the situation is much more complicated because of additional factors such as RNA-binding proteins and several intracellular compartments. Thus, in spite of the apparent simplicity of antisense approaches, many problems such as intracellular stability of the antisense molecule, target site selection, or colocalization of antisense molecule and target RNA remain to be solved. This review will address several of these problems and will discuss possible solutions. Due to space limitations, it is restricted to aspects of ribozyme applications. The reader interested in antisense RNA or oligonucleotides is referred to several reviews covering those topics (Stein and Cheng, 1993; Sczakiel, 1994; Heidenreich et al., 1995b; Nellen and Sczakiel, 1996).

2. Natural antisense RNAs and ribozymes

In nature, antisense RNA controls both eukaryotic and prokaryotic gene expression (Nellen and Sczakiel, 1996). In prokaryotes, antisense RNAs are usually short RNAs with specific stem-loop structures (Wagner and Simons, 1994). Extensive studies particularly of the CopA/CopT system in *E.coli* contributed to our understanding of antisense mechanisms. CopA RNA controls the replication of plasmid RI, thereby regulating its copy number in the bacterial cell. It binds to the CopT region of the repA mRNA and inhibits RepA expression post-transcriptionally. Since RepA is limiting for the initiation of plasmid replication, its CopA-mediated reduction lowers the plasmid copy number. Both CopA and CopT form stem-loop structures with two complementary loops. Base pairing is initiated between these two loops. However, the stems are not opened after formation of a short double strand (Malmgren et al., 1997). Instead, in a second step single-stranded regions adjacent to the stems form a stretch of some 35 base pairs. The formation of the perfect doublestrand by opening the stems is a very slow process *in vitro* and may not be important for the inhibition of RepA expression. Thus, the CopA/CopT complex consists of a discontinous double-stranded RNA interupted by the two opposing stems. Formation of such imperfect double-stranded regions between sense and antisense RNA may be a rule and not an exception in prokaryotes.

In eukaryotes, antisense RNAs belong to two different classes: Short RNAs of 20-100 nucleotides and long transcripts of one to several kb in length. An example for the first class is the lin-4 transcript from *Caenorhabditis elegans* (Lee et al., 1993; Wightman et al., 1993). Two different lin-4 transcripts of 20 and 45 nucleotides have been identified. Its target RNA encodes the Lin-14 protein which specifies the normal temporal sequence of all cell lineages during development. Downregulation of Lin-14 does not coincide with a decrease in its mRNA level. This indicates a posttranscriptional mechanism that does not involve the degradation of the lin-14 mRNA. The 3'-untranslated region (UTR) of the lin-14 mRNA contains nine short sequence stretches with partial complementarity to the lin-4 RNAs. The lin-4 transcripts can bind to these regions and form complexes with the lin-14 mRNA, which consists of two successive 5-9bp long double-strands, and which are interupted by bulges of similar length. A possible explanation for this is the avoidance of gene silencing which has been observed in *C. elegans* with long double-stranded RNAs (Fire et al., 1998).

The antisense RNA controlling EB4 expression in *Dictyostelium discoideum* belongs to the class of long antisense transcripts with perfect homology to their target RNAs (Hildebrandt and Nellen, 1992). The EB4 protein is associated with the membrane of prespore vesicles and is encoded by a 2.2 kb long mRNA which accumulates only during the aggregation stage. EB4 accumulation is regulated by a 1.8 kb long antisense RNA transcribed from a promoter probably located in an intron of the EB4 gene. The antisense transcript is short-lived; when it is not expressed, EB4 mRNA levels increase. This accumulation is not due to a competition between sense and antisense promoter, since both RNAs are transcribed simultaneously. The downregulation of EB4 mRNA is dependent on the synthesis of new antisense RNA due to the low stability of the antisense transcript. Since the inhibition of antisense RNA transcription prolongs the half-life of the mRNA, the antisense transcript induces the degradation of the sense transcript, probably by activating a double-strand-specific RNase activity. Similar regulation modes have been described e.g. for mammalian thymidine kinase expression (Sutterluety et al., 1998) and early polyoma virus gene expression (Liu et al., 1994).

Ribozymes are RNAs with inherent catalytic properties. RNA catalysis was first described for the RNA component of RNaseP (ribonuclease P) and the group I and group II introns. The self-splicing of group I introns, which is dependent on a guanosine cofactor and magnesium, was first described for the intron of the nuclear 26S rRNA gene in *Tetrahymena thermophila* (Cech, 1990). The excised intron can be converted into a true enzyme which acts *in trans* on its substrates. Group I introns have been manipulated to repair mutated RNAs by transsplicing (Lan et al., 1998).

Group II introns are also able to undergo self-splicing (Schmelzer and Schweyen, 1986). In contrast to group I introns, such introns do not need a nucleoside cofactor. Instead, they use an internal adenosine for catalysis. The reaction products are the ligated exons and the excised intron which forms a lariat.

RNase P is a ubiquitous endoribonuclease that processes the 5'-end of precursor tRNA molecules in the presence of a divalent metal ion as cofactor. RNase P consists of a protein and an RNA with the latter being the catalytic component (Altman, 1990). RNase P can be directed to cleave any RNA when the target is complexed to a short complementary oligonucleotide called external guide sequence (EGS) (Yuan et al., 1992).

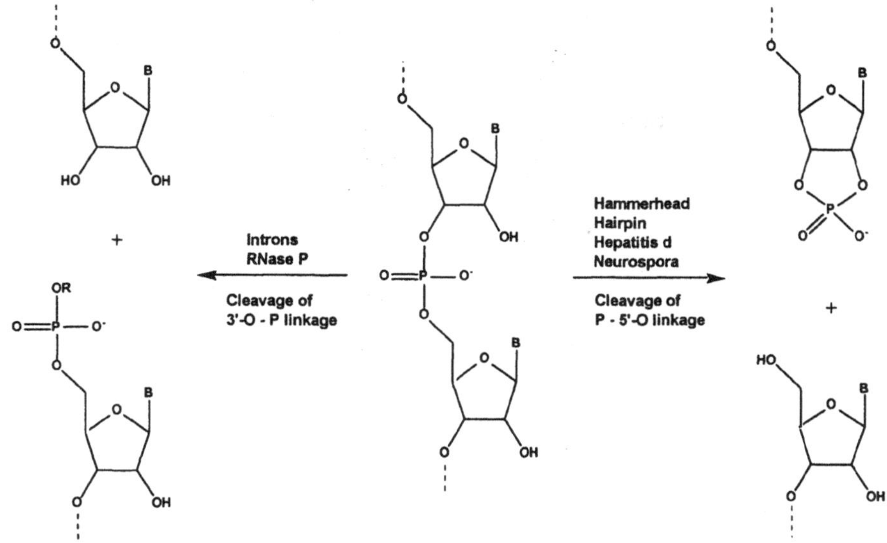

Figure 1. Ribozyme-mediated RNA cleavage
group I and II introns and RNase P cleave the linkage between the 3'-oxygen and the phosphorus. The cleavage products are a 3'-hydroxygroup and a 5'-phosphoester. R is for RNase P-mediated RNA cleavage a hydrogen, and the 3'- or 2'-C in the case of group I- or group II-mediated RNA cleavage, respectively. The small ribozymes cleave the linkage between phosphate and the 5'-oxygen yielding a 2'-3'-cyclophosphate and a 5'-hydroxy group.

Most of the ribozymes cleave and ligate RNA molecules. Due to their size and cleavage products, they can be divided into two classes. The first class contains the large group I and group II introns as well as the RNA moiety of RNase P. These ribozymes always cleave the linkage between phosphorus and the 3'-oxygen leading to the formation of a 3'-hydroxyl group (Cech, 1990; Altman, 1990; Schmelzer and Schweyen, 1986). In contrast, the small ribozymes of the hammerhead, hairpin and HdV type (fig. 2) as well as the VS RNA found in *Neurospora* mitochondria cleave the linkage between phosphorus and the 5'-oxygen (Symons, 1992). Members of this ribozyme class catalyse an attack of the 2'-oxygen on the phosphorus. The products of the RNA cleavage are a 5'-hydroxyl group and a 2',3'-cyclophosphate. Both the hairpin and the hammerhead ribozymes also catalyse the reverse reaction: They ligate the two RNA cleavage products by using the energy of the cyclic phosphate. These ribozyme structures are crucial for the replication of some viroids or viroid satellite RNAs (Symons, 1992). These plant pathogens are naked, singlestranded circular RNAs, which replicate by a "rolling-circle" mechanism. The circular (+) strand serves as a template for the synthesis of a linear concatemeric (-) strand RNA. Formation of the monomeric form proceeds by autocatalytic cleavage catalyzed by the ribozyme structure followed by self-ligation. The ribozyme is always part of the processed RNA. Thus, cleavage and ligation proceed in cis. The resulting circular (-) strand serves again as a template for the rolling circle replication and the entire process is repeated. The HdV ribozymes fulfill similar functions in the replication of this satellite virus of the Hepatitis B virus (Symons, 1992).

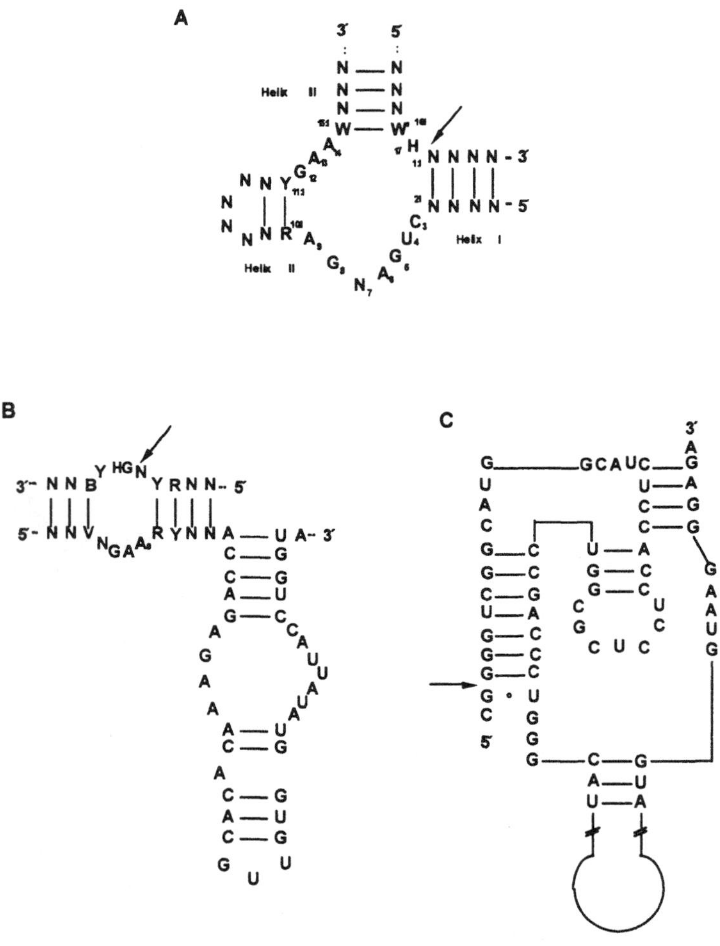

Figure 2. Small ribozymes. The secondary structure of the hammerhead (A), hairpin (B) and hepatitis δ ribozyme (C) is shown. The cleavage sites are indicated by arrows. The hammerhead core is numbered according to Hertel et al. (Hertel et al., 1992). The stem-loop of the hepatitis δ ribozyme is not conserved and only drawn as a line. A conserved GU basepair 5' of the cleavage site is indicated by a circle. R stands for A or G; Y for C or U; H for A, C or U; W for A or U, W' for the nucleotide complementary to W.

2. Ribozyme synthesis

Ribozymes are synthesised either by enzymatic synthesis using SP6, T3 or T7 RNA polymerase or, like oligodeoxynucleotides, by chemical synthesis using a DNA synthesizer. For enzymatic synthesis the ribozyme gene has to be placed behind the corresponding phage promoter, which itself has a length of some 20 nucleotides. Thus, the complete transcription unit can be as short as 60 nucleotides. Either linearised plasmid DNA or double-stranded oligonucleotides serve as templates for the transcription. The 3'-end of the ribozyme RNA is defined by the end of the oligonucleotide, or by the restriction cut in the plasmid DNA. Since the RNA polymerase will drop off the template at this site, this synthesis is also known as "run-off transcription". The protocols are very simple and can be performed in any molecular biology laboratory (Hoffman and Johnson, 1994). Run-off transcription allows the synthesis of very long RNAs of up to several kb in length. The yields are often in milligram quantities. One limitation of this method is that chemically modified nucleosides cannot be site-selectively incorporated. Moreover, only few modified nucleoside triphosphates such as 2'-amino- and 2'-fluoronucleoside triphosphates are efficiently incorporated, whereas others, such as 2'-deoxy- or 2'-alkoxynucleoside triphosphates, are very poor substrates for phage RNA polymerases (Aurup et al., 1992).

Ribozymes can be chemically synthesised as an alternative to run-off transcription. Using solid phase phosphoramidite chemistry, oligoribonucleotides of up to 50 nucleotides in length can be synthesised with reasonable yield (Gait et al., 1991; Wincott et al., 1995). However, compared to conventional DNA synthesis the 2'-hydroxyl group of the ribose moiety poses an additional problem for RNA synthesis. During synthesis this group must be blocked to prevent formation of 2'-5'-linkages or strand cleavage. After completion of the synthesis, the 2'-hydroxygroup must be deprotected under mild conditions to prevent RNA degradation. Because of these additional reaction steps, automated RNA synthesis is more expensive and achieves lower yields than DNA synthesis. In contrast to enzymatic ribozyme synthesis, chemical synthesis permits the site-selective incorporation of modified building blocks such as nucleoside moieties with sugar or base modifications, or altered internucleotydic linkages such as phosphorothioate or 3'-3' linkages. Because of these advantages, chemical synthesis has been widely used to investigate ribozyme cleavage mechanisms (Tuschl et al., 1995). Moreover, in this way hammerhead ribozymes can be protected against nucleases without affecting their catalytic activity, which is important for the exogenous delivery of ribozymes to cells (Heidenreich and Eckstein, 1997).

3. Structure and sequence requirements of hammerhead ribozymes

The hammerhead ribozyme consists of three stems connected by a core region (fig. 2). All core nucleotides except the one at position 7 are conserved. Introduction of any mutation severely impairs RNA cleavage. This core region contains two G-A non-Watson-Crick basepairs extending stem II, and, in addition, a U turn formed by the CUGA sequence. The resulting tertiary structure resembles a wishbone; the smallest angle is between stem I and II, and helix II stacks on helix III (Pley et al., 1994; Tuschl et al., 1994; Scott et al., 1995). Conventional trans cleaving ribozymes form stems I and III by binding to their target sequences (Haseloff and Gerlach, 1988). Until recently, it was assumed that the hammerhead ribozyme cleaves RNA only 3' of an NUH (H=A, C, U). Naturally occurring cleavage triplets are GUC, GUA and AUA with GUC being the most efficiently cleaved triplet. Changing GUC to any other cleavable triplet affects the cleavage rate k_{cat} and/or the Michaelis-Menten constant K_m (Shimayama et al., 1995; Zoumadakis and Tabler, 1995). For example, triplets with an A at the first position are cleaved with a dramatically increased K_m but with a similar k_{cat} compared to GUC. Uridine at the first position does not affect the K_m but decreases k_{cat}. With respect to k_{cat}, that is under substrate-saturating conditions, GUC, AUC and AUA belong to the more efficiently cleaved triplets. Under k_{cat}/K_m conditions, when the substrate concentration is much smaller than the K_m, the naturally occuring triplets AUA and GUA are only poorly cleaved. However, since hammerhead ribozymes are optimized by evolution to cleave RNA in cis, which corresponds to k_{cat} conditions, these two triplets should be cleaved reasonably well during viroid replication.

Several studies demonstrated that even the uridine of the NUH is not essential for hammerhead ribozyme-mediated RNA cleavage. GAC and GAA are also cleaved when the nucleoside at position 15.1 is a uridine instead of an adenosine (Kore et al., 1998). In addition, an inosine at this position changes the substrate specificity from NUH to NCH (Ludwig et al., 1998). Inosine-containing hammerhead ribozymes cannot be transcribed inside the cell, but can only be chemically synthesised. Therefore, the consensus sequence for hammerhead-catalysed RNA cleavage can be extended from NUH to NHH (Kore et al., 1998).

Recently, the group of Eckstein reported a ribozyme-catalysed cleavage 3' of AUG and of related triplets containing a 3'-terminal purine nucleoside (fig. 3) (Vaish et al., 1998). They obtained this ribozyme by *in vitro* selection starting with a randomised hammerhead sequence. The secondary structure of this purine nucleoside-specific ribozyme resembles the hammerhead structure. Despite this similarity, the ribozyme core sequence differs at several positions from the hammerhead core.

Furthermore, several positions in the stem-loop II seem to be essential for AUG cleavage. This is not the case for the hammerhead ribozyme, where only the orientation of the base pair closing stem II is conserved with a purine nucleoside at position 10.1 and a pyrimidine nucleoside at position 11.1 (fig. 2; Tuschl and Eckstein, 1993). This AUG cleaving ribozyme permits targeting of the starting codon of any mRNA. In conclusion, since the majority of triplets can be cleaved by ribozymes of the hammerhead type, the selection of target sites for hammerhead ribozymes is, at worst, hardly impaired by the presence or absence of cleavable triplets.

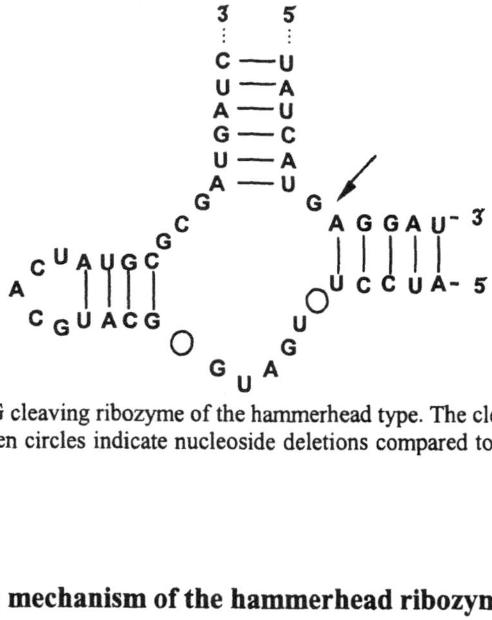

Figure 3. AUG cleaving ribozyme of the hammerhead type. The cleavage site is marked by an arrow. The open circles indicate nucleoside deletions compared to the wildtype hammerhead core.

4. Catalytic mechanism of the hammerhead ribozyme

The hammerhead ribozyme activity depends strongly on the presence of divalent metal ions such as Mg^{2+} or Mn^{2+} (Dahm and Uhlenbeck, 1991). Zn^{2+} supports RNA cleavage only in the presence of spermine as an additional counterion, suggesting that divalent metal ions are required for both RNA folding and catalysis (Dahm and Uhlenbeck, 1991). The concentration of magnesium ions influences the global structure of the hammerhead ribozyme. Electrophoretic data suggest a two step folding scheme. In the absence of magnesium ions the hammerhead is fully unfolded with the arms extended (Bassi et al., 1995). Addition of magnesium leads to a folding transition, in which stems II and III associate coaxially. At magnesium concentrations below 1 mM the smallest angle lies between stem

I and III. Increasing the magnesium concentration results in a conformational change with stem I approximating stem II. This conformation is in agreement with crystallography (Pley et al., 1994; Scott et al., 1995) and fluorescence resonance energy transfer studies (Tuschl et al., 1994) and, thus, very likely represents the active form. Furthermore, studies of the effect of pH changes on the apparent reaction rate showed a striking correlation between the pH optimum and the pK_a of the corresponding hydrated metal ion in water (Dahm et al., 1993). This finding suggests that a metal hydroxide ion is involved in hammerhead ribozyme catalysis. In this form, it triggers the nucleophilic attack of the 2'-oxygen on the phosphorus.

Since RNA cleavage results in the inversion of the phosphate configuration, it very likely proceeds according to an S_N2 mechanism (van Tol et al., 1990; Slim and Gait, 1991). The reaction intermediate is a pentacoordinated phosphate with the 2'-oxygen attacking from one side and the 5'-oxygen as the leaving group on the opposite side. Thus, the linkage between phosphorus and 5'-oxygen is cleaved at the same time as the new linkage between 2'-oxygen and phosphorus is formed, resulting in the 2',3'-cyclic phosphate and the 5'-hydroxyl group as reaction products.

The hairpin ribozyme also cleaves RNA by an S_N2 mechanism (van Tol et al., 1990). However, divalent metal ions are unlikely to participate directly in the reaction (Young et al., 1997). Instead, they mainly have a structural role. Interestingly, a recent report demonstrates that unphysiologically high concentrations of monovalent metal ions are sufficient for hammerhead self-cleavage, thereby confining the requirement of divalent metal ions for hammerhead self cleavage (Murray et al., 1998).

5. *In vitro* duplex formation between sense and antisense RNA

Inhibiting gene expression by ribozymes includes several crucial steps: colocalisation of the ribozyme and its target RNA, formation of an RNA duplex and cleavage of the target RNA. Selection of target sequences as well as choice of delivery and expression systems for ribozymes will affect these steps and, therefore, will contribute to the success or failure of a ribozyme strategy. So far, the best understood step is the formation of RNA duplices between sense and antisense RNAs, since it can be easily investigated under *in vitro* conditions. Hybridisation of target RNA and ribozyme may contain several reaction steps as shown in figure 4. First, one or both RNAs may have to perform a conformational change from a "closed" to a more open tertiary structure to permit complex formation. To start hybridisation, an initiation complex must form involving a few initiating base pairs. Starting from these base pairs, the double-stranded region must be extended to yield a stable ribozyme-target RNA complex. All three steps contribute to the apparent hybridisation or association rate k_{on}. Hybridisation is followed by RNA cleavage, which is again associated with conformational changes of the ribozyme-target RNA complex (see chapter 3). Particularly hybridisation has been extensively studied with antisense RNAs. Since those results are also of relevance for ribozymes they will be discussed here.

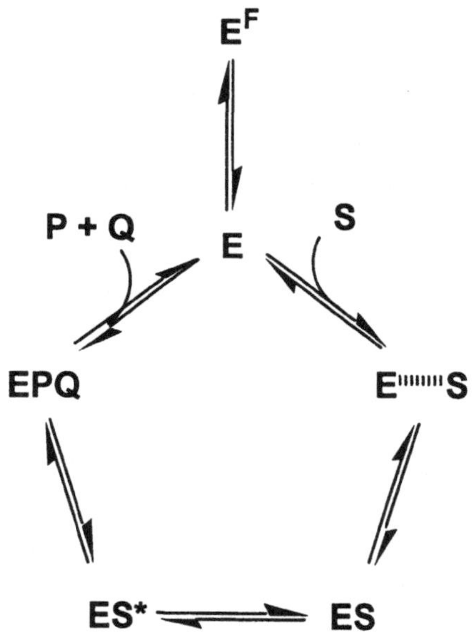

Figure 4. Ribozyme reaction scheme

A conformational change from a closed structure (E^F) to an open structure (E) allows the ribozyme to form with the substrate RNA S an initiation complex indicated by the ladder. After complete formation of the RNA duplices, the ribozyme-substrate complex ES undergoes another conformational change to reach the transition state indicated by ES*. RNA cleavage results in a ribozyme-product complex EPQ. After dissociation of the cleavage products P and Q, the ribozyme E is ready for another reaction round.

Several studies suggest that the apparent association rate k_{on} is crucial for an efficient inhibition of gene expression by an antisense RNA (Rittner et al., 1993; Patzel and Sczakiel, 1998). The fastest hybridisation rates - in this case with oligonucleotides - measured so far *in vitro* are in the range of 10^6-10^7 M^{-1} s^{-1} (Pörschke and Eigen, 1971). These apparent association rates are two orders of magnitude lower than those observed for the diffusion-controlled encounter between small substrates and protein enzymes (Fersht,

1985). An explanation for this relatively slow association rate is the formation of the first correct base pairs. This nucleation step involves fidelity, which results in a slower apparent k_{on}. Plainly spoken, fidelity costs time. Therefore, 10^6-10^7 M^{-1} s^{-1} may represent a theoretical threshold for the hybridisation of nucleic acids (Pörschke and Eigen, 1971).

In the CopA/CopT system both transcripts associate *in vitro* with a k_{on} of 10^6 M^{-1} s^{-1} thereby reaching this threshold (Persson et al., 1988). This is not surprising, since this system evolved by selection for optimal association between sense and antisense transcripts. Artificial antisense RNAs, however, bind often with much lower association rates to their target RNAs. For an antisense RNA targeted against the gag gene of HIV-1 an apparent k_{on} of $3x10^4$ M^{-1} s^{-1} was measured (Homann et al., 1993a). Despite this low value, a significant inhibition of HIV replication was observed in cell culture. Comparison of *in vitro* association rates with inhibition of HIV revealed that efficiently inhibiting antisense RNAs often have higher k_{on} values than less efficient ones (Rittner et al., 1993) suggesting that *in vitro* association rates may predict *ex vivo* (in cell culture) or *in vivo* antisense efficacy. For longer antisense RNAs the duplex formation is dependent on complementary loops in the sense and the antisense RNA, since introduction of point mutations into these loops significantly reduced the association rate (Homann et al., 1993a). Thus, the first interactions between sense and antisense transcripts may take place in these loop regions. This finding is reminiscent of the first interaction between CopA and CopT, where also two complementary loops make the initial contact (Wagner and Simons, 1994). Therefore, the tertiary structures of sense and antisense RNAs determine their association rates.

In enzyme reactions, the quotient of the reaction rate over the Michaelis-Menten constant, also called specificity constant k_{cat}/K_m, is a measure for the catalytic efficiency of an enzyme and represents a lower limit for the association of enzyme and substrate (Fersht, 1985). Thus, for ribozymes, association rates can be estimated by determining their catalytic efficiencies. Hammerhead ribozymes cleave short RNA substrates of 15-20 nucleotides in length with k_{cat}/K_m of up to 10^6 M^{-1} s^{-1} (Heidenreich and Eckstein, 1992). However, longer target RNAs are less efficiently cleaved. Published catalytic efficiencies of 500 to 10^4 M^{-1} s^{-1} are in a similar range as association rates reported for longer antisense RNAs (Heidenreich and Eckstein, 1992; Pace et al., 1994; Birikh et al., 1997).

6. Selection of efficient ribozymes

As discussed in the previous paragraphs, the choice of binding and cleavage sites on target RNAs for antisense RNAs or ribozymes is crucial for antisense efficacy. Several different strategies for site selection have been used with mixed success. One of the more common approaches is the use of computer programs, such as mfold which predict RNA secondary structure. However, particularly with ribozymes this approach often fails to predict RNA sequences accessible for binding by an antisense molecule. Sequences suggested to be single-stranded by mfold turned out to be almost uncleavable by ribozymes, whereas efficiently cleaved sites were predicted to be part of almost perfect double-strands (Birikh et al., 1997). A possible reason for these failures is the limitation of such programs to RNA secondary structure. Tertiary interactions of RNA structural elements are not taken into account. In spite of these limitations, a recent report used computer analysis based on mfold to design a set of fast-binding antisense RNAs targeted against the gag region of HIV-1 mRNAs (Patzel and Sczakiel, 1998). The authors searched in this study not for open structures in the target RNA but for relatively short antisense RNAs of 100 nucleotides in length containing large loops and single-stranded termini. Antisense sequences identified by this approach exhibited association rates of up to 10^6 M^{-1} s^{-1} and inhibited HIV replication by more than 90% (Patzel and Sczakiel, 1998). Therefore, RNA structure prediction by computer programs may be useful, at least for the design of shorter antisense RNAs and ribozymes (Denman, 1993).

An alternative approach for the identification of efficient antisense RNAs or ribozymes is the *in vitro* selection of such molecules from libraries. For such an approach no detailed knowledge is necessary about the target or antisense RNA structure. Several protocols have been published up to now ranging from identification of accessible sites on the target RNA to direct isolation of efficiently binding and cleaving antisense RNAs and ribozymes. One approach takes advantage of RNaseH-mediated RNA cleavage (Birikh et al., 1997). This RNase specifically degrades the RNA moiety of a DNA-RNA double-strand.The target RNA is incubated with a library consisting of 10 nucleotides long oligodeoxynucleotides of random sequence followed by digestion of the heteroduplices by RNase H. The cleavage products are separated on sequencing gels, and the cleavage sites are identified according to their lengthes. Accessible sites can then be targeted by ribozymes. Using this protocol, hammerhead ribozymes have been designed which cleave their target RNA with a cleavage efficiency of 10^4 M^{-1} s^{-1} (Birikh et al., 1997). In a variation of this approach, Scherr and Rossi (1998) targeted the endogenous DNA methyltransferase mRNA present in cell extracts with an oligodeoxynucleotide library. This selection protocol takes the role of RNA-binding proteins in RNA structure into consideration. In a somewhat more complicated approach, a hammerhead ribozyme library was incubated with the target RNA. The 5'-cleavage product was ligated to a oligo C tail which allowed the amplification of the cleavage product by PCR and its identification by sequencing of the amplified product. An anti-HGH ribozyme targeted against a site identified by this protocol reduced HGH expression by up to 90% in cell culture, as well as in a transgenic mouse system. A similar protocol using a hairpin ribozyme library has been published by the group of Hampel (Yu et al., 1998).

While the selection protocols described so far identify accessible sequences on the target RNA, they give only indirect information about the antisense molecule, in these cases the cleaving ribozyme. The actual sequence of the ribozyme can only be inferred from the target sequences flanking the cleavage site. Crucial information such as the lengths of the ribozyme antisense arms is not directly provided. A better approach would be the direct isolation of efficient antisense molecules by *in vitro* selection. The group of Sczakiel created an antisense RNA library by limited alkaline hydrolysis of an antisense RNA (Rittner et al., 1993). Since RNAs of different lengths will adopt different structures, the library contains RNA molecules with varying hybridisation capabilities. The target RNA was incubated with the library, and double-stranded RNA was separated from non-hybridised RNA by nondenaturing gel electrophoresis. Both the double-stranded and the single-stranded fractions were eluted and analysed by denaturing gel electrophoresis. Distinct clusters of fast and slow binding antisense lengthes were identified. Interestingly, all rapidly binding antisense RNAs examined were efficient inhibitors of HIV replication, whereas only very few slow binders exhibited a significant effect on virus proliferation (Rittner et al., 1993). A similar protocol has been applied to antisense RNAs containing ribozyme moieties (Kronenwett et al., 1996).

7. Specificity of ribozymes

The human genome consists of $3x10^9$ 4^{16} base pairs. Thus, to bind to a target sequence in average once, the antisense molecule needs a complementary stretch of 16 nucleotides. One might think, that longer antisense stretches increase the specificity, and, in addition, one has more options to select an efficiently binding antisense molecule. However, this is not necessarily the case. On the one hand, a longer antisense sequence may rapidly form a very stable complex with its target RNA. On the other hand, such stable complexes may well tolerate base mismatches possibly leading to the inhibition of genes related to the true target gene. One should keep in mind that "more isn't always better" (Herschlag, 1991). However, this somewhat over-simplifying remark does not consider potential differences in association of an antisense RNA with closely related target RNAs. Furthermore, RNA cleavage by e.g. hammerhead ribozymes is very sensitive to base mismatches directly flanking the cleavage site (Zoumadakis et al., 1994; Werner and Uhlenbeck, 1995), since such mismatches probably disturb the formation of a catalytically competent ribozyme-substrate complex. However, base mismatches located more distantly from the cleavage site affect RNA cleavage to a lesser extent (Zoumadakis et al., 1994; Werner and Uhlenbeck, 1995).

To achieve a reasonable compromise between efficiency and specificity it is necessary to know whether the target RNA has some closely related RNAs or not. If a viral RNA, such as HIV mRNAs, shall be targeted by an antisense approach, longer antisense sequences may be considered due to the probable intracellular singularity of the target transcript. In this case, the only *in vitro* parameters of interest are the association rate and the stability of the RNA duplex. In contrast, if a fusion gene product such as bcr-abl, or an mRNA carrying a point mutation such as Ha-ras mRNA is the target RNA, specificity may become a problem. In this case, short antisense RNAs or ribozymes should be applied to reduce or prevent inhibition of nonmutated allelic gene products.

8. RNA-binding proteins

Inside the cell, RNAs including ribozymes are complexed with proteins (Luzi et al., 1997). Since interaction with such proteins may change the RNA structure and interaction, it is not surprising that RNA-binding proteins have sometimes pronounced effects on antisense and ribozyme action. For quite a while it was assumed that RNA-binding proteins inhibit the binding of antisense molecules by competing with them for the RNA binding sites thereby decreasing the target RNA accessibility. However, recent reports suggest that at least some of these proteins such as HIV NCp7 (Tsuchihashi et al., 1993), hnRNP A1 (Bertrand and Rossi, 1994; Herschlag et al., 1994), glyceraldehyde 3-phosphate dehydrogenase (Sioud and Jespersen, 1996) or p53 (Nedbal et al., 1997) enhance the binding of antisense RNAs to, or the ribozyme-mediated cleavage of target RNAs. These enhancing effects seem to be due to either destabilisation of intramolecular RNA structure (RNA chaperone) or an increase in the probability of encounter between sense and antisense RNA by stabilisation of the initial pre-complex. Thus, these proteins facilitate the association of antisense and sense RNA (Bertrand and Rossi, 1994; Herschlag et al., 1994). In the case of ribozymes, the dissociation of ribozyme-substrate complexes is also enhanced, thereby increasing the ribozyme turnover (Herschlag et al., 1994). Interestingly, the same group described for hnRNP A1, an abundant protein involved in mRNA transport, a chaperone activity which reactivates misfolded ribozyme-substrate complexes. In addition, imperfectly paired ribozyme-substrate complexes due to single base substitutions in the substrate RNA were destabilised by hnRNP A1, thereby enhancing the fidelity of the ribozyme-mediated RNA cleavage. Enhancing the ribozyme catalysis by hnRNP A1 and by NCp7 is dependent on their concentrations. At higher protein concentrations these proteins inhibit ribozyme-mediated RNA cleavage. Thus, it is of interest to know, whether enhancement takes place inside the cell. We addressed this question by comparing ribozyme activity in cell nuclei with that in the absence of proteins (Heidenreich et al., 1995a). The ribozyme-mediated cleavage of an endogenous RNA proceeded more than thirtyfold faster than without protein. Thus, RNA binding proteins very likely support intracellular ribozyme catalysis.

Besides complex formation between antisense and target RNA, the degradation or modification of this complex is also dependent on RNA binding proteins. Double-stranded RNA is thought to be degraded by RNaseIII-like enzymes, and deamination of adenosines in RNA duplices is catalysed by adenosine deaminase. In addition to these direct effects on antisense and ribozyme action, RNA binding proteins will also exert more indirect effects affecting the stability and the intracellular localisation of antisense RNAs and ribozymes. For example, an unmodified anti-TNFα ribozyme is protected against degradation by intracellular nucleases by forming a ribozyme-protein complex which is still catalytically competent (Sioud, 1994). Introduction of protein recognition sites has also been employed for enhancing ribozyme catalysis. A hairpin ribozyme containing the binding site for the coat protein of bacteriophage R17 increased the RNA cleavage reaction twofold and the ligation tenfold (Sargueil et al., 1995).

9. Colocalization of ribozyme and target RNA

To inhibit gene expression, a ribozyme must be present in the same cellular compartments as its target RNA. Sullenger and Cech (1993) were the first to demonstrate the necessity for colocalisation of ribozyme and target RNA. A packaging cell line was first transduced with a lacZ encoding retroviral vector followed by a second transfection with a vector containing a hammerhead ribozyme targeted against *lacZ* mRNA. The lacZ transcript could be either translated in the cytosol, or, since it contained the Ψ packaging signal, packaged yielding recombinant virus. The authors assumed that the presence of the retroviral packaging signal on the ribozyme containing retroviral transcript results in colocalisation of the ribozyme with the lacZ RNA in the newly formed virus. They found that the viral titres released from cells containing the active ribozyme were reduced by 90% with no effect on intracellular β-galactosidase levels. Although packaged and translated mRNA may not have the same accessibility for ribozyme-mediated cleavage, this study strongly suggests that colocalisation of ribozyme with its substrate RNA is essential for the successfull inhibition of gene expression.

The Ψ element does not only control the encapsidation, but also causes dimerisation of retroviral RNA. Thus, due to its colocalising effect, a ribozyme transcript carrying the same Ψ site as the viral target RNA should be more effective in inhibiting retroviral replication than a ribozyme without this sequence. To prove this assumption, Pal et al. (1998) cotransfected cells with two retroviral MMLV vectors. One vector carried the neo gene, the second contained the hygro gene in combination with an anti-neo ribozyme. Both transcripts contained the MMLV Ψ element. To measure ribozyme efficacy, the number of hygromycin- and geneticin-resistant colonies was counted. In both packaging and non-packaging cell lines, the Ψ containing ribozyme RNA reduced the colony formation by up to 90%, whereas the ribozyme without the Ψ element caused only a twofold reduction. A catalytically inactive ribozyme with Ψ element also reduced colony formation twofold. Therefore, the reduction of geneticin-resistant colonys was dependent on both the Ψ site and a catalytically competent ribozyme.

In conclusion, optimisation of colocalisation of ribozyme and target RNA should improve intracellular ribozyme efficacy. Moreover, since it may not be necessary to flood the whole cell with ribozymes, such approaches may reduce or prevent unspecific effects unrelated to ribozyme-mediated RNA cleavage. Therefore, colocalisation strategies will contribute to the development of specific and efficient "gene shears".

10. Chemically modified ribozymes

Ribozymes are very susceptible towards degradation by nucleases. In analogy to antisense oligonucleotides, introduction of chemical modifications improves ribozyme stability against nucleases (fig. 5). However, such modifications may also impair ribozyme-mediated RNA cleavage. Introduction of modified nucleosides at certain positions in the hammerhead core diminishes its catalytic activity. Moreover, several modifications such as 2'-aminonucleosides destabilise RNA duplices thereby impairing binding of the ribozyme to its complementary RNA (Aurup et al., 1994). On the other hand, 2'-fluoro and 2'-methoxy groups have the opposite effect (Monia et al., 1993).

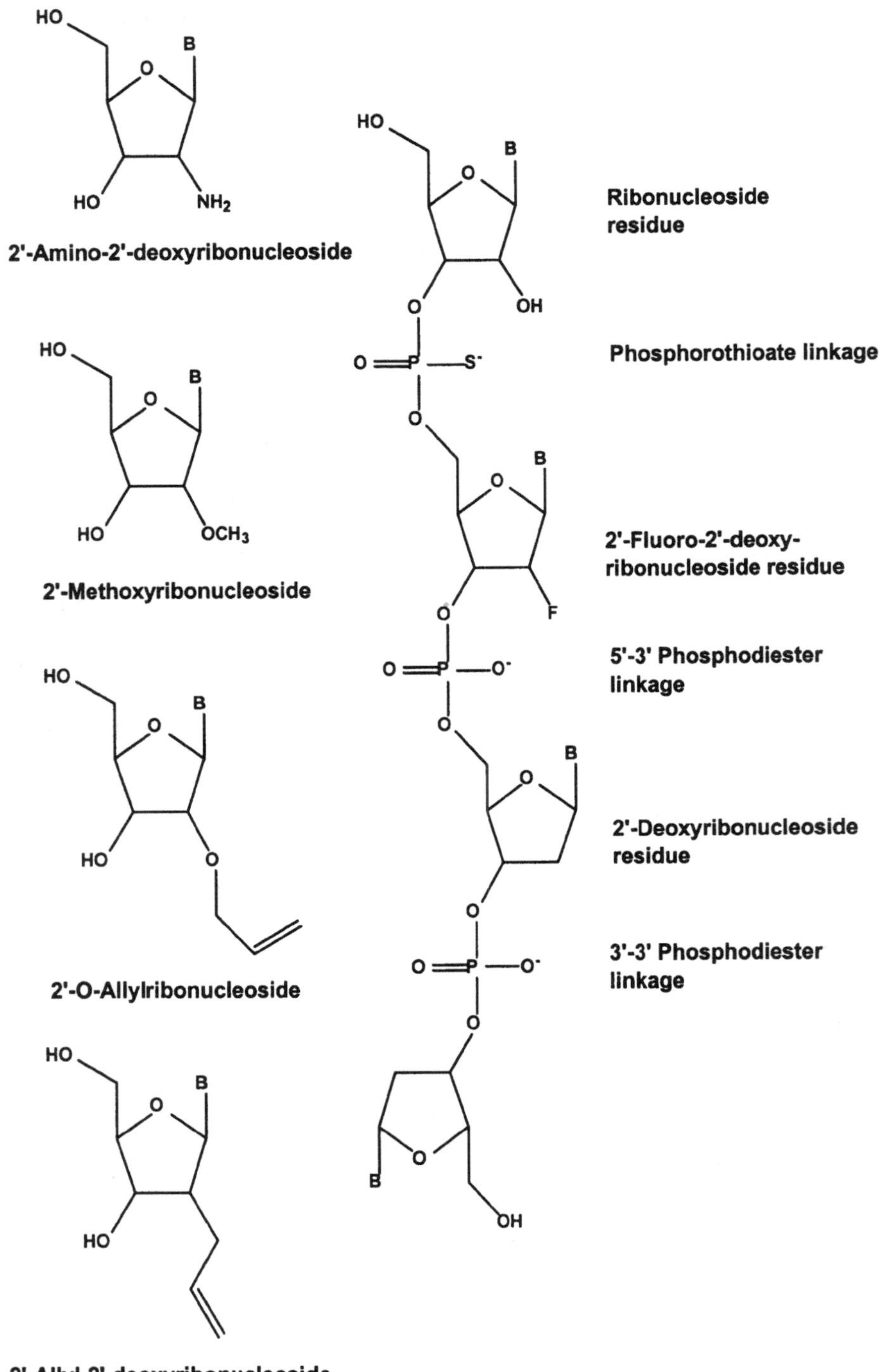

Figure 5. Chemical modifications for the stabilisation of ribozymes against nucleases.

Thus, in the extreme case, a "ribozyme" may be very stable but is incapable of cleaving RNA. Moreover, modifications such as phosphorothioates can cause unspecific and even toxic effects in cells (Heidenreich et al., 1995b). For these reasons, it is necessary to keep the extent of modifications at a minimum.

To avoid unnecessary modifications, the nucleases to be encountered by the ribozyme should be considered. The blood serum, for example, contains mainly pyrimidine-specific RNases and a 3'-exonuclease activity (Heidenreich et al., 1994). In cell nuclei, 5'- and 3'-exonuclease activities have been found (Heidenreich et al., 1996). The incorporation of terminal phosphorothioate linkages or inverted 3'-3'-linkages protects nucleic acids against degradation by exonucleases (Heidenreich et al., 1994; Heidenreich et al., 1996; Beigelman et al., 1995). In addition, when only incorporated at the termini, they do not affect the catalytic activity of the hammerhead ribozyme (Heidenreich and Eckstein, 1992).

Since the serum RNases need the 2'-hydroxyl group for RNA cleavage, replacement of this group by an alkoxy-, alkyl-, amino group, a fluorine or a hydrogen abolishes RNase-mediated degradation at this site. Selective introduction of sugar modifications in combination with terminal modifications increases the half-life of hammerhead ribozymes in blood serum from a few seconds to several days without seriously affecting their catalytic efficiency (Heidenreich et al., 1994; Beigelman et al., 1995; Scherr et al., 1997). Such modified ribozymes also have a superior stability in cell nuclei (Heidenreich et al., 1995a; Heidenreich et al., 1996). These results demonstrate the possibility to synthesise nuclease-resistant ribozymes without diminishing their catalytic efficiency.

11. Ribozyme delivery

The delivery of ribozymes to cells or tissues is crucial for the success of a ribozyme approach. Two basic strategies for ribozyme delivery can be distinguished: Exogenous delivery of preformed ribozymes and endogenous delivery of ribozyme genes using a vector system. Both strategies have similar problems such as low stability or unfavourable cellular localisation, but different tools dependent on the delivery mode can be applied to overcome them. The successfull delivery of ribozymes will result in the efficient cleavage of the intracellular target RNA.

11.1. EXOGENOUS RIBOZYME DELIVERY

To function as therapeutic agents, synthesised ribozymes must gain access to their target RNAs within the appropriate cellular compartment.

Both enzymatically and chemically synthesized ribozymes can be directly delivered to the cells and tissues (Marschall et al., 1994). The advantage of this method is that ribozymes can be chemically modified which dramatically increases their resistance against nucleases without compromising catalytic activity. Furthermore, exogenous delivery of ribozymes allows a transient downmodulation of gene expression without causing an immune response or insertional mutagenesis.

The cellular uptake of "naked" oligonucleotides including ribozymes by endocytosis is very low because of the negatively charged phosphate groups. In addition, the oligonucleotides are still enclosed in membrane vesicels that hinder the ribozymes to reach their target RNAs in the cytoplasm or the nucleus (Stein and Cheng, 1993). Consequently, the delivered oligonucleotide concentration has to be very high. Nevertheless, direct injection of chemically modified ribozymes has been successfully applied for transient inhibition of gene expression.

In most cases, ribozymes have been delivered using cationic lipids such as lipofectin (Sioud et al., 1992), DOTAP (Lange et al., 1993) or transfectam (Snyder et al., 1993). These cationic lipids contain quaternary amines or spermines as positively charged groups, which bind to the phosphate groups of e.g. the ribozyme. When fluorescently tagged ribozymes were delivered with cationic lipids to leukemic cells, the fluorescence was found in intracellular vesicles reminiscent of endosomes; faint cytoplasmic and nuclear staining was also observed (Lange et al., 1994). Thus, similar to oligodeoxynucleotides (Zabner et al., 1995), the ribozyme-lipid complexes are taken up by endocytosis with the majority of the complexes being trapped in endosomes. Only a small fraction of the ribozyme-lipid complex leaks into the cytoplasm and can find its substrate RNA. Nevertheless, cationic lipids as ribozyme carriers have already been successfully tested in cell culture as well as *in vivo*. However, while these reports demonstrate the feasibility of cationic lipids for delivering ribozymes to cells, there are several problems to solve. At high concentrations, cationic lipids are toxic to cells, so that application conditions must be carefully controlled and optimized (Felgner et al., 1994). In addition, many cationic lipid formulations show a significantly decreased efficacy in the presence of serum. Nevertheless, they are promising and fast-developing tools for oligonucleotides delivery (Gregoriadis, 1995; Lewis et al., 1996). Future generations of cationic lipids with improved stability and decreased cell toxicity may greatly facilitate ribozyme applications in therapeutic settings.

11.2. ENDOGENOUS RIBOZYME DELIVERY

Endogenous delivery of ribozymes requires a vector containing a ribozyme expression cassette. Such cassettes contain, in addition to the ribozyme-encoding sequence, promoters and termination sequences such as polyadenylation sites or Pol III termination sites. The promotor is a very important part of an expression cassette since it determines the tissue specificity and the level of ribozyme expression. Commonly, Pol II and Pol III promotors are used for ribozyme transcription.

Pol II promotors, such as the SV40 early promotor, the CMV, the β-actin promotor or viral LTRs derived from MMLV or RSV are commonly used for ribozyme expression. One disadvantage of conventional Pol II promotors is that transcripts not containing a functional open reading frame are rather unstable resulting in low intracellular ribozyme levels (Jacobson and Peltz, 1996). Alternatively, a U1 cassette may be used for ribozyme expression. The U1 promoter is also Pol II dependent and controls the expression of the U1 snRNA, a short transcript of 165 nucleotides in length. Several reports demonstrate the feasibility of U1 expression cassettes for ribozyme expression (De Young et al., 1994; Michienzi et al., 1996).

Pol III promotors, which naturally control the expression of e.g. tRNAs, U6 snRNA or the adenoviral VA1RNA, allow high-level and non-tissue-specific expression of short RNAs. They need only relatively short sequences to produce properly initiated and terminated transcripts. VA1 and tRNA genes contain internal promoters, whereas the U6 promoter is externally located like a conventional Pol II promoter. Another difference is the localisation of the transcripts with tRNAs and VA1 RNA accumulating in the cytoplasm and the U6 snRNA remaining in the nucleus (Bertrand et al., 1997). The nuclear localisation of the U6 transcript is dependent on the γ-methylated 5'-terminus. In contrast, U6 cassettes with certain mutations in their 5' termini are not methylated and are also found in the cytoplasm (Westaway et al., 1998).

The vector containing the ribozyme expression cassette can be either a plasmid, a retroviral vector, or a viral vector such as recombinant adenovirus or adeno-associated virus (AAV). In cell culture, transfection using plasmids is the easiest way for the endogenous delivery of ribozymes. Several methods, such as electroporation, calcium phosphate precipitation, or lipofection with cationic lipids permit the transfection of almost any cell line and of many primary cell types. For stable expression one can select for plasmid integration into the genome. However, integration proceeds randomly and is a rare event. If only a minor fraction of cells can be transfected, the selection of stably transfected cells can be extremely difficult and frustrating. Moreover, the expression levels will vary from clone to clone because of different integration sites. An alternative to this random integration approach is the use of plasmids which contain the epstein barr nuclear antigen 1 (EBNA 1) and the corresponding origin of replication. These vectors remain episomal; their copy number can be controlled by the selection pressure. Episomal vectors may permit selection of transfected cells at low levels of ribozyme expression. After initial selection, ribozyme expression may be increased by increasing selection pressure. Such an approach is very usefull for delivering ribozymes which target gene products essential for cell growth (Nason-Burchenal et al., 1998b).

Several gene therapy reviews discuss the advantages and disadvantages of the various viral vectors used for the delivery of genes to cells (Miller, 1992; Morgan and Anderson, 1993; Mulligan, 1993). Presently, the most frequently used vectors for *ex vivo* gene transfer are retroviral vectors. Ribozyme expression can be either controlled directly by the 5' long terminal repeat (LTR) (Zhou et al., 1994) or by an internal promoter (Yu et al., 1993). However, in the latter case promoter interference and silencing is frequently observed (Bertrand et al., 1997), whereas in the first case the long flanking sequences may diminish the ribozyme activity. Alternatively, ribozyme expression cassettes using a Pol III dependent promoter have been inserted into the downstream U3 region. Such cassettes produce ribozymes with short flanking sequences. Since during viral replication this region is copied to generate a functional 5'-LTR, such double-copy vectors yield high levels of ribozyme expression (Gervaix et al., 1997). For the production of infectious recombinant virions, a packaging cell line is transfected with the vector. The target cells are then transduced with the virions obtained from the packaging cell culture supernatant. The proviral sequence is always integrated into the host genome. This stable integration may result in long lasting ribozyme expression, but promoter silencing is frequently observed (Salmons and Günzburg, 1997). In addition, integration may also cause insertional mutagenesis or other genetic changes. Moreover, cell proliferation is essential for successful retroviral transduction based on MMLV-derived vectors. This poses a problem for many primary cells such as hematopoietic stem cells, where differentiation is linked to proliferation resulting in the loss of pluripotency, or neurons which do not divide anymore. Lentiviruses such as HIV can infect non-dividing cells. Currently, lentiviral vectors are developed to circumvent the problem of proliferation dependence (Zufferey et al., 1997).

Other viral vectors, such as adenoviruses and adeno-associated viruses (AAV) have been employed for gene therapy including the delivery of ribozyme genes (Feng et al., 1995; Lewin et al., 1998). Particularly adenoviral vectors are increasingly used for the endogenous delivery of ribozymes. In contrast to retroviral or AAV vectors, adenoviral DNA does not integrate into the host genome. Thus, the ribozyme is only transiently expressed. Hematopoietic tissues are very inefficiently infected because of the absence of receptors. One interesting exception are EBV transformed B cells which are susceptible to AdV infection (Huang et al., 1997). Another limitation for the *in vivo* application of recombinant AdV vectors is the rapid elimination of infected cells. The reason for this is the presence of viral genes on the vector whose expression causes immunogenic responses.

However, this problem may be solved by the recent development of adenoviral vectors deleted of all viral genes (Fisher et al., 1996).

12. Ribozyme effect, antisense effect or artefact?

At first sight, inhibition of gene expression by a ribozyme seems to be straightforward. The most often assumed case is that the ribozyme binds to its complementary RNA and cleaves it followed by the complete degradation of the target RNA by intracellular nucleases. This, however, may not be the case for several reasons. As already stated above, hammerhead ribozymes are metalloenzymes; they are dependent on divalent metal ions such as magnesium ions. The standard concentration of magnesium in cell-free cleavage assays is 10 mM. This value is in the range of the total cellular magnesium concentration. However, the majority of magnesium is complexed to proteins or phosphate groups of nucleotides or nucleic acids and is therefore not directly available to the ribozyme. The concentration of free, uncomplexed magnesium has been determined to be only 200 to 500 μM (Romani and Scarpa, 1992). Thus, a ribozyme may cleave its target RNA inefficiently inside the cell. Furthermore, ribozymes interact like other RNAs or oligodeoxynucleotides with many proteins (Sioud, 1994; Heidenreich et al., 1995b). Particularly, when biological effects are observed, one should consider a decoy function of the ribozyme in addition to a ribozyme-target RNA interaction. Finally, the observed reduction of target RNA may not be due to an intracellular cleavage, but instead may take place during RNA preparation (Beck and Nassal, 1995; Heidenreich et al., 1996). Since such artefacts have been described for RNA preparation protocols using magnesium-containing buffers, such protocols should be avoided. In addition, the copurified ribozyme may also cleave the target RNA during treatment with DNaseI or during reverse transcription, since these reactions are performed in the presence of magnesium (Scherr et al., 1997). Thus, apparent target RNA levels determined by RT-PCR should be interpreted with caution.

Which controls are necessary to examine a ribozyme function? In the case of endogenous delivery, one obvious control is the parental vector not containing the ribozyme sequence. In addition, ribozymes with scrambled arms and inactive ribozymes control for a decoy effect and the contribution of the RNA cleavage to the observed inhibition, respectively. Alternatively, the straight antisense sequence without the core region can be used. Notably, Kawasaki et al. (1996) reported, that a ribozyme inhibited the expression of its target gene significantly better than an inactive control ribozyme. However, the antisense RNA was as efficient as the catalytically competent ribozyme. Similar results have been reported by Gavin and Gupta (1997). The discrepancy between the antisense RNA and the inactive ribozyme is unclear, but may be due to different binding kinetics. These findings demonstrate, that ribozymes are not necessarily superior to simple antisense RNAs for the inhibition of gene expression. Interestingly, asymmetric ribozymes containing a long antisense arm have been shown to be fivefold more efficient in inhibiting HIV replication than the corresponding antisense RNA (Homann et al., 1993b). In this study, similar binding kinetics for both antisense and ribozyme RNA were measured. This result is somewhat surprising givien the fact that such long antisense RNAs bind tightly to their complementary RNA and inhibit efficiently HIV even by themselves. To explain this result, Szakiel and Goody (1994) suggested a duplex-destabilising helicase activity which reduces the antisense effect. In such a scenario the irreversible RNA cleavage by the ribozyme moiety could significantly improve the inhibition of gene expression.

13. Ribozyme Applications in Cell Culture

Ribozymes are increasingly used to establish gene functions and to inhibit the expression of pathogenic gene products. However, for reasons discussed in the previous chapter, particularly for functional studies, the specificity and efficacy of a given ribozyme must be carefully tested both at the RNA and the protein level. Checking a ribozyme effect on the amount of synthesized target protein is dependent on the availability of sufficiently specific antiserum. If no antibody is available, this becomes a difficult task. One option is the use of a fusion gene, in which the ribozyme target sequence is connected to reporter genes, such as the luciferase gene or the green fluorescent protein gene. If the target site locates in an open reading frame, it can be cloned in frame with respect to the reporter. Thus, the intracellular ribozyme efficacy can be easily followed by measuring the reporter activity, for example as light units in the case of luciferase. The wildtype reporter can serve as a control for ribozyme specificity. However, it should be kept in mind, that the overall RNA structure of such fusion constructs may differ from that of the endogenous target RNA. Therefore, such fusion genes should only be used for either rapidly comparing the intracellular activity of several ribozymes or, as an additional proof of activity, in combination with data showing reduction of endogenous target RNA levels. Several reports demonstrate the feasibility of such an approach. The *ex vivo* activity of chemically modified hammerhead ribozymes targeted against N-ras was examined using an N-ras-luciferase fusion gene (Scherr et al., 1997). A 2'-fluoropyrimidine nucleoside substituted ribozyme reduced the luciferase activity by almost 60%. In this case, the degradation of N-ras RNA was confirmed by a semi-quantitative RT-PCR (Scherr et al., 1997). To investigate the intracellular activity of anti-hepatitis C virus ribozymes, the 5'- untranslated region and part of the core region was fused in frame with the luciferase gene (Sakamoto et al., 1996). The ribozyme expression was controlled by a CMV promoter. The ribozymes reduced the luciferase activity by 40-50%, when cotransfected with the reporter plasmid. In addition, transfection of the reporter plasmid into cells stably expressing a ribozyme yielded similarly reduced luciferase activities compared to non-expressing control cells (Sakamoto et al., 1996).

An excellent example of a ribozyme approach to establish a gene function has been recently published by Kawasaki et al. (1998). CBP and p300 are transcriptional coactivators, which are involved in cell growth and differentiation. Both proteins contain a histone acetylase activity thereby facilitating transcription. However, functional differences between CBP and p300 were unknown. This question was addressed by targeting both genes with specific hammerhead ribozymes. The ribozymes were endogenously expressed with a tRNA expression vector, where ribozyme sequences were cloned between the RNA polymerase III-dependent $tRNA^{Val}$ promoter and the termination signal. For an initial intracellular screening, the authors used a luciferase assay as discussed in the previous paragraph (Kawasaki et al., 1996). Catalytically disabled ribozymes served as controls. In addition, mutated p300 and CBP cDNAs carrying a translationally silent point mutation in the cleavage triplet, which prevented ribozyme-mediated RNA cleavage, were used for rescue experiments (Kawasaki et al., 1998). This study established for the first time different functions for p300 and CBP in cell differentiation, apoptosis and cell cycle control.

Promising targets for ribozymes are genes essential for oncogenesis. These potential target genes can be divided into two groups. Oncogenes resulting from point mutations or gene fusions within the mRNA provide unique targets for a ribozyme approach. In this case, the presence of ribozymes in nonpathogenic cells should not matter. Examples for such oncogenes are the various ras forms or fusion genes such as bcr-abl. However, when the target gene is also expressed in normal cells, ribozyme delivery and expression should be restricted to the pathogenic tissue. For example, protein kinase C (PKC) is involved in many biological effects such as cell proliferation, differentiation and tumour-promotion. Targeting a ribozyme against PKCα suggested, that this kinase isoform is critical for glioma cell proliferation and survival (Sioud and Sørensen, 1998). In this study, an anti-PKCα-ribozyme complexed to the cationic lipid DOTAP was directly injected into the tumour. In contrast to untreated tumours, a single ribozyme dose caused an almost complete block of tumour growth. Interestingly, an inactive ribozyme control was less efficient than the active ribozyme. Thus, targeted delivery to pathogenic tissue by, for example, direct injection may provide a new opportunity of modulating gene expression of genes essential for both normal cell function and oncogenesis.

Ras oncogenes contain point mutations leading to a constitutively active status of the corresponding Ras protein. Ras is essential for cell proliferation and its oncogenic forms can contribute to cell transformation. Ha-ras contain a point mutation in codon 12 changing a GGC triplet to a GUC, which can be cleaved by a conventional hammerhead ribozyme. Cells carrying Ha-ras were either transfected with a ribozyme-encoding plasmid or were infected with adenoviral vectors (Kashani-Sabet et al., 1994; Feng et al., 1995). In both approaches ribozyme expression was driven by a β-actin promoter. Transfection of Ha-ras transformed NIH3T3 with anti-Ha-ras ribozyme plasmid reduced both Ha-ras mRNA and protein which was not the case with an inactive ribozyme control (Kashani-Sabet et al., 1994). Since Ki-ras mRNA and protein levels were not affected by the ribozyme, the observed inhibition was sequence-specific. Phenotypically, the ribozyme-expressing cells had a similar morphology and proliferation rate as untransformed NIH3T3 cells. In an extension of this study, the ribozyme cassette was cloned into an adenoviral vector, and human bladder carcinoma cells were infected with the corresponding recombinant virus (Feng et al., 1995). Infection with ribozyme encoding virus strongly reduced Ha-ras mRNA levels and abolished cell proliferation, whereas several control viruses had no effect on either mRNA levels or proliferation. Injection of ribozyme-infected cells into nude mice did not form tumours in contrast to uninfected cells or cells infected with control constructs. N-ras carrying point mutations at codon 12 or 13 is the major oncogenic form of Ras found in leukemia. A GC transversion in codon 13 changes a GUG to a GUC, which can be cleaved by hammerhead ribozymes (Scherr et al., 1997). Since generally leukemic cells cannot be efficiently transduced with recombinant adenoviruses, a retroviral approach was chosen for delivery of anti-N^{13}-ras ribozymes (Scherr et al., 1998). The long terminal repeat of the retroviral vector controlled the ribozyme expression. Thus, the ribozyme sequence was part of the complete viral transcript. However, the long flanking sequences did not affect the catalytic efficiency of this long ribozyme in comparison with the short, parental ribozyme. Infection of N-ras transformed hematopoietic cells with the ribozyme-encoding retrovirus reduced N-ras mRNA levels tenfold, restored IL3 dependent cell proliferation and, at least partially, the erythropoietin dependent cell differentiation. Parental vector or scrambled ribozyme controls did not affect any of these parameters. Therefore, such a ribozyme-encoding retroviral vector may be a suitable gene therapy approach against N-ras-dependent oncogenesis.

Leukemias are frequently associated with chromosomal rearrangements such as translocations. The translocation t(9;22) is detected in 95% of all cases of chronic myeloid leukemia (CML) yielding the Philadelphia chromosome. This translocation fuses the 5'-region of bcr encoding a phosphotransferase to the abl gene, which codes for a tyrosine kinase. Two bcr-abl fusion sites occur dependent on the chromosomal break point. Either, the second or the third bcr exon is fused to the second exon of abl leading to the B2-A2 or the B3-A2 form, respectively. These fusion RNAs are only expressed in leukemic cells and are therefore an ideal target for a ribozyme. Particularly the B3-A2 form has been frequently targeted with hammerhead ribozymes, since a cleavable GUU is in close proximity to the fusion site. Both endogenously, as well as exogenously delivered ribozymes have been tested. One group examined the effect of ribozymes containing a 5'-cap structure as sole modification (Lange et al., 1993; Lange et al., 1994). A second group transfected chimeric ribozymes containing 2'-deoxynucleotides in the stems and ribonucleotides in the core region (Snyder et al., 1993). Transfection with active ribozymes resulted in a decrease in cell proliferation, bcr-abl mRNA levels and bcr-abl protein and kinase activity, respectively. Inactive control ribozymes had only minimal effects on these parameters. Similar results were obtained with a multi-unit ribozyme targeting three neighboured cleavage sites surrounding the B3-A2 fusion site (Leopold et al., 1995). The B3-A2 transformed cell line K562 was also infected with retroviral vectors encoding anti-B3-A2 ribozyme. The β-actin, the thymidine kinase and the $tRNA^{Met}$ promoter were used to express the ribozyme. The first two expression cassettes were located between the LTRs, whereas the tRNA ribozyme cassette was cloned into the U3 region of the LTR resulting in a double copy vector. Only the latter construct caused a reduction of kinase activity in a majority of cell clones. Thus, at least in this study, the double copy vector containing a pol III promoter was superior to the internal cassette driven by pol II promoters.

The translocation t(15;17) associated with acute promyelocytic leukemia (APL) yields the PML/RARα fusion protein. The function of this fusion protein in leukemogenesis was addressed by targeting a hammerhead ribozyme against the corresponding fusion message (Nason-Burchenal et al., 1998b; Nason-Burchenal et al., 1998a). To express the ribozyme in a leukemic cell line, it was cloned into an episomal vector carrying the nuclear antigen-gene EBNA 1 and the Epstein Barr virus ori. Ribozyme-expressing cells were selected at a low concentration of hygromycin, since with increasing hygromycin concentrations cell growth was suppressed (Nason-Burchenal et al., 1998b). The growth suppression was attributed to an increase in apoptosis. The growth suppression was paralleled by a decreased level of PML/RARα fusion protein. Interestingly, the inactive control ribozyme had no effects on protein levels, cell proliferation or apoptosis. However, RARα levels expressed from the nontranslocated allele were also reduced. Since the cleavage site is located in the RARa part of the fusion message some 4 nucleotides downstream of the fusion site, the normal RARα mRNA may also be cleaved by the ribozyme (Pace et al., 1994). Therefore, the observed antileukemic effects can be attributed to the ribozyme function, but it remains open, whether these effects are solely due to the inhibition of PML/RARα.

Pleiotrophin is a polypeptide growth factor which contributes to the development and maintenance of normal tissue. It is essential as an autocrine factor for tumour growth and metastasis since it supports angiogenesis in expanding tumours. In two studies, Czubayko and coworkers inhibited pleiotrophin expression using a hammerhead ribozyme (Czubayko et al., 1994; Czubayko et al., 1996). Ribozyme expression was driven by the CMV promoter. Stable transfection of two melanoma cell lines yielded clones with strongly reduced pleiotrophin mRNA and protein levels. In addition, one melanoma line showed in reduced soft agar colony formation (Czubayko et al., 1994). *Ex vivo* colony formation of the second line was not affected indicating that this line is not dependent on pleiotrophin as a growth factor (Czubayko et al., 1996). However, when ribozyme-expressing clones of the second line were injected into nude mice, tumour growth and angiogenesis were decreased and apoptosis was increased. Concomitantly, metastasis in the lung was prevented. In an extension of these two studies, transduction of several cancer cell lines with ribozyme-encoding adenoviral vectors caused a strongly reduced proliferation (Czubayko et al., 1997).

Multidrug resistance is a great challenge for the chemotherapy of cancer, which can be often attributed to an increased expression of a glycosylated transporter protein, MDR1. In several approaches, endogenously expressed ribozymes were used to downmodulate MDR expression (Holm et al., 1994; Scanlon et al., 1994; Matsushita et al., 1998). For ribozyme expression, all groups used the β-actin promoter. The ribozyme inhibited MDR1 expression and increased the sensitivity of carcinoma cells towards cytostatic drugs such as daunorubicin or actinomycin D (Holm et al., 1994). Interestingly, an anti-fos ribozym also restored drug sensitivity in these cells (Scanlon et al., 1994). Prolonged treatment of both anti-mdr- and anti-fos-ribozyme expressing cells resulted again in an increase of drug resistance. However, whereas with anti-mdr ribozyme this increase was eightfold, anti-fos-ribozyme expressing cells were only threefold less sensitive towards actinomycin D. Thus, the anti-fos ribozyme may be more efficient in reducing multiple drug resistance than the anti-mdr-ribozyme chosen in this study. In a recent study, the effect of an anti-mdr-ribozyme on the resistance of leukemic myeloid cells against all-trans retinoic acid was examined (Matsushita et al., 1998). After complete remission caused by treatment with retinoic acid, APL patients sometimes relapse and again develop APL, this time resistant against retinoic acid treatment. To examine the possible role of MDR1 in this resistance, leukemic myeloid cells resistant against retinoic acid were stably transfected with an anti-mdr-ribozyme encoding vector (Matsushita et al., 1998). The ribozyme yielded a restoration of retinoic acid sensitivity. The addition of retinoic acid inhibited cellular proliferation and induced differentiation in ribozyme-expressing cells. Thus, MDR1 seems to be, at least in part, responsible for the resistance of leukemic cells against retinoic acid. In addition, this ribozyme has therapeutic potential for patients with retinoic-acid resistant APL (Matsushita et al., 1998).

Ribozymes are particularly promising in the treatment and prevention of viral diseases. Since viral sequences are usually not present in the host genome, they provide unique targets for ribozyme-mediated RNA destruction. Therefore, ribozymes may have a higher specificity and thereby lower toxicity than conventional antiviral drugs, such as nucleoside analogues. In addition, malignant cell transformation induced by viral proteins may be combated by ribozymes directed against the corresponding transcripts. One interesting example for this has been recently published by Huang et al. (1997). Epstein-Barr virus (EBV), which can persist in latently infected B cells, is usually not associated with the development of fatal diseases. However, in immunocompromised patients such as AIDS patients or transplant recipients EBV infection can contribute to malignant transformation resulting in lymphomas or leukemias. The Epstein-Barr nuclear antigen 1 (EBNA-1) is required for viral replication and gene expression. It was therefore chosen as a target for a ribozyme approach. For ribozyme delivery, the authors took advantage of the notion, that unlike uninfected B cells, EBV infected cells express the αv integrin, the adenovirus internalisation receptor. Thus, in contrast to uninfected B cells, EBV infected cells are susceptible to adenovirus-mediated gene delivery. The ribozyme expression cassette contained the RSV LTR as promoter. Infection of EBV-immortalised B lymphoctes with recombinant adenovirus carrying the ribozyme gene suppressed EBNA-1 expression, reduced the EBV genome number and inhibited cell proliferation. Furthermore, an inactive ribozyme construct did not affect EBNA expression and reduced the number of EBV genomes to a lesser extent than the active ribozyme. Therefore, adenovirus-encoded ribozymes may have great potential for the treatment of EBV-associated malignancies.

The lymphocytic choriomeningitis virus (LCMV) belongs to the arenaviruses and has a bisegmented single-stranded RNA genome. Its natural host is the mouse, but it can also cause choriomeningitis in humans. Cells stably expressing an anti-LCMV ribozyme under the control of the MMTV LTR show lower viral RNA levels and a 100-fold lower production of infectious virus (Xing and Whitton, 1993). To control for the LCMV specificity of the ribozyme, ribozyme-expressing cells were also infected with pichinde virus, which is closely related with LCMV, but contains several point mutations in the ribozyme target site. The yield of this arenavirus is not affected by the ribozyme strongly suggesting that an interaction of the ribozyme with the LCMV RNA is responsible for the observed effects (Xing and Whitton, 1993).

Human immunodeficiency virus type 1 (HIV-1) is probably the most often chosen ribozyme target. In fact, at least two groups are presently performing phase 1 clinical studies to investigate ribozyme safety and antiviral effects in patients (Wong-Staal et al., 1998). Several regions of the HIV genome, such as the U5 region (Dropulic et al., 1992; Ojwang et al., 1992), the gag region (Sarver et al., 1990; Sun et al., 1994) or tat/rev region (Zhou et al., 1994; Yamada et al., 1994), have been successfully targeted by ribozymes. So far, ribozyme delivery has been mainly achieved by transduction with amphotropic retroviruses carrying ribozyme genes. Different ribozyme expression strategies have been employed: directly by the retroviral long terminal repeat (LTR) (Zhou et al., 1994), by an internal Pol II promoter such as the β-actin promoter (Yu et al., 1993), by an internal Pol III promoter such as $tRNA^{Val}$ or the VA1 promoter (Yu et al., 1993), or as a double copy construct with the ribozyme cassette located in the U3 regions of the LTR (Gervaix et al., 1997). HIV is one of the few examples, for which hairpin ribozymes are also intensively studied for antiviral gene therapy.

A major concern for a ribozyme approach is the genetic variability of HIV. Two reports could demonstrate that HIV strains carrying point mutations either in or flanking the cleavage triplet are not inhibited by ribozymes targeted against the unmutated sequence (Yamada et al., 1994; Sun et al., 1995b). One the one hand, this finding proves the necessity of ribozyme-mediated RNA cleavage for the efficient inhibition of HIV, on the other, however, it points to the problem of escape mutants which will not be inhibited by the ribozyme anylonger. Furthermore, such ribozymes will only protect against a subset of HIV strains. Several strategies have been suggested to circumvent these problems. One possibility is the choice of target sequences being conserved in a majority of HIV strains, such as certain regions in the U5 part of the HIV LTR (Yu et al., 1993). Alternatively, multimeric ribozyme transcripts targeting several locations in the HIV genome will affect the majority of HIV strains and, furthermore, will minimise the likelihood for generation of escape mutants (Chen et al., 1992). Finally, combination of ribozymes with other approaches, such as RNA decoys specific for the HIV regulator protein Rev, have been shown to inhibit HIV replication more potently than a ribozyme or the decoy alone (Yamada et al., 1996). In an extension of this work, inclusion of a second ribozyme into the retroviral vector increased the inhibition of the replication of several HIV clades (Gervaix et al., 1997).

HIV infects both CD4 positive T helper cells and cells of the monocyte/macrophage lineage. Therefore, transduction of $CD34^+$ early hematopoietic progenitor cells by ribozyme-encoding amphotropic retroviruses would protect both cell lineages against HIV infection. Several studies claim that up to 100% growth factor-stimulated stem cell populations can be transduced with such retroviruses (Yu et al., 1995; Bauer et al., 1997; Rosenzweig et al., 1998). The transduction does not significantly affect hematopoietic differentiation. Moreover, monocytes (Bauer et al., 1997; Yu et al., 1995) and T cells (Rosenzweig et al., 1998) differentiated from transduced stem cells still express the ribozyme and are protected against HIV, or in another study SIV (Rosenzweig et al., 1998), infection. Thus, a gene therapy approach seems to be feasible, in which hematopoietic stem cells isolated from the patient are transduced *ex vivo* with a ribozyme encoding retrovirus followed by reinfusion into the patient. Alternatively, peripheral blood lymphocytes can be transduced with such retroviruses without affecting their proliferation (Sun et al., 1995a). In this case, however, newly differentiated T lymphocytes are not protected against HIV. As already mentioned above, such trials are already in progress.

14. Ribozyme Approaches in Animal Models

The use of ribozymes under "real" *in vivo* condition, i.e. in whole organisms, is currently the most challenging ribozyme application. Ribozymes have been exogenously and endogenously delivered to target tissues, and transgenic mice have been generated as an alternative to homologous recombination. Furthermore, ribozymes are increasingly used to establish RNA or protein function in embryologic development (Heinrich et al., 1993; Zhao and Pick, 1993; Steinbach et al., 1997; Xie et al., 1997; Zhao and Lemke, 1998).

Exogenously delivered ribozymes have been examined for their therapeutical potential in mouse and rabbit model systems. Amelogenin, the major translational product in mammalian teeth during enamel synthesis, is involved in the enamel mineralisation. Lyngstadaas et al. (1995) injected 2'-O-allylmodified ribozymes directly into the mandibles close to the developing molar teeth of newborn mice. As controls they used an inactive ribozyme, an antisense and a scrambled oligonucleotide. Whereas the scrambled oligonucleotide had no effect on amelogenin synthesis, both the mutated ribozyme and the antisense oligonucleotide delayed amelogenin synthesis by one day. The catalytically competent ribozyme was the most potent oligonucleotide with a delay of 3.5 days. Furthermore, this pronounced delay was paralleled by a failure of normal enamel mineralisation. This report demonstrated for the first time the successful application of exogenously delivered ribozymes in an animal model.

Stromelysin is a matrix metalloproteinase which degrades extracellular matrix components such as collagens, gelatins, proteoglycans and fibronectin. Matrix metalloproteinases are important for embryogenesis and tissue remodeling, but they contribute also, when abnormally expressed, to diseases such as atherosclerosis, arthritis and cancer. In a rabbit model, expression of stromelysin is induced by injection of interleukin 1 into the knee joint. To modulate stromelysin expression, chemically modified hammerhead ribozymes containing 2'-aminouridines in combination with 2'-allylpyrimidine nucleosides and a terminal 3'-3' linkage were examined (Flory et al., 1996). Intra-articularly injected ribozymes were taken up by synovial cells with full length ribozyme being visible for at least 7 days. When injected 24 hours prior to stimulation with interleukin 1, stromelysin mRNA levels were reduced by up to 60%. Inactive control ribozymes were ineffective. Notably, as in the previous study, no carrier such as cationic lipids have been used for ribozyme delivery.

Lieber and Kay (1996) employed adenovirus-mediated gene transfer to express ribozymes directed against human growth hormone (hGH) in transgenic mice. The ribozyme gene was cloned into a VA1 cassette. The natural VA1 gene served as a control. In addition, the authors tested whether combining this basal ribozyme cassette with either the pol II dependent RSV promoter or with a T7 RNA polymerase expression system improves the efficacy of ribozyme delivery and action. Injection of any of the ribozyme-containing recombinant adenoviruses into the tail vein caused a transient reduction of serum hGH levels by up to 50%. Reduction was maximal between one and two weeks and lasted up to one month. In parallel, the hGH mRNA levels in the liver were reduced by up to 90%. Notably, ribozyme cassettes downstream of the RSV LTR or transcribed by T7 RNA polymerase were somewhat more efficient than the parental VAI ribozyme cassette. In contrast, the control virus containing only the VAI cassette without the ribozyme sequence did not affect either serum hGH or hepatic hGH mRNA levels. In conclusion, this study suggests, that ribozyme-encoding adenoviruses may be useful tools for transient somatic gene therapy.

In 12% of patients suffering from autosomal dominant retinitis pigmentosa, a histididine for proline substitution at codon 23 of the rhodopsin gene results in photoreceptor cell death and leads finally to blindness. Ribozymes targeted against this mutation were delivered to photoreceptors in a rat model by injecting recombinant adeno-associated virus into the eye (Lewin et al., 1998). Both a hairpin and a hammerhead ribozyme delayed photoreceptor degeneration for at least three months. Inactive control ribozymes had a less pronounced effect. Furthermore, the ribozymes caused a 10-15% reduction of the mutated transcript. The authors stated, that this reduction may be significant, since only a quarter of the photoreceptors could be transduced. Unfortunately, the mRNA levels were estimated by semiquantitative RT-PCR. Thus, an arteficial RNA reduction cannot be excluded.

Transgenic mice carrying a ribozyme gene permit the investigation of gene function by downmodulating its expression. In contrast to knockout approaches using homologous recombination ribozymes usually do not completely shut off the gene of interest, but reduce it to a certain extent. Therefore, the function of gene products essential for an organism can be examined by such an approach. Furthermore, if the reduced gene expression causes a predisposition for disease, such transgenic lines are valuable model systems to examine genetic and enviromental factors affecting disease development. Mutations in the glucokinase gene have been linked to the development of a special type of diabetes named maturity-onset diabetes of the young (MODY), which is of autosomal dominant inheritance. To obtain an animal model for MODY, transgenic mice were generated with a ribozyme expression cassette containing the rat insulin promoter (Efrat et al., 1994). In initial cell culture experiments the ribozyme decreased both the mRNA and protein levels of glucokinase. Transgenic mice had reduced levels of glucokinase activity in the pankreas whereas hexokinase activity was not affected. In addition, glucose-induced insulin secretion was impaired in perfused pancreas. However, serum levels of glucose and insulin were normal. The authors concluded that these mice may be predisposed to type II diabetes with increased susceptibility to additional diabetic factors.

L'Huillier et al. (1996) generated transgenic mice which expressed a ribozyme targeted against bovine α-lactalbumin. The ribozyme chosen for these experiments had been characterized in cell culture experiments (L'Huillier et al., 1992). Ribozyme expression was controlled by MMTV LTR, which is very active in mammary glands (L'Huillier et al., 1996). Crossbreeding with transgenic mice expressing bovine α-lactalbumin yielded mice with reduced levels of the transgenic α-lactalbumin. Expression of murine α-lactalbumin was not reduced in these double-transgenic mice. This example shows the feasibility of using ribozymes for controlling the composition of milk and underlines the potential of ribozymes in biotechnology.

15. Outlook

The study of ribozyme structure and mechanisms of catalysis are currently strongly promoted at the academic and commercial level. It can therefore be expected that many new developments will contribute to a rapid progress of this technology. However, there are several obstacles to overcome, such as efficient delivery to a sufficient percentage of the cell population, efficient expression of active ribozymes from vectors, colocalisation of the ribozyme with the target, specifity of ribozymes for the targeted RNA and improvement of the intracellular activity of ribozymes. As the knowledge of RNA secondary and tertiary structure increases, it will be possible to target the RNA more rationally, thereby achieving a high specificity of the ribozyme for its substate RNA. At the same time, the understanding of RNA transport and localisation will also help to achieve colocalization of the ribozyme and its target RNA. Finally, the development of efficient and specific delivery systems is a very important step towards the application of ribozymes in therapy and biotechnology.

Results obtained with ribozymes so far look very promising. This is particularly the case for HIV-1, where currently several clinical trials for ribozymes are ongoing. The human clinical trials, as well as animal studies will provide the information about ribozyme safety necessary for subsequent efficacy trials. In conclusion, ribozymes and related antisense molecules may mark the beginning of a new era in medicine and biotechnolgy in which the drugs selectively block the flow of genetic information. Thus, the dream of the molecular scissors may come true after all.

16. Acknowledgements

We thank L. Peterson and S. Blumenthal for critically reading the manuscript. Work in the authors laboratory has been supported by the Deutsche Krebshilfe (10-1217 He1).

17. References

Altman, S. (1990) Enzymatic cleavage of RNA by RNA. *Angew. Chem. Int. Ed. Engl.* **29,** 749-758.

Aurup, H., Tuschl, T., Benseler, F., Ludwig, J. and Eckstein, F. (1994) Oligonucleotide duplexes containing 2'-amino-2'-deoxycytidines: thermal stability and chemical reactivity. *Nucleic Acids Res.* **22,** 20-24.

Aurup, H., Williams, D.M. and Eckstein, F. (1992) 2'-Fluoro- and 2'-amino-2'-deoxynucleoside 5'-triphosphates as substrates for T7 RNA polymerase. *Biochemistry* **31,** 9636-9641.

Bassi, G.S., Møllegard, N.-E., Murchie, A.I.H., von Kitzing, E. and Lilley, D.M.J. (1995) Ionic interactions and the global conformations of the hammerhead ribozyme. *Nature Struct. Biol.* **2,** 45-55.

Bauer, G., Valdez, P., Kearns, K., Bahner, I., Wen, S.F., Zaia, J.A. and Kohn, D.B. (1997) Inhibition of human immunodeficiency virus-1 (HIV-1) replication after transduction of granulocyte colony-stimulating factor-mobilized CD34+ cells from HIV-1-infected donors using retroviral vectors containing anti-HIV-1 genes. *Blood* **89,** 2259-2267.

Beck, J. and Nassal, M. (1995) Efficient hammerhead ribozyme-mediated cleavage of the structured hepatitis B virus encapsidation signal *in vitro* and in cell extracts, but not in intact cells. *Nucleic Acids Res.* **23,** 4954-4962.

Beigelman, L., McSwiggen, J.A., Draper, K.G., Gonzalez, C., Jensen, K., Karpeisky, A.M., Modak, A.S., Matulic-Adamic, J., DiRenzo, A.B., Haeberli, P., Sweedler, D., Tracz, D., Grimm, S., Wincott, F.E., Thackray, V.G. and Usman, N. (1995) Chemical modification of hammerhead ribozymes. *J. Biol. Chem.* **270,** 25702-25708.

Bertrand, E., Castanotto, D., Zhou, C., Carbonnelle, C., Lee, N.S., Good, P., Chatterjee, S., Grange, T., Pictet, R., Kohn, D., Engelke, D. and Rossi, J.J. (1997) The expression cassette determines the functional activity of ribozymes in mammalian cells by controlling their intracellular localization. *RNA* **3,** 75-88.

Bertrand, E.L. and Rossi, J.J. (1994) Facilitation of hammerhead ribozyme catalysis by the nucleocapsid protein of HIV-1 and the heterogeneous nuclear protein A1. *EMBO J.* **13,** 2904-2912.

Birikh, K.R., Berlin, Y.A., Soreq, H. and Eckstein, F. (1997) Probing accessible sites for ribozymes on human acetylcholinesterase RNA. *RNA* **3,** 429-437.

Cech, T.R. (1990) Self-splicing and enzymatic activity of an intervening sequence RNA from tetrahymena. *Angew. Chem. Int. Ed. Engl.* **29,** 759-768.

Chen, C.J., Banerjea, A.C., Harmison, G.G., Haglund, K. and Schubert, M. (1992) Multitarget-ribozyme directed to cleave at up to nine highly conserved HIV-1 env RNA regions inhibits HIV-1 replication - potential effectiveness against most presently sequenced HIV-1 isolates. *Nucleic Acids Res.* **20,** 4581-4589.

Czubayko, F., Downing, S.G., Hsieh, S.S., Goldstein, D.J., Lu, P.Y., Trapnell, B.C. and Wellstein, A. (1997) Adenovirus-mediated transduction of ribozymes abrogates HER-2/neu and pleiotrophin expression and inhbits tumor cell proliferation. *Gene Ther.* **4,** 943-949.

Czubayko, F., Riegel, A.T. and Wellstein, A. (1994) Ribozyme-targeting elucidates a direct role of pleiotrophin in tumor growth. *J. Biol. Chem.* **269,** 21358-21363.

Czubayko, F., Schulte, A.M., Berchem, G.J. and Wellstein, A. (1996) Melanoma angiogenesis and metastasis modulated by ribozyme targeting of the secreted growth factor pleiotrophin. *Proc. Natl. Acad. Sci. USA* **93,** 14753-14758.

Dahm, S.C., Derrick, W.B. and Uhlenbeck, O.C. (1993) Evidence for the role of solvated metal hydroxide in the hammerhead cleavage mechanism. *Biochemistry* **32,** 13040-13045.

Dahm, S.C. and Uhlenbeck, O.C. (1991) Role of divalent metal ions in the hammerhead RNA cleavage reaction. *Biochemistry* **30,** 9464-9469.

De Young, M.B., Kincade-Denker, J., Boehm, C.A., Riek, R.P., Mamone, J.A., McSwiggen, J.A. and Graham, R.M. (1994) Functional characterization of ribozymes expressed using U1 and T7 vectors for the intracellular cleavage of ANF mRNA. *Biochemistry* **33,** 12127-12138.

Denman, R.B. (1993) Using RNAFOLD to predict the activity of small catalytic RNAs. *Biotechniques* **15,** 1090-1094.

Dropulic, B., Lin, N.H., Martin, M. and Jeang, K.-T. (1992) Functional characterization of a U5 ribozyme: Intracellular suppression of human immunodeficiency virus type 1 expression. *J. Virol.* **66,** 1432-1441.

Efrat, S., Leiser, M., Wu, Y.-J., Fusco-DeMaine, D., Emran, O.A., Surana, M., Jetton, T.L., Magnuson, M.A., Weir, G. and Fleischer, N. (1994) Ribozyme-mediated attenuation of pancreatic β-cell glucokinase expression in transgenic mice results in impaired glucose-induced insulin secretion. *Proc. Natl. Acad. Sci. USA* **91,** 2051-2055.

Felgner, J.H., Kumar, R., Sridhar, C.N., Wheller, C.J., Tsai, Y.J., Border, R., Ramsey, P., Martin, M. and Felgner, P.L. (1994) Enhanced gene delivery and mechanism studies with a novel series of cationic lipid formulations. *J. Biol. Chem.* **269,** 2550-2561.

Feng, M., Cabrera, G., Deshane, J., Scanlon, K.J. and Curiel, D.T. (1995) Neoplastic reversion accomplished by highe efficiency adenoviral-mediated delivery of an anti-ras ribozyme. *Cancer Res.* **55,** 2024-2-28.

Fersht, A. (1985). Enzyme structure and mchanism. New York, W.H. Freeman.

Fire, A., Xu, S., Montgomery, M.K., Kostas, S.A., Driver, S.E. and Mello, C.C. (1998) Potent and specific genetic interference by double-stranded RNA in Caenorhabditis elegans. *Nature* **391,** 806-811.

Fisher, K.J., Choi, H., Burda, J., Chen, S.J. and Wilson, J.M. (1996) Recombinant adenovirus deleted of all viral genes for gene therapy in cystic fibrosis. *Virology* **217,** 11-22.

Flory, C.M., Pavco, P.A., Jarvis, T.C., Lesch, M.E., Wincott, F.E., Beigelman, L., Hunt III, S.W. and Schrier, D.J. (1996) Nuclease-resistant ribozymes decrease stromelysin mRNA levels in rabbit synovium following exogenous delivery to the knee joint. *Proc. Natl. Acad. Sci. USA* **93,** 754-758.

Gait, M.J., Pritchard, C. and Slim, G. (1991). Oligoribonucleotide synthesis. in *Oligonucleotides and analogues*. Eckstein, F., Ed. Oxford, IRL press. 25-48.

Gavin, D.K. and Gupta, K.C. (1997) Efficient hammerhead ribozymes targeted to the polycistronic sendai virus P/C mRNA. *J. Biol. Chem.* **272,** 1461-1472.

Gervaix, A., Li, X., Kraus, G. and Wong-Staal, F. (1997) Multigene antiviral vectors inhibit divers human immunodeficiency virus type 1 clades. *J. Virol.* **71,** 3048-3053.

Gregoriadis, G. (1995) Engineering liposomes for drug delivery: progress and problems. *Trends Biotechnol.* **13527-537,** 527-537.

Haseloff, J. and Gerlach, W.L. (1988) Simple RNA enzymes with new and highly specific endoribonuclease activities. *Nature* **334,** 585-591.

Heidenreich, O., Benseler, F., Fahrenholz, A. and Eckstein, F. (1994) High activity and stability of hammerhead ribozymes containing 2'-modified pyrimidine nucleosides and phosphorothioates. *J. Biol. Chem.* **269,** 2131-2138.

Heidenreich, O. and Eckstein, F. (1992) Hammerhead ribozyme-mediated cleavage of the long terminal repeat RNA of human immunodeficiency virus type 1. *J. Biol. Chem.* **267,** 1904-1909.

Heidenreich, O. and Eckstein, F. (1997). Synthetic ribozymes. in *Concepts in Gene Therapy*. Strauss, M. and Barranger, J., Ed. Berlin, Walter de Gruyter.

Heidenreich, O., Kang, S.-H., Brown, D.A., Xu, X., Swiderski, P., Rossi, J.J., Eckstein, F. and Nerenberg, M. (1995a) Ribozyme-mediated RNA degradation in nuclei suspension. *Nucleic Acids Res.* **23,** 2223-2228.

Heidenreich, O., Kang, S.-H., Xu, X. and Nerenberg, M. (1995b) Application of antisense technology to therapeutics. *Mol. Medicine Today* **1,** 128-133.

Heidenreich, O., Xu, X. and Nerenberg, M. (1996) A hammerhead ribozyme cleaves its target RNA during RNA preparation. *Antisense Nucleic Acids Drug Dev.* **6,** 141-144.

Heidenreich, O., Xu, X., Swiderski, P., Rossi, J.J. and Nerenberg, M. (1996) Correlation of activity with stability of chemically modified ribozymes in nuclei suspension. *Antisense Nucleic Acid Drug Dev.* **6,** 111-118.

Heinrich, J.C., Tabler, M. and Louis, C. (1993) Attenuation of white gene expression in transgenic Drosophila melanogaster: possible role of a catalytic antisense RNA. *Dev. Genetics* **14,** 258-265.

Herschlag, D. (1991) Implications of ribozyme kinetics for targeting the cleavage of specific RNA molecules *in vivo*: More isn't always better. *Proc. Natl. Acad. Sci. USA* **88,** 6921-6925.

Herschlag, D., Khosla, M., Tsuchihashi, Z. and Karpel, R.L. (1994) An RNA chaperone activity of non-specific RNA binding proteins in hammerhead ribozyme catalysis. *EMBO J.* **13,** 2913-2924.

Hertel, K.J., Pardi, A., Uhlenbeck, O.C., Koizumi, M., Ohtsuka, E., Uesugi, S., Cedergren, R., Eckstein, F., Gerlach, W.L., Hodgson, R. and Symons, R.H. (1992) Numbering system for the hammerhead. *Nucleic Acids Res.* **20,** 3252.

Hildebrandt, M. and Nellen, W. (1992) Differential antisense transcription from the Dictyostelium EB4 gene locus: Implications on antisense-mediated regulation of mRNA stability. *Cell* **69,** 197-204.

Hoffman, L.M. and Johnson, M.G. (1994) Enzymatic synthesis of milligram quantities of ribozymes in small volumes. *Biotechniques* **17,** 372-375.

Holm, P.S., Scanlon, K.J. and Dietel, M. (1994) Reversion of multidrug resistance in the P-glycoprotein-positive human pancreatic cell line (EPP85-181RDB) by introduction of a hammerhead ribozyme. *Brit. J. Cancer* **70,** 239-243.

Homann, M., Rittner, K. and Sczakiel, G. (1993a) Complementary large loops determine the rate of RNA duplex formation *in vitro* in the case of an effective antisense RNA directed against the human immunodeficieny virus type 1. *J. Mol. Biol.* **233,** 7-15.

Homann, M., Tzortzakaki, S., Rittner, K., Sczakiel, G. and Tabler, M. (1993b) Incorporation of the catalytic domain of a hammerhead ribozyme into antisense RNA enhances its inhibitory effect on the replication of human immunodeficiency virus type 1. *Nucleic Acids Res.* **21,** 2809-2814.

Huang, S., Stupack, D., Mathias, P., Wang, Y. and Nemerow, G. (1997) Growth arrest of epstein-barr virus immortalized B lymphocytes by adenovirus-delivered ribozymes. *Proc. Natl. Acad. Sci. USA* **94,** 8156-8161.

Jacobson, A. and Peltz, S.W. (1996) Interrelationships of the pathways of mRNA decay and translation in eukaryotic cells. *Ann. Rev. Biochem.* **65,** 693-739.

Kashani-Sabet, M., Funato, T., Florenes, V.A., Fodstad, O. and Scanlon, K.J. (1994) Suppression of the neoplastic phenotype *in vivo* by an anti-ras ribozyme. *Cancer Res.* **54,** 900-902.

Kawasaki, H., Eckner, R., Yao, T.-P., Taira, K., Chiu, R., Livingston, D.M. and Yokoyama, K.K. (1998) Distinct roles of the co-activators p300 and CBP in retinoic-acid-induced F9-cell differentiation. *Nature* **393,** 284-289.

Kawasaki, H., Ohkawa, J., Tanishige, N., Yoshinari, K., Murata, T., Yokoyama, K.K. and Taira, K. (1996) Selection of the best target site for ribozyme-mediated cleavage within a fusion gene for adenovirus E1A-associated 300 kDa protein (p300) and luciferase. *Nucleic Acids Res.* **24,** 3010-3016.

Kore, A.R., Vaish, N.K., Kutzke, U. and Eckstein, F. (1998) Sequence specificity of the hammerhead ribozyme revisited; the NHH rule. *Nucleic Acids Res.* **26,** 4116-4120.

Kronenwett, R., Haas, R. and Sczakiel, G. (1996) Kinetic selectivity of complementary nucleic acids: bcr-abl-directed antisense RNA and ribozymes. *J. Mol. Biol.* **259,** 632-644.

L'Huillier, P., Soulier, S., Stinnakre, M.-G., Lepourry, L., Davis, S.R., Mercier, J.-C. and Vilotte, J.-L. (1996) Efficient and specific ribozyme-mediated reduction of bovine a-lactalbumin expression in double transgenic mice. *Proc. Natl. Acad. Sci. USA* **93,** 6698-6703.

L'Huillier, P.J., Davis, S.R. and Bellamy, A.R. (1992) Cytoplasmic delivery of ribozymes leads to efficient reduction in α-lactalbumin mRNA levels in C127I mouse cells. *EMBO J.* **11,** 4411-4418.

Lan, N., Howrey, R.P., Lee, S.-W., Smith, C.A. and Sullenger, B.A. (1998) Ribozyme-mediated repair of sickle b-globin mRNAs in erythrocyte precursors. *Science* **280,** 1593-1596.

Lange, W., Cantin, E.M., Finke, J. and Dölken, G. (1993) *In vitro* and *in vivo* effects of synthetic ribozymes targeted against BCR/ABL mRNA. *Leukemia* **7,** 1786-1794.

Lange, W., Daskalatis, M., Finke, J. and Dölken, G. (1994) Comparison of different ribozymes for efficient and specific cleavage of BCR/ABL related mRNAs. *FEBS Lett.* **338,** 175-178.

Lee, R.C., Feinbaum, R.L. and Ambros, V. (1993) The C. elegans heterochronic gene lin-4 encodes small RNAs with antisense complementary to lin-14. *Cell* **75,** 843-854.

Leopold, L.H., Shore, S.K., Newkirk, T.A., Reddy, R.M.V. and Reddy, E.P. (1995) Multi-unit ribozyme-mediated cleavage of bcr-abl mRNA in myeloid leukemias. *Blood* **85,** 2162.

Lewin, A.S., Drenser, K.A., Hauswirth, W.W., Nishikawa, S., Yasumura, D., Flannery, J.G. and LaVail, M.M. (1998) Ribozyme rescue of photoreceptor cells in a transgenic rat model of autosomal dominant retinitis pigmentosa. *Nature Med.* **4,** 967-971.

Lewis, J.G., Lin, K.J., Kothavale, A., Flanagan, W.M., Matteucci, M.D., DePrince, R.B., Mook, R.A., Hendren, R.W. and Wagner, R.W. (1996) A serum-resistent cytofectin for cellular delivery of antisense oligodeoxynucleotides and plasmid DNA. *Proc. Natl. Acad. Sci. USA* **93,** 3176-3181.

Lieber, A. and Kay, M.A. (1996) Adenovirus-mediated expression of ribozymes in mice. *J. Virol.* **70,** 3153-3158.

Liu, Z., Batt, D.B. and Carmichael, G.G. (1994) Targeted nuclear antisense RNA mimics natural antisense-induced degradation of polyoma virus early RNA. *Proc. Natl. Acad. Sci. USA* **91,** 4258-4262.

Ludwig, J., Blaschke, M. and Sproat, B.S. (1998) Extending the cleavage rules for the hammerhead ribozyme:mutating adenosine$^{15.1}$ to inosine$^{15.1}$ change the cleavage site specificity from $N^{16.2}U^{16.1}H^{17}$ to $N^{16.2}C^{16.1}H^{17}$. *Nucleic Acids Res.* **26,** 2279-2285.

Luzi, E., Eckstein, F. and Barsacchi, G. (1997) The newt ribozyme is part of a riboprotein complex. *Proc. Natl. Acad. Sci. USA* **94,** 9711-9716.

Lyngstadaas, S.P., Risnes, S., Sproat, B.S., Thrane, P.S. and Prydz, H.P. (1995) A synthetic, chemically modified ribozyme eliminates amelogenin, the major translation product in developing mouse enamel *in vivo*. *EMBO J.* **14,** 5224-5229.

Malmgren, C., Wagner, E.G.H., Ehresmann, C., Ehresmann, B. and Romby, P. (1997) Antisense RNA control of plasmid R1 replication. *J. Biol. Chem.* **272,** 12508-12512.

Marschall, P., Thomson, J.B. and Eckstein, F. (1994) Inhibition of gene expression with ribozymes. *Cell. Mol. Neurobiol.* **14,** 523-538.

Matsushita, H., Kizaki, M., Kobayashi, H., Ueno, H., Muto, A., Takayama, N., Awaya, N., Kinjo, K., Hattori, Y. and Ikeda, Y. (1998) Restoration of retinoid sensitivity by MDR1 ribozymes in retinoic acid-resistant myeloid leukemic cells. *Blood* **91,** 2452-2458.

Michienzi, A., Prislei, S. and Bozzoni, I. (1996) U1 small nuclear RNA chimeric ribozymes with substrate specificity for the Rev pre-mRNA of human immunodeficiency virus. *Proc. Natl. Acad. Sci.* **93,** 7219-7224.

Miller, A.D. (1992) Human gene therapy comes of age. *Nature* **357,** 455-460.

Monia, B.P., Lesnik, E.A., Gonzalez, C., Lima, W.F., McGee, D., Guinosso, C.J., Kawasaki, A.M., Cook, P.D. and Freier, S.M. (1993) Evaluation of 2'-modified oligonucleotides containing 2'-deoxy gaps as antisense inhibitors of gene expression. *J. Biol. Chem.* **268,** 14514-14522.

Morgan, R.A. and Anderson, W.F. (1993) Human gene therapy. *Ann. Rev. Biochem.* **62,** 191-217.

Mulligan, R.C. (1993) The basic science of gene therapy. *Science* **260,** 926-932.

Murray, J.B., Seyhan, A.A., Walter, N.G., Burke, J.M. and Scott, W.G. (1998) The hammerhead, hairpin and VS ribozymes are catalytically proficient in monovalent cations alone. *Chem. Biol.* **5,** 587-595.

Nason-Burchenal, K., Allopenna, J., Begue, A., Stehelin, D., Dmitrovsky, E. and Martin, P. (1998a) Targeting of PML/RARα is lethal to retinoic acid-resistant promyelocytic leukemia cells. *Blood* **92,** 1758-1767.

Nason-Burchenal, K., Takle, G., Pace, U., Flynn, S., Allopenna, J., Martin, P., George, S.T., Goldberg, A.R. and Dmitrovsky, E. (1998b) Targeting the PML/RARα translocation product triggers apoptosis in promyelocytic leuemia cells. *Oncogene* **17,** 1759-1768.

Nedbal, W., Frey, M., Willemann, B., Zentgraf, H. and Sczakiel, G. (1997) Mechanistic insights into p53-promoted RNA-RNA annealing. *J. Mol. Biol.* **266,** 677-687.

Nellen, W. and Sczakiel, G. (1996) *In vitro* and *in vivo* action of antisense RNA. *Mol. Biotechnol.* **6,** 7-15.

Ojwang, J.O., Hampel, A., Looney, D.J., Wong-Staal, F. and Rappaport, J. (1992) Inhibition of human immunodeficiency virus type 1 expression by a hairpin ribozyme. *Proc. Natl. Acad. Sci. USA* **89,** 10802-10806.

Pace, U., Bockmann, J.M., MacKay, B.J., Miller, J., W.H., Dmitrovsky, E. and Goldberg, A.R. (1994) A ribozyme which discriminates *in vitro* between PML/RARα, the t(15;17)-associated fusion RNA of acute promyelocytic leukemia, and PML and RARα, the transcripts from the nonrearranged alleles. *Cancer Res.* **54,** 6365-6369.

Pal, B.K., Scherer, L., Zelby, L., Bertrand, E. and Rossi, J.J. (1998) Monitoring retroviral RNA dimerization *in vivo* via hammerhead ribozyme cleavage. *J. Virol.* **72,** 8349-8355.

Patzel, V. and Sczakiel, G. (1998) Theoretical design of antisense RNA structures substantially improves annealing kinetics and efficacy in human cells. *Nat. Biotech.* **16,** 64-68.

Persson, C., Wagner, E.G.H. and Nordström, K. (1988) Control of replication of plasmid R1: Kinetics of *in vitro* interaction between the antisense RNA, CopA, and its target, CopT. *EMBO J.* **7,** 3279-3288.

Pley, H.W., Flaherty, K.M. and McKay, D.B. (1994) Three-dimensional structure of a hammerhead ribozyme. *Nature* **372,** 68-74.

Pörschke, D. and Eigen, M. (1971) Cooperative non-enzymatic base recognition. *J. Mol. Biol.* **62,** 361-381.

Rittner, K., Burmester, C. and Sczakiel, G. (1993) *In vitro* selection of fast-hybridizing and effective antisense RNAs directed against the human immunodeficiency virus type 1. *Nucleic Acids Res.* **21,** 1381-1387.

Romani, A. and Scarpa, A. (1992) Regulation of cell magnesium. *Arch. Biochem. Biophys.* **298,** 1-12.

Rosenzweig, M., Marks, D.F., Hempel, D., Heusch, M., Kraus, G., Wong-Staal, F. and Johnson, R.P. (1998) Intracellular immunization of rhesus CD34+ hematopoietic progenitor cells with a hairpin ribozyme protects T cells and macrophages from simian immunodeficiency virus infection. *Blood* **90,** 4822-4831.

Sakamoto, N., Wu, C.H. and Wu, G.Y. (1996) Intracellular cleavage of hepatitis C virus RNA and inhibition of viral protein translation by hammerhead ribozymes. *J. Clin. Invest.* **98,** 2720-2728.

Salmons, B. and Günzburg, W.H. (1997). Retroviral vectors. in *Concepts in gene therapy*. Strauss, M. and Barranger, J. A., Ed. Berlin, Walter de Gruyter. 3-24.

Sargueil, B., Pecchia, D.B. and Burke, J.M. (1995) An improved version of the hairpin ribozyme functions as a ribonucleoprotein complex. *Biochemistry* **34,** 7739-7748.

Sarver, N., Cantin, E.M., Chang, P.S., Zaia, J.A., Ladne, P.A., Stephens, D.A. and Rossi, J.J. (1990) Ribozymes as potential anti-HIV-1 therapeutic agents. *Science* **247,** 1222-1225.

Scanlon, K.J., Ishida, H. and Kashani-Sabet, M. (1994) Ribozyme-mediated reversal of the multidrug-resistant phenotype. *Proc. Natl. Acad. Sci. USA* **91,** 11123-11127.

Scherr, M., Grez, M., Ganser, A. and Engels, J.W. (1997) Specific hammerhead ribozyme-mediated cleavage of mutant N-ras mRNA *in vitro* and *ex vivo*. *J. Biol. Chem.* **272,** 14304-14313.

Scherr, M., Maurer, A.B., Klein, S., Ganser, A., Engels, J.W. and Grez, M. (1998) Effecive reversal of a transformed phenotype by retrovirus-mediated transfer of a ribozyme directed againt mutant N-ras. *Gene Ther.* **5,** 1227-1234.

Scherr, M. and Rossi, J.J. (1998) Rapid determination and quantitation of the accessibility to native RNAs by antisense oligodeoxynucleotides in murine cell extracts. *Nucleic Acids Res.* **26,** 5079-5085.

Schmelzer, C. and Schweyen, R.J. (1986) Self-splicing of group II introns *in vitro*: mapping ofthe branch point and mutational inhibition of lariat formation. *Cell* **46,** 557-565.

Scott, W.G., Finch, J.T. and Klug, A. (1995) The crystal structure of an all-RNA hammerhead ribozyme: a proposed mechanism for RNA catalytic cleavage. *Cell* **81,** 991-1002.

Sczakiel, G. (1994) Antisense strategies for the control of aberrant gene expression. *J. Hematother.* **3,** 305-313.

Sczakiel, G. and Goody, R.S. (1994) Antisense principle or ribozyme action? *Biol. Chem. Hoppe-Seyler* **375,** 745-746.

Shimayama, T., Nishikawa, S. and Taira, K. (1995) Generality of the NUX rule: Kinetic analysis of the results of systematic mutations in tthe trinucleotide at the cleavage site of the hammerhead ribozyme. *Biochemistry* **34,** 3649-3654.

Sioud, M. (1994) Interaction between tumour necrosis factor α ribozyme and cellular proteins. *J. Mol. Biol.* **242,** 619-629.

Sioud, M. and Jespersen, L. (1996) Enhancement of hammerhead ribozyme catalysis by glyceraldehyde-3-phosphate dehydrogenase. *J. Mol. Biol.* **257,** 775-789.

Sioud, M., Natvig, J.B. and Førre, Ø. (1992) Preformed ribozyme destroys tumour necrosis factor mRNA in human cells. *J. Mol. Biol.* **223,** 831-835.

Sioud, M. and Sørensen, D.R. (1998) A nuclease-resistant protein kinase Ca ribozyme blocks glioma cell growth. *Nat. Biotech.* **16,** 556-561.

Slim, G. and Gait, M.J. (1991) Configurationally defined phosphorothioate-containing oligoribonucleotides in the study of the mechanism of cleavage of hammerhead ribozymes. *Nucleic Acids Res.* **19,** 1183-1188.

Snyder, D.S., Wu, Y., Wang, J.L., Rossi, J.J., Swiderski, P., Kaplan, B.E. and Forman, S.J. (1993) Ribozyme-mediated Inhibition of bcr-abl gene expression in a Philadelphia chromosome-positive cell line. *Blood* **82,** 600-605.

Stein, C.A. and Cheng, Y.-C. (1993) Antisense oligonucleotides as therapeutic agents - Is the bullet really magical? *Science* **261,** 1004-1012.

Steinbach, O.C., Wolffe, A.P. and Rupp, R.A.W. (1997) Somatic linker histones cause loss of mesodermal competence in Xenopus. *Nature* **389,** 395-399.

Sullenger, B.A. and Cech, T.R. (1993) Tethering ribozymes to a retroviral packaging signal for destruction of viral RNA. *Science* **262,** 1566-1569.

Sun, L.-Q., Warrilow, D., Wang, L., Witherington, C., MacPherson, J. and Symonds, G. (1994) Ribozyme-mediated suppression of moloney murine leukemia virus and human immunodeficiency virus type I replication in permissive cell lines. *Proc. Natl. Acad. Sci. USA* **91,** 9715-9719.

Sun, L.Q., Pyati, J., Smythe, J., Wang, L., MacPherson, J., Gerlach, W. and Symonds, G. (1995a) Resistance to human immunodeficiency virus type 1 infeciton conferred by transduction of human peripheral blood lymphocytes with ribozyme, antisense, or polymeric trans-activation response element constructs. *Proc. Natl. Acad. Sci. USA* **92,** 7272-7276.

Sun, L.Q., Wang, L., Gerlach, W.L. and Symonds, G. (1995b) Target sequence-specific inhibition of HIV-1 replication by ribozymes directed to tat RNA. *Nucleic Acids Res.* **23,** 2909-2913.

Sutterluety, H., Bartl, S., Doetzlhofer, A., Khier, H., Wintersberger, E. and Seiser, C. (1998) Growth-regulated antisense transcription of the mouse thymidine kinase gene. *Nucleic Acids Res.* **26,** 4989-4995.

Symons, R.H. (1992) Small catalytic RNAs. *Annu. Rev. Biochem.* **61,** 641-671.

Tsuchihashi, Z., Khosla, M. and Herschlag, D. (1993) Protein enhancement of hammerhead ribozyme catalysis. *Science* **262,** 99-102.

Tuschl, T. and Eckstein, F. (1993) Hammerhead ribozymes: importance o stem-loop II for activity. *Proc. Natl. Acad. Sci. USA* **90,** 6991-6994.

Tuschl, T., Gohlke, C., Jovin, T.M., Westhof, E. and Eckstein, F. (1994) A three-dimensional model for the hammerhead ribozyme based on fluorescence measurements. *Science* **266,** 785-789.

Tuschl, T., Thomson, J.B. and Eckstein, F. (1995) RNA cleavage by small catalytic RNAs. *Curr. Opin. Struct. Biol.* **5,** 296-302.

Vaish, N.K., Heaton, P.A., Fedorova, O. and Eckstein, F. (1998) *In vitro* selection of a purine nucleotide-specific hammerhead-like ribozyme. *Proc. Natl. Acad. Sci. USA* **95,** 2158-2162.

van Tol, H., Buzayan, J.M., Feldstein, P.A., Eckstein, F. and Bruening, G. (1990) Two autolytic processing reactions of a satellite RNA proceed with inversion of configuration. *Nucleic Acids Res.* **18,** 1971-1975.

Wagner, E.G.H. and Simons, R.W. (1994) Antisense RNA control in bacteria, phages, and plasmids. *Ann. Rev. Microbiol.* **48,**

Werner, M. and Uhlenbeck, O.C. (1995) The effect of base mismatches in the substrate recognition helices of hammerhead ribozymes on binding and catalysis. *Nucleic Acids Res.* **23,** 2092-2096.

Westaway, S.K., Cagnon, L., Chang, Z., Li, S., Li, H., Larson, G.P., Zaia, J.A. and Rossi, J.J. (1998) Virion encapsidation of tRNA3Lys-ribozyme chimeric RNAs inhibits HIV infection. *Antisense Nucleic Acid Drug Dev.* **8,** 185-197.

Wightman, B., Ha, I. and Ruvkun, G. (1993) Posttranscriptional regulation of the heterochronic gene lin-14 by lin-4 mediates temporal pattern formation in C. elegans. *Cell* **75,** 855-862.

Wincott, F., DiRenzo, A., Shaffer, C., Grimm, S., Tracz, D., Workman, C., Sweedler, D., Gonzalez, C., Scaringe, S. and Usman, N. (1995) Synthesis, deprotection, analysis and purification of RNA and ribozymes. *Nucleic Acids Res.* **23,** 2677-2684.

Wong-Staal, F., Poeschla, E.M. and Looney, D.J. (1998) A controlled, phase 1 clinical trial to evaluate the safety and effects in HIV-1 infected humans of autologous lymphoctes transduced with a ribozyme that cleaves HIV-1 RNA. *Hum. Gene Ther.* **9,** 2407-2425.

Xie, Y., Chen, X. and Wagner, T.E. (1997) A ribozyme-mediated, gene "knockdown" strategy for the identification of gene function in zebrafish. *Proc. Natl. Acad. Sci. USA* **94,** 13777-13781.

Xing, Z. and Whitton, J.L. (1993) An anti-lymphocytic choriomeningitis virus ribozyme expressed in tissue culture cells diminishes viral RNA levels and leads to a reduction in infectious virus yield. *J. Virol.* **67,** 1840-1847.

Yamada, O., Kraus, G., Leavitt, M.C., Yu, M. and Wong-Staal, F. (1994) Activity and cleavage site specificity of an anti-HIV-1 hairpin ribozyme in human T-cells. *Virology* **205,** 121-126.

Yamada, O., Kraus, G., Luznik, L., Yu, M. and Wong-Staal, F. (1996) A chimeric human immunodeficiency virus type 1 (HIV-1) minimal Rev response element-ribozyme molecule exhibits dual antiviral function and inhibits cell-cell transmission of HIV-1. *J. Virol.* **70,** 1596-1601.

Young, K.J., Gill, F. and Grasby, J.A. (1997) Metal ions play a passive role in the hairpin ribozyme catalysed reaction. *Nucleic Acids Res.* **25,** 3760-3766.

Yu, M., Leavitt, M.C., Maruyama, M., Yamada, O., Young, D., Ho, A.D. and Wong-Staal, F. (1995) Intracellular immunization of human fetal cord blood stem/progenitor cells with a ribozyme against human immunodeficiency virus type 1. *Proc. Natl. Acad. Sci. USA* **92,** 699-703.

Yu, M., Ojwang, J., Yamada, O., Hampel, A., Rapapport, J., Looney, D. and Wong-Staal, F. (1993) A hairpin ribozyme inhibits expression of diverse strains of human immunodeficiency virus type 1. *Proc. Natl. Acad. Sci. USA* **90,** 6340-6344.

Yu, Q., Pecchia, D.B., Kingsley, S.L., Heckman, J.E. and Burke, J.M. (1998) Cleavage of highly structured viral RNA molecules by combinatorial libraries of hairpin ribozymes. *J. Biol. Chem.* **273,** 23524-23533.

Yuan, Y., Hwang, E.-S. and Altman, S. (1992) Targeted cleavage of mRNA by human RNase P. *Proc. Natl. Acad. Sci. USA* **89,** 8006-8010.

Zabner, J., Fasbender, A.J., Moninger, T., Poellinger, K.A. and Welsh, M.J. (1995) Cellular and molecular barriers to gene transfer by a cationic lipid. *J. Biol. Chem.* **270,** 18997-19007.

Zhao, J.J. and Lemke, G. (1998) Selective disruption of neuregulin-1 function in vertebrate embryos using ribozyme-tRNA transgenes. *Development* **125,** 1899-1907.

Zhao, J.J. and Pick, L. (1993) Generating loss-of-function phenotypes of the fushi tarazu gene with a targeted ribozyme in Drosophila. *Nature* **365,** 448-451.

Zhou, C., Bahner, I.C., Larson, G.P., Zaia, J.A., Rossi, J.J. and Kohn, D.B. (1994) Inhibition of HIV-1 in human T-Lymphocytes by retrovirally transduced anti-tat and rev hammerhead ribozymes. *Gene* **149,** 33-39.

Zoumadakis, M., Neubert, W.J. and Tabler, M. (1994) The influence of imperfectly paired helices I and III on the catalytic of hammerhead ribozymes. *Nucleic Acids Res.* **22,** 5271-5278.

Zoumadakis, M. and Tabler, M. (1995) Comparative analysis of cleavage rates after systematic permutation of the NUX consensus target motif for hammerhead ribozymes. *Nucleic Acids Res.* **23,** 1192-1196.

Zufferey, R., Nagy, D., Mandel, R.J., Naldini, L. and Trono, D. (1997) Multiply attenuated lentiviral vector achieves efficient gene delivery *in vivo*. *Nat. Biotechnol.* **15,** 871-875.

HIGH YIELD EXPRESSION OF RECOMBINANT PROTEINS REQUIRING PROTEOLYTIC MATURATION: USE OF THE ENDOPROTEASE FURIN

M. HIMMELSPACH, U. SCHLOKAT*, B. PLAIMAUER, F. G. FALKNER and F. DORNER

Biomedical Research Center, Hyland Immuno, Division of Baxter Inc., Uferstrasse 15, 2304 Orth/Donau, Austria
* *Correspondence: fax +43-1-20100-4000, e-mail schloku@baxter.com*

1. Introduction

During biogenesis, secreted eucaryotic proteins may undergo multiple post-translational modifications. The most prominent modifications include glycosylation, phosphorylation, acetylation, sulfation, β-hydroxylation, γ-carboxylation and disulphide-bond formation. In addition, the generation of biologically active molecules may require post-translational proteolytic processing. In this case the protein is synthesised as an inactive precursor molecule which, beside the removal of the signal peptide upon translocation into the endoplasmic reticulum, undergoes additional proteolytic cleavage(s) before secretion. Such endoproteolytic processing events typically remove N-terminal propeptides and/or convert a single chain precursor into a mature heterodimeric form. Secretion of proteins is commonly accomplished by two alternate routes. Precursor cleavage of proteins secreted via the regulated pathway ('regulated secretory proteins') usually occurs C-terminal to paired basic amino acid residues, while precursor processing of secreted proteins which transit the constitutive secretory pathway ('constitutive secretory proteins') requires a more complex basic motif. In either case, the enzymes responsible for precursor cleavage at these motifs were found to be homologous to the bacterial subtilisin/yeast kexin endoproteases and termed the family of pro-protein convertases. Furin, the first mammalian member of this family, is ubiquitously expressed in all cell types examined and was found to process the precursor molecules of a wide variety of physiologically important proteins, such as hormones, growth factors, receptors, plasma proteins, viral envelope proteins and bacterial toxins.

The ability to produce proteins recombinantly in eucaryotic expression systems, whose use is imperative for complex proteins, has resulted in the development of pharmaceutical proteins, used, e.g., for the treatment of patients suffering from haemophilia, heart attack or anemia. While high expression levels sufficient for industrial purposes are fairly routinely achieved by amplification of the heterologous coding sequence in these systems, post-translational modifications, often a prerequisite for the biological activity of the recombinant protein, frequently become incomplete at overexpression. Particularly, endoproteolytic conversion of precursor molecules into their mature biologically active forms is often severely affected.

This report will review the usefulness of furin and derivatives for the maturation of potentially therapeutic recombinant proteins, i.e. the plasma factors protein C, factor IX, factor X and von Willebrand factor, all of which require post-translational cleavage events for their conversion into biologically active biopharmaceuticals. Different strategies chosen to ensure complete precursor processing of proteins, expressed either in mammalian cell lines or in transgenic animals, will be described. Amino acids are indicated by the three letter code throughout this chapter and numbered relative to the translational initiator methionine (Met), to which position 1 has been assigned.

2. Constitutive and regulated secretion

Newly synthesised secretory proteins are routed from the rough endoplasmic reticulum (RER) through the Golgi complex (GC) to the trans Golgi network (TGN) where they are sorted into the constitutive or regulated secretory pathway (fig. 1; see Urbé et al., 1997, for review). In the constitutive pathway, common to all eucaryotic cell types, proteins are transported in secretory vesicles (SV) and continuously released from there into the environment. In the regulated pathway, coexistent with constitutive secretion in specialised cell types, e.g. endocrine, exocrine or neuroendocrine cells, proteins are stored in mature secretory granules (MSG) from which they are delivered upon stimulation by an appropriate trigger. Sorting of the proteins into their corresponding pathway is believed to occur in the TGN and/or the immature secretory granules (ISG). The extent of sorting at these two levels varies between cell types. In the TGN, regulated secretory proteins have been proposed to be segregated from other proteins by being selectively directed into ISGs. The second sorting event occurs during maturation of ISGs to MSGs. Proteins, having entered the ISGs but not destined to the MSGs, are actively removed in vesicles budding from the ISGs, while regulated secretory proteins are retained. The proteins retrieved from the ISGs may then be targeted to the endosomal pathway, recycled to the TGN or secreted via a constitutive-like pathway. At acidic pH and in the calcium environment typical to the TGN, regulated secretory proteins form membrane-associated aggregates. This particular feature of regulated secretory proteins has been proposed to allow their segregation from other proteins (Bauerfeind et al., 1993). An alternative mechanism for selective targeting of secretory proteins to their correct destination suggests a direct or indirect interaction of the cargo proteins with specific 'sorting receptors', implying the recognition of particular sequences or structural elements in the secretory proteins by the sorting machinery. The evidence that the N-terminal disulphide-linked loop structure common to chromogranin B (Chanat et al., 1993) and pro-opiomelanocortin (POMC; Cool et al., 1995) represents a molecular targeting signal for the regulated pathway supports the existence of a receptor-mediated sorting of pro-hormones in the secretory pathway. More recently, carboxypeptidase E, which is involved in the proteolytic maturation of proteins in secretory granules, has been suggested to act as a sorting receptor for POMC, pro-insulin and pro-enkephalin (Cool et al., 1997; Normant et al., 1998). Furthermore, evidence that N-linked carbohydrate moieties mediate Golgi to cell membrane transport as well as apical targeting of membrane and secretory proteins (Scheiffele et al., 1995; Gut et al., 1998) suggests that lectins may act as sorting receptors for proteins leaving the TGN.

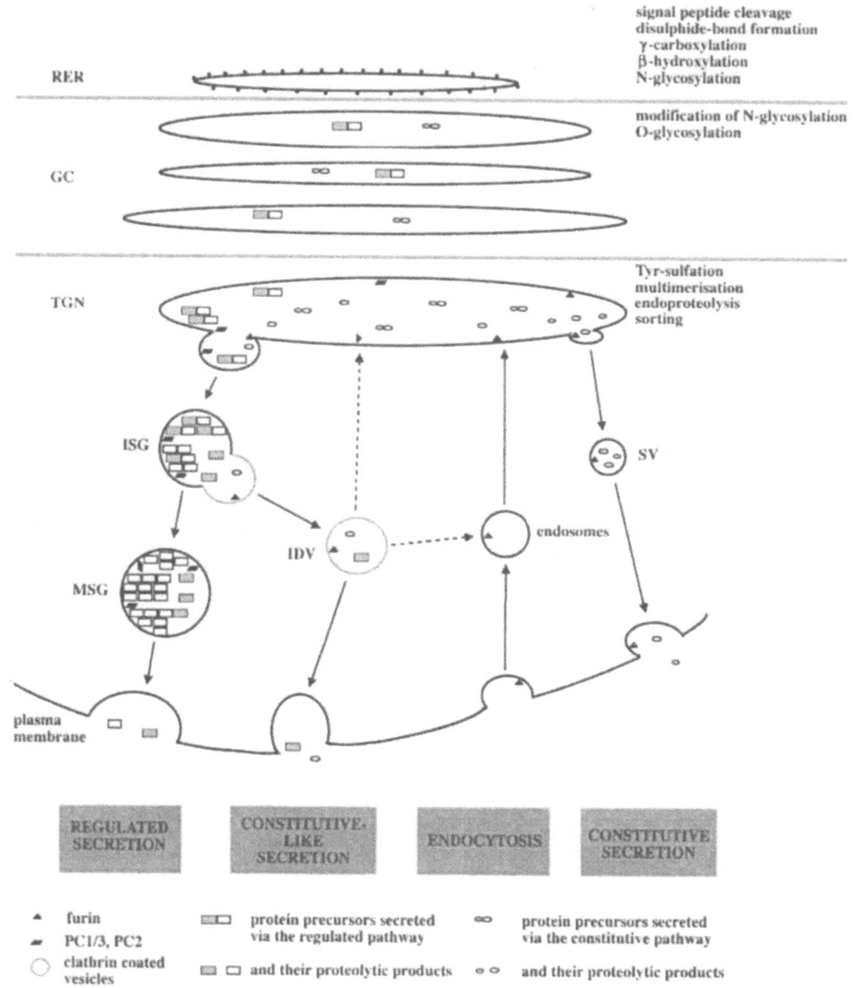

Figure 1. Schematic representation of the post-translational modifications of proteins during their transit through the secretory pathway(s). Post-translational modifications that occur within the rough endoplasmic reticulum (RER), the Golgi complex (GC) and the trans Golgi network (TGN) are indicated. The endoproteolytic maturation of constitutive secretory proteins mainly occurs in the TGN and is performed by endoproteases like furin. Furin delivered at the plasma membrane recycles to the TGN via the endocytic pathway. The maturation of regulated secretory proteins occurs in post-TGN compartments, i.e. in the immature secretory granules (ISG) and the mature secretory granules (MGS), by membrane-associated endoproteases such as PC1/3 and PC2. Constitutive secretory proteins, which have mistakenly entered the ISGs, or soluble processing products are removed through ISG-derived clathrin coated vesicles (IDV) and secreted via a constitutive-like pathway.

During their transit through the secretory pathway, proteins are subjected to multiple post-translational modifications. Some modifications such as disulphide-bond formation, γ-carboxylation, β-hydroxylation or N-glycosylation occur in the rough endoplasmic reticulum, while others such as O-glycosylation and sulfation are specific to the Golgi complex and the TGN, respectively. In addition to the removal of the signal peptide in the lumen of the rough endoplasmic reticulum (for review see Walter et al., 1994), additional endoproteolysis is often required in order to render a secretory protein mature and biologically active. Cleavage of regulated secretory proteins occurs predominantly within newly formed secretory granules (Orci et al., 1985), while it has been suggested that constitutive secretory proteins are cleaved mainly within the TGN (Nagahama et al., 1991). Generally, cleavage is performed C-terminal to specific sites consisting of single, paired or multiple basic amino acid motifs. Proteins secreted via the regulated pathway are frequently cleaved after the paired dibasic amino acids Arg-Arg and Lys-Arg. The cleavage site of proteins released from the constitutive pathway requires a more complex basic motif following the rule: an Arg immediately preceding the cleavage site, i.e. at amino acid position -1, and commonly, two additional basic residues at positions -2, -4 or -6 (Watanabe et al., 1993).

3. The family of convertases mediating protein precursor processing

The first eucaryotic endoprotease involved in protein precursor cleavage at basic amino acid motifs was identified as the translational product of the *kex* gene from the lower eucaryote *Saccharomyces cerevisiae* and found to resemble the procaryotic subtilisin. The corresponding protein, kexin, mediates the cleavage of the precursors of the α-mating factor and the killer toxin at the dibasic Lys-Arg motif (Fuller et al., 1989). Subsequently, several homologous endoproteases were identified in higher eucaryotes also. These proteases, which are involved in the processing of pro-proteins and pro-hormones, are collectively called subtilisin/kexin-like **pro**-protein **c**onvertases (SPCs) or **pro**-hormone **c**onvertases (PCs). The mammalian convertases are currently comprised of 7 members: furin which is also called PACE (**pa**ired basic **a**mino acid **c**leavage **e**nzyme), PC1/3, PC2, PC4, PACE4, PC5/6 and, finally, PC7 which is also known as LPC, PC8 or SPC7 (Nakayama, 1997; Gensberg et al., 1998; Creemers et al., 1998; Seidah et al., 1998). In addition, isoforms of PACE4 (Zhong et al., 1996; Tsuji et al., 1997; Mori et al., 1997), PC4 (Seidah et al., 1992) and PC5/6 (Nakagawa et al., 1993; De Bie et al., 1996), produced by alternative splicing, have been identified.

Each member of the convertase family exhibits a unique tissue distribution; different cell types were found to express individual combinations of these enzymes (Seidah et al., 1994). PC4 was found to be restricted to testicular germ cells (Seidah et al., 1992; 1994). PC1/3, PC2, and PC5A were localised to the secretory granules of regulated secreting cells, e.g. endocrine and neuroendocrine cells (Smeekens et al., 1990; Smeekens et al., 1991; Malide et al., 1995; Tanaka et al., 1996; De Bie et al., 1996). PC1/3 and PC2 were shown to process regulated secretory proteins, e.g. neuroendocrine-specific pro-insulin (Smeekens et al., 1992), POMC (Benjannet et al., 1991), pro-glucagon (Rouillé et al., 1994) and pro-somatostatin (Xu et al., 1994), at the dibasic amino acid motifs Arg-Arg and Lys-Arg (see Rouillé et al., 1995, for review). PACE4, PC5B, and PC7/LPC/PC8 are present in a variety of different tissues, both in regulated and constitutively secreting cells (Seidah et al., 1994). Furin is expressed ubiquitously. The widespread distribution and the sublocalisation of furin, PC5B, and PC7 to the TGN suggests that these endoproteases are involved in the processing of pro-proteins in the constitutive pathway (Molloy et al., 1994; Schäfer et al., 1995; De Bie et al., 1996; Seidah et al., 1996; Van de Loo et al., 1997).

4. The endoprotease furin

The transcription unit of the *fur* gene was originally identified as the open reading frame located immediately upstream of the fes proto-oncogene (**f**es **u**pstream **r**egion; Roebroek et al., 1986). The human *fur* cDNA, isolated from a liver cell line cDNA library (Wise et al., 1990; Van den Ouweland et al., 1990; Barr et al., 1991), encodes a primary translation product of 794 amino acids. The structural organisation of furin reveals different domains which exhibit significant homology with other members of the family of convertases, as schematically shown in figure 2.

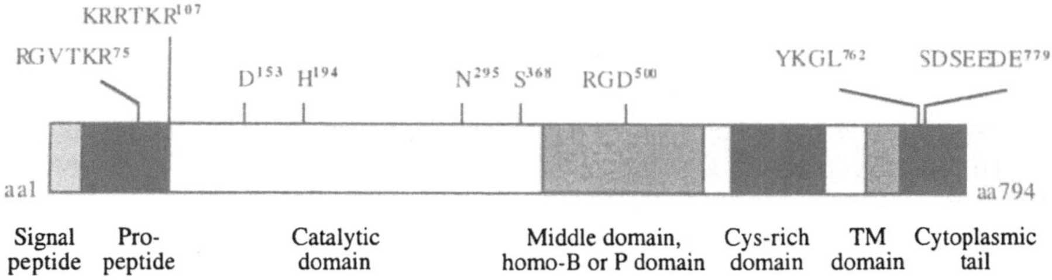

Figure 2. Schematic representation of the structural domains of furin. The amino acid motifs important for maturation, catalytic activity and intracellular localisation are indicated, and discussed in the text. The numbers give the position of individual amino acids within the furin molecule, relative to the translational initiator amino acid methionine.

An N-terminal signal peptide indicates its targeting to the secretory pathway. The adjacent propeptide, which has been suggested to be crucial to proper folding of the protease during its synthesis in the ER, must be removed in order to render furin functionally active (Rehemtulla et al., 1992). Precursor cleavage occurs in the ER by an autocatalytic intramolecular process past the sequence Arg-Thr-Lys-Arg^{107} (Leduc et al., 1992; Creemers et al., 1993). This cleavage is required for the translocation of furin to the TGN and is essential to, but not sufficient for, full activation (Molloy et al., 1994; Vey et al., 1994; Creemers et al., 1995; Anderson et al., 1997). In fact, the propeptide remains associated with the protease in the neutral environment of the ER where it autoinhibits the enzymatic function (Anderson et al., 1997). The release of the propeptide and the resulting activation of furin was demonstrated to be a pH and calcium-dependent process, requiring a TGN characteristic environment (pH 6, calcium in the millimolar range), which ultimately led to an additional cleavage within the propeptide C-terminal to the sequence Arg-Gly-Val-Thr-Lys-Arg[75]. The propeptide may thus regulate furin activity in different cellular compartments.

The catalytic domain is located at the N-terminus of the mature protein and consists of approximately 300 amino acids. It contains the conserved catalytic triad Asp, His and Ser and the oxyanion hole Asn residue. The catalytic triad is essential for autocatalytic maturation and substrate processing. The Asn residue is involved in the latter also (Creemers et al., 1993).

The homo-B domain, also called P or middle domain, is about 150 amino acids in length and located adjacent to the catalytic domain. It contains the Arg-Gly-Asp motif (RGD), an integrin binding sequence. Partial deletion (Hatsuzawa et al., 1992; Creemers et al., 1993) of or mutation (Takahashi et al., 1995A; Spence et al., 1995) within the homo-B domain of furin retained the unprocessed pro-protease in the ER, suggesting a role of the homo-B domain in folding of furin into a conformation appropriate for efficient autocatalytic propeptide removal. For PC1/3, the integrity of the Arg-Arg-Gly-Asp-Leu (RRGDL) motif has been found to be critical for correct sorting of the protease to the regulated secretory pathway, where it is subsequently converted into a more active form by a C-terminal autocatalytic truncation (Lusson et al., 1997). Recently, evidence has been gathered for a role of the homo-B domain in calcium and pH dependence as well as substrate specificity of the individual convertases (Zhou et al., 1998).

The C-termini of the individual convertases vary from one another. Furin contains a cysteine-rich region followed by a transmembrane domain and, finally, a cytoplasmic tail. To date, the role of the cysteine-rich region remains unclear. The transmembrane domain anchors furin in the membranes of the TGN/endosomal system, while the cytoplasmic tail contains targeting signals essential for compartmentalisation and intracellular trafficking between TGN, cell surface and endosomes. Deletion of the cytoplasmic tail results in mislocalisation of the protease (Bosshart et al., 1994; Molloy et al., 1994; Schäfer et al., 1995; Jones et al., 1995; Voorhees et al., 1995; Creemers et al., 1996; Dittié et al., 1997). One of these targeting determinants, the YKGL[765] motif, has been found to be essential for the retrieval of furin from the cell surface to the TGN by promoting the internalisation of the endoprotease from the plasma membrane (Takahashi et al., 1995). This sequence resembles the YXXØ consensus sequence (where Ø is an amino acid with a bulky hydrophobic side chain, i.e. Leu, Ile, Phe, Met or Val). Motifs conforming to this consensus were shown to direct internalisation of receptors from the cell surface, as well as to target transmembrane proteins to the TGN or to lysosomal compartments. This motif has also been shown to interact with the 'medium' subunits of the adaptor complexes AP-1 and AP-2, which are components of the clathrin coated vesicles (Ohno et al., 1996). A second targeting signal, the acidic cluster SDSEEDE[783], contains two Ser residues both of which can be phosphorylated by casein kinase II (Jones at al., 1995; Takahashi et al., 1995). This motif has been demonstrated to mediate the localisation of furin to the TGN (Jones et al., 1995; Schäfer et al., 1995; Takahashi et al., 1995; Voorhees et al., 1995). In addition, the phosphorylation-dependent interaction of furin with AP-1, the TGN/ISG specific adaptor complex, is essential for the removal of furin from the regulated pathway and, hence, for its correct localisation to the TGN and endosomes (Dittie et al., 1997). The cytoskeletal protein ABP-280 was also shown to directly interact with the cytoplasmic domain of furin and to direct furin sorting from the endosomes to the TGN, thereby modulating the processing activity of furin in the endocytic pathway (Liu et al., 1997). Thus, the C-terminal domain of furin is essential for its individual localisation in specific compartments, where it exerts its biological function.

For functional activity, the C-terminus is dispensable. Deletion mutants of furin, lacking the transmembrane domain and the cytoplasmic tail, were found to exhibit enzymatic properties similar to the ones shown by the full length protein (Thomas et al., 1992; Rehemtulla et al., 1992; Molloy et al., 1992; Oda et al., 1992; Takahashi et al., 1994; Bravo et al., 1994; Preininger et al., 1998). Furthermore, overexpression of full length furin results in the secretion of a naturally truncated form termed 'shed' furin, lacking the transmembrane and cytosolic domains (Vidricaire et al., 1993; Vey et al., 1994; Schäfer et al., 1995). It remains to be determined whether 'shedding' occurs only upon expression of furin beyond a certain threshold level, or, rather, is part of a physiological phenomenon involved in furin trafficking between the TGN and the cell surface.

The substrate specificity of furin has been extensively analysed by coexpression experiments (Hosaka et al., 1991; Watanabe et al., 1993; Takahashi et al., 1994) and by in vitro studies (Molloy et al., 1992; Hatsuzawa et al., 1992; Oda et al., 1992), the latter of which employed purified furin or derivatives thereof. Although furin has been found to preferentially cleave C-terminal to the consensus sequence Arg-X-Arg/Lys-Arg, it is also able to process cleavage sites containing Arg at position -1 and two additional amino acid residues, either Arg or Lys, at positions -2, -4 or -6. The amino acid at position +1 may modulate the cleavage efficiency; amino acids with a hydrophobic aliphatic side chain such as Val, Leu or Ile significantly impair cleavage. In a naturally occurring pro-albumin mutant, pro-albumin Bleinheim, which harbors a Val residue rather than the wildtype Asp at position +1, cleavage is entirely abolished (Oda et al., 1992; Brennan et al., 1994). The substrate specificity of furin has been proposed to be governed by negatively charged amino acid residues located in the vicinity of the substrate binding region (Siezen et al., 1994). Mutation of these negatively charged amino acids alters the substrate specificity (Creemers et al., 1993; Roebroek et al., 1994).

At present, furin has experimentally been shown to process the precursors of a wide variety of proteins which transit the secretory pathway (table 1). These substrates include hormone peptides, growth factors, receptors, coagulation factors, complement proteins and matrix metalloproteinases, as well as several viral envelope proteins and bacterial toxins. Apart from the precursors listed in table 1, several other pro-proteins harbour cleavage sites also matching the furin consensus sequence; however, their actual cleavage by furin needs to be confirmed experimentally. Quite a few of these furin substrates represent attractive candidates for the development of potential therapeutic agents, particularly those proteins which play a crucial role in coagulation/fibrinolysis.

5. Biotechnological exploitation of furin

Blood coagulation and fibrinolysis is highly complex and strictly controlled physiologically. Genetic and acquired malfunction or absence of individual proteins involved generally results in severe clinical symptoms of bleeding and thrombosis, respectively. Patients suffering from the corresponding diseases have conventionally been treated with components purified from human plasma. The potential of transmitting blood borne adventitious agents, such as immunodeficiency or hepatitis viruses, has led to the industrial development of recombinant plasma proteins, e.g. recombinant factors VIII, IX, and activated VII (Recombinate®, Kogenate®, BeneFIX®, Novoseven®).

The complex nature of these proteins requires eucaryotic cells as hosts for production, since only the presence of intact and complete post-translational modifications ensures a high degree of functionality and efficacy of the desired recombinant protein. The most prominent modifications include vitamin K-dependent γ-carboxylation of individual glutamic acid residues, essential for interaction of the corresponding factors with phospholipid surfaces, complex N- and O-glycosylation potentially regulating secretion, half-life and biological activity, tyrosine sulfation involved in protein-protein interaction, phosphorylation, β-hydroxylation and proteolytic processing (see Kaufmann, 1998, for a review).

The production of large amounts of recombinant protein in mammalian tissue culture systems is routinely achieved by the amplification of the target protein gene using an amplification marker (Schlokat et al., 1997). For production of therapeutic proteins in the milk of transgenic animals, typically multiple copies of the gene of interest are introduced into fertilised eggs by microinjection. The gene of interest is fused to regulatory sequences targeting the expression to the mammary gland (Lubon et al., 1996). High concentration of the desired protein in the milk of the resulting animals is typically accomplished by the densely packed mammary tissue. Unfortunately, in either system, the cellular machinery involved in post-translational modifications often becomes rapidly insufficient when a foreign protein is overexpressed, resulting in the secretion of only partially active molecules which must subsequently be removed by additional purification steps.

The limiting step to be discussed here concerns the endoproteolytic maturation of recombinant proteins. Incomplete propeptide removal and/or conversion of single chain precursors to mature proteins generally abolishes the function of the protein. A variety of strategies, employing furin and derivatives thereof in coexpression or proteinchemical downstream processing procedures, have demonstrated the potential to ensure complete proteolytic cleavage even at large scale production. Currently, biotechnological processes employing furin or derivatives have been developed for the production of the human plasma proteins von Willebrand factor, factor IX, factor X and protein C.

The necessity of propeptide removal became evident early in 1986 by the characterisation of fIX mutant molecules from haemophilic patients. Amino acid alterations at positions -1, -2 and -4 preceding the cleavage site caused the secretion of a propeptide containing non-functional fIX molecules into plasma (Bentley et al., 1986; Diuguid et al., 1986; Ware et al., 1989). The importance of basic amino acid residues

TABLE 1. Protein precursors shown to be susceptible to the endoprotease furin. Listed are the precursors of proteins which have been found to be cleaved by the endoprotease furin either in coexpression experiments or in vitro using partially purified components. The six amino acid residues directly preceding the cleavage site are indicated. Amino acids are given in the one letter code. Lys and Arg residues, which are throught to represent the major determinants of furin specificity, are indicated in bold type.

	-6	-5	-4	-3	-2	-1
Furin consensus sequence	**R**	X	**R**	X	**R/K**	**R**
Growth factors and hormones						
• mouse pro-β-nerve growth factor	T	H	**R**	S	**K**	**R**
• porcine pro-brain-derived neutrophic factor	S	M	**R**	V	**R**	**R**
• human pro-neutrophin-3	T	S	**R**	**R**	**K**	**R**
• human pro-transforming growth factor β-1	S	S	**R**	H	**R**	**R**
• rat pro-Müllerian inhibiting substance	**R**	G	**R**	A	G	**R**
• human pro-insulin-like growth factor 1	P	A	**K**	S	A	**R**
• human pro-endothelin-1	L	**R**	**R**	S	**K**	**R**
• human pro-parathyroid hormone related-peptide	L	**R**	**R**	L	**K**	**R**
• human pro-parathyroid hormone	**K**	S	V	**K**	**K**	**R**
Receptors						
• human insulin pro-receptor	P	S	**R**	**K**	**R**	**R**
• human hepatocyte growth factor pro-receptor	E	**K**	**R**	**K**	**K**	**R**
• human pro-LRP	S	N	**R**	H	**R**	**R**
• human integrin α3-chain	P	Q	**R**	**R**	**R**	**R**
• human integrin α6-chain	N	S	**R**	**K**	**K**	**R**
Plasma proteins						
• human pro-albumin	**R**	G	V	F	**R**	**R**
• human pro-factor IX	L	N	**R**	P	**K**	**R**
• human pro-factor X	L	A	**R**	V	T	**R**
• human factor X single chain precursor	L	E	**R**	**R**	**K**	**R**
• human pro-protein C	L	**R**	I	**R**	**K**	**R**
• human protein C single chain precursor	**R**	S	H	L	**K**	**R**
• human pro-von Willebrand factor	S	H	**R**	S	**K**	**R**
• rat complement pro-C3	A	A	**R**	**R**	**R**	**R**
Matrix metalloproteinases						
• human stromelysin-3	**R**	N	**R**	Q	**K**	**R**
• human MT-MMP1	N	V	**R**	**R**	**K**	**R**
Endoproteases						
• human pro-furin	**K**	**R**	**R**	T	**K**	**R**
• rat endopeptidase 3.4.24.18.	P	S	**R**	P	**K**	**R**
• mouse pro-7B2	Q	**R**	**R**	**K**	**R**	**R**
Viral envelope glycoproteins						
• HIV gp 160	V	Q	**R**	E	**K**	**R**
• human CMV glycoprotein B	H	N	**R**	T	**K**	**R**
• mouse mammary tumor virus-7 superantigen	E	N	**R**	**K**	**R**	**R**
• avian influenza virus A hemagglutinin	**K**	**K**	**R**	E	**K**	**R**
• measles virus F0	S	**R**	**R**	H	**K**	**R**
• newcastle disease virus F0	G	**R**	**R**	Q	**R**	**R**
• sindbis virus gp E2	S	G	**R**	S	**K**	**R**
• human parainfluenza virus type 3 F0	D	P	**R**	T	**K**	**R**
• ebola virus glycoprotein	G	**R**	**R**	T	**R**	**R**
Bacterial exotoxins						
• anthrax toxin protective antigen	N	S	**R**	**K**	**K**	**R**
• diphteria toxin	G	N	**R**	V	**R**	**R**
• pseudomonas exotoxin A	**R**	H	**R**	Q	P	**R**
• shiga toxin	A	S	**R**	V	A	**R**

for successful propeptide removal at these positions was subsequently confirmed by mutagenesis of the corresponding positions in protein C, factor IX, and von Willebrand factor (Foster et al., 1987; Galeffi et al., 1987; Rehemtulla et al., 1992A). Apart from propeptide removal, protein C and factor X require additional proteolysis of the polypeptide chain precursor for the conversion into a biologically active heterodimer.

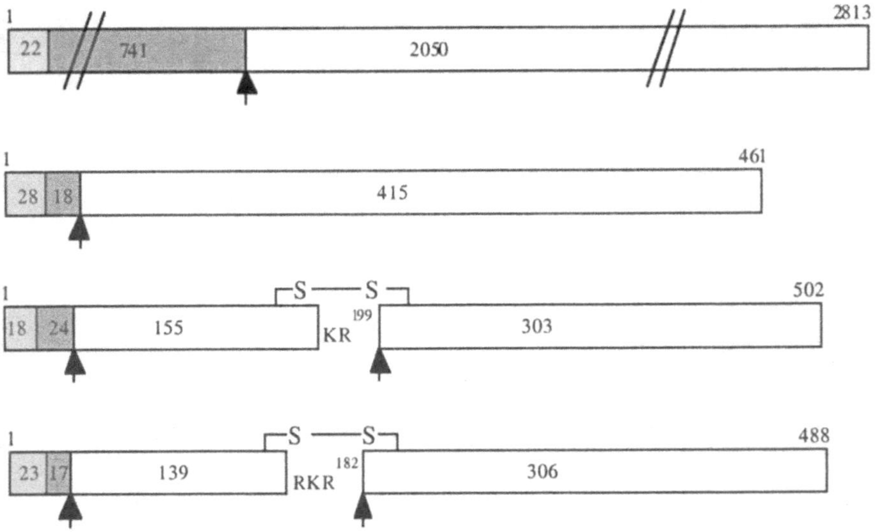

Amino acid sequences at the cleavage sites:

	-6	-5	-4	-3	-2	-1	+1
hvWF propeptide	S	H	**R**	S	**K**	**R**	S
hfIX propeptide	L	N	**R**	P	**K**	**R**	Y
hPC propeptide	L	**R**	I	**R**	**K**	**R**	A
hPC single chain	**R**	S	H	L	**K**	**R**	D
hfX propeptide	L	A	**R**	V	T	**R**	A
hfX single chain	L	E	**R**	**R**	**K**	**R**	S
Consensus cleavage sequence	**R**	X	**R**	X	**R/K**	**R**	X

Figure 3. Schematic representation of the structural organisation of the von Willebrand factor, factor IX, protein C, and factor X precursor molecules. The signal peptide is depicted in light, the propeptide in dark grey and the mature molecule in white. The numbers indicate the size of the respective protein moiety in amino acids. Endoproteolytic processing sites during biosynthesis which are potentially used by furin are marked by arrows. The connecting dipeptide Lys-Arg^{199} and tripeptide Arg-Lys-Arg^{182} in protein C and factor X, respectively, are removed after endoproteolysis by exoproteolytic trimming.

5.1. VON WILLEBRAND FACTOR PROPEPTIDE REMOVAL BY FULL LENGTH FURIN

Von Willebrand factor (vWF) is a multimeric plasma glycoprotein that mediates platelet adhesion to the subendothelium at the site of vascular injury, promotes platelet aggregation and stabilises factor VIII in circulation. Recombinant vWF (rvWF) is currently being developed as a potential therapeutic agent for the treatment of patients suffering from von Willebrand disease (Lethagen, 1995). vWF is synthesised in megakaryocytes and endothelial cells as a large molecule containing a signal peptide 22 amino acids in length, a propeptide harbouring 741 amino acids, and the mature monomer consisting of 2050 amino acids (fig. 3). After removal of the signal peptide, coinciding with translocation to the endoplasmic reticulum, the precursor molecules undergo dimerization through intermolecular disulphide linkage between the C-termini of two monomers. Subsequently, the resulting dimers multimerise in order to form high molecular weight forms up to 20000 kDa in size. These multimers are released constitutively as well as stored in specialised secretory granules called Weibel-Palade bodies and α-granules. The release from these storage organelles is triggered by appropriate stimuli. It is throught that, prior to secretion, propeptides are removed and mature vWF multimers and free propeptides are released (see Furlan, 1996, for review). Despite the requirement of the propeptide for the intracellular multimerisation of vWF (Leyte et al., 1991; Wise et al., 1991), which in turn is believed to be essential for functional activity, its subsequent removal is necessary to enable vWF to interact with factor VIII.

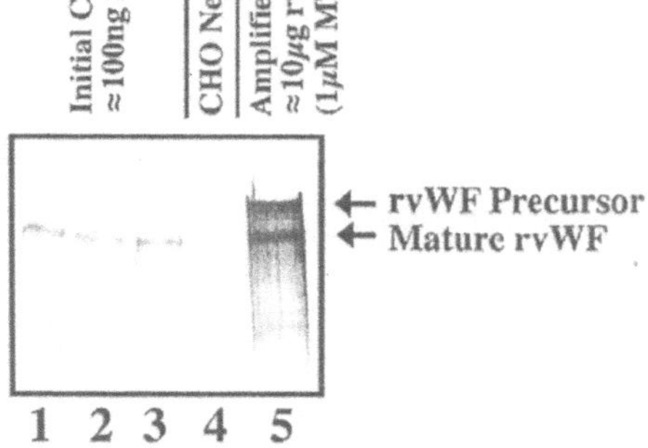

Figure 4. Insufficient rvWF precursor processing by endogenous furin upon overexpression. rvWF molecules in conditioned media derived from three low and one high yield CHO cell clones were visualised by Western blotting under reducing and denaturing conditions. Lanes 1 to 3, low yield CHO-rvWF clones; lane 4, unmanipulated CHO cells; lane 5, amplified CHO-rvWF cell clone. From Preininger et al. (1998), with kind permission from Kluwer Academic Publishers.

Human vWF was expressed in a variety of heterologous cells such as monkey kidney derived COS-1 and CV1 cells, baby hamster kidney cells BHK, human and mouse fibroblasts, as well as Chinese hamster ovary cells CHO (Bonthron et al., 1986; Verweij et al., 1987; Meulien et al., 1992). Analysis of secreted rvWF molecules derived from permanent CHO-rvWF cell clones demonstrated that even cells devoid of the regulated secretory pathway are capable of removing the propeptide and of assembling high molecular weight vWF multimers. Depending on the expression level achieved, significant amounts of the recombinant molecules were, however, found to still contain propeptide (fig. 4). In conditioned media from cell clones mediating low expression from 50 to 200ng rvWF/10^6 cells x day, only mature rvWF molecules were detectable. In supernatants from amplified CHO/rvWF cells with 100 fold increased expression levels (10μg/10^6 cells x day) approximately 50% of the secreted rvWF molecules were found to still contain the propeptide moiety (Preininger et al., 1998). Propeptide removal thus had become severely insufficient upon overexpression. rvWF multimer analysis with intermediate yield CHO-rvWF cell derived conditioned medium demonstrated that mature monomers and precursor molecules formed high molecular weight heteromultimers (Fischer et al., 1994). rvWF precursor was the first substrate shown to be properly cleaved by furin upon transient coexpression from furin and vWF cDNAs in COS-1 cells (Wise et al., 1990; Van de Ven et al., 1990). In COS-1 cells, processing of rvWF has been postulated to occur exclusively intracellularly when expressed with recombinant furin (Rehemtulla et al., 1992, 1992A).

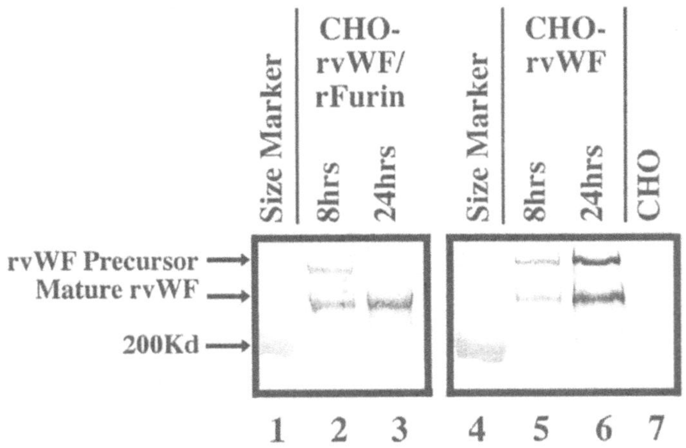

Figure 5. Proteolytic processing of rvWF precursor molecules alone and upon coexpression with full length rfurin. rvWF from a stable intermediate yield CHO-rvWF cell clone (lanes 5 and 6) and from a successor cell clone CHO-rvWF/rfurin, additionally transfected in order to express full length rfurin (lanes 2 and 3), was visualised by Western blotting. The conditioned media were collected after 8 (lanes 2 and 5) or 24 hours (lanes 3 and 6). Lanes 1 and 4, 200 kilodalton size marker; lane 7, 24 hours conditioned medium derived from unmanipulated CHO cells. From Preininger et al. (1998), with kind permission from Kluwer Academic Publishers.

In order to ensure complete rvWF propeptide removal in stably transfected higher yield CHO-rvWF cells also, the cells were additionally transfected with the entire furin cDNA (Fischer et al., 1995; Schlokat et al., 1996). In the conditioned medium of the resulting CHO-rvWF/rfurin cells, producing intermediate rvWF levels of about 2μg/10^6 cells x day, only propeptide free rvWF molecules were detectable. Coexpression of rvWF and rfurin in BHK cells (Lankhof et al., 1996) yielded similar results, demonstrating that propeptide removal of rvWF is improved by the expression of full length furin in stably transfected cells. rvWF multimers derived from such a CHO-rvWF/rfurin clone, when produced at industrial scale in a high cell density perfusion bio-reactor, were found to be composed exclusively of entirely propeptide free rvWF monomers and to exhibit an extraordinary integrity, significantly surpassing the one shown by plasma derived vWF (Fischer et al., 1995; Fischer et al., 1997).

Interestingly, when the supernatant of the CHO-rvWF/rfurin cells was collected more frequently, i.e. every 8 hours, significant amounts of rvWF precursor molecules were again detectable, and the ratio between precursor and mature rvWF molecules was found to closely resemble that found in the conditioned medium of CHO cells solely expressing rvWF (fig. 5). This finding suggests that, despite overexpression, intracellular recombinant full length furin does not significantly contribute to rvWF precursor processing, but, rather, that rvWF propeptide removal occurs largely, or even exclusively, in the conditioned medium. A smaller form of furin was subsequently found to be present in the cell culture supernatant. This form likely is identical to 'shed' furin. It is this form of furin that accomplishes processing of the rvWF precursor molecules extracellularly, as demonstrated experimentally by mixing conditioned media containing 'shed' furin and incompletely processed rvWF, respectively (fig. 6; Schlokat et al., 1996).

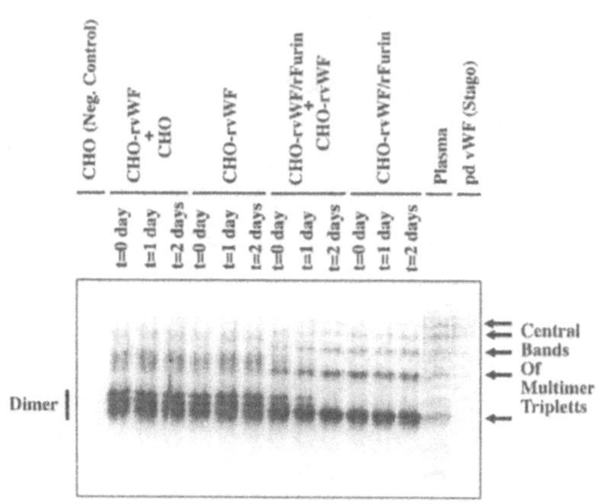

Figure 6. rvWF multimer pattern upon 'shed' rfurin mediated in vitro processing. High resolution multimer analysis was performed using aliquots of conditioned medium from CHO, CHO-rvWF and CHO-rvWF/rfurin cells, incubated alone or after mixing, as indicated. Plasma and plasma derived vWF, as well as CHO cell derived supernatant, were used as controls. The primary rvWF dimer bands, as well as the central bands of plasma derived vWF dimer, tetramer, hexamer, octamer and decamer triplets are indicated. Incubation time (t) is given in days. From Schlokat et al. (1996), with kind permission from Portland Press, Ltd.

These results contrast earlier data according to which rvWF precursor processing in vitro could not be achieved, employing an experimentally truncated furin derivative called PACE SOL (Rehemtulla et al., 1992; 1992A). These contradictions remain to be resolved. Different host cells used, the size of PACE SOL (being slightly larger than 'shed' furin) and other factors may explain these discrepancies. Neither 'shed' furin nor rvWF precursors, when derived from CHO-cells, require a cell-membrane association for processing (Schlokat et al., 1996).

5.2. PRODUCTION OF RECOMBINANT FACTOR IX USING A TRUNCATED SOLUBLE FURIN DERIVATIVE

The blood coagulation factor IX is a vitamin K-dependent plasma glycoprotein. A functional deficiency of factor IX results in haemophilia B. Factor IX (fIX) is synthesised as a precursor polypeptide containing a signal peptide, 28 amino acids in length, an 18 amino acids propeptide and a mature protein of 415 amino acids. The propeptide directs the post-translational vitamin K-dependent γ-carboxylation to the 12 glutamic acid residues of the N-terminus of the mature protein. This modification is required for the molecule's calcium-dependent interaction with phospholipid surfaces (Vermeer, 1990). Removal of the propeptide is crucial to fIX functionality; fIX is able to acquire its correct conformational structure only after propeptide removal. This process is performed by cleavage after the amino acid sequence Leu-Asn-Arg-Pro-Lys-Arg^{46}, which represents the C-terminus of the propeptide (fig. 3).

Since 1985, in a variety of attempts to produce biologically active fIX, several different cell lines expressing recombinant human fIX were established. In all the systems used (Anson et al., 1985; De la Salle et al., 1985; Busby et al., 1985; Kaufman et al., 1986) only partially active rfIX has been obtained, mainly due to insufficient γ-carboxylation and incomplete endoproteolytic propeptide removal. Transient transfection of furin, but not PACE4, into CHO-rfIX cells improved the degree of rfIX processing, thereby increasing the specific activity of the secreted fIX 2-3 fold (Wasley et al., 1993). Subsequently, CHO-rfIX cell clones were established, which permanently coexpressed full length furin or PACE SOL. PACE SOL retains the 715 N-terminal amino acids and lacks the transmembrane and cytosolic domain. Cell clones yielding approximately 4.2μg and 30μg completely processed rfIX/ml were achieved by coexpression of fIX with full length PACE and PACE SOL, respectively (Wasley et al., 1993). This demonstrated that full length furin as well as its truncated form could mediate complete rfIX processing in stably transfected cells. Processing of the rfIX precursors by the truncated, soluble form of furin was reported to occur intracellularly as well as extracellularly, contrasting the situation reported by this group for rvWF propeptide removal.

The commercially available rfIX derives from stably transfected CHO cells co-expressing PACE SOL. Current findings indicate that this expression system is highly reliable with regard to the propeptide removal of rfIX. Extracellular precursor molecules, which had escaped intracellular endoproteolysis, were shown to be efficiently cleaved in the culture medium, during fermentation, and in the downstream manufacturing process, during cell removal and medium concentration (Hamilton et al., 1997).

5.3. PROCESSING OF RECOMBINANT FACTOR X PRECURSORS USING FURIN DERIVATIVES IN VITRO

Factor X (fX) is a vitamin K-dependent glycoprotein which also plays a key role in haemostasis. FX is present in plasma as a heterodimer composed of a light chain and heavy chain linked by a disulphide-bond. During coagulation, it becomes activated to fXa either by the intrinsic fVIIIa/fIXa complex or by the extrinsic fVIIa/tissue factor complex. fXa in combination with factor Va and phospholipids is the physiological activator of prothrombin. FX deficiency results in an enhanced bleeding tendency (see Watzke et al., 1995, for review). FX is synthesised as a precursor molecule consisting of a signal peptide of 23 amino acids in length, a propeptide of 17 amino acids and a 139 amino acids light chain linked to a 306 amino acid heavy chain by the tripeptide Arg-Lys-Arg^{182}. After signal peptide cleavage, fX precursor molecules require two additional endoproteolytic processing steps to reach their biologically active form. Cleavage past the amino acid sequence Arg-Val-Thr-Arg^{40} releases the propeptide, while proteolysis past the amino acid sequence Arg-Arg-Lys-Arg^{182} generates the heterodimer consisting of a 142 amino acids light chain and the heavy chain (fig. 3). Subsequent exoproteolytic trimming of the connecting tripeptide yields the mature light chain C-terminus.

Even at an expression level as modest as 1-2µg rfX/10^6 cells x day in stably transfected HEK 293 cells, a fraction of secreted rfX molecules was shown to already retain covalently linked propeptide (Rudolph et al., 1997). In stable CHO cells at similar expression levels, significant amounts of rfX were found to be secreted even as single chain molecules (Wolf et al., 1991). The difference of rfX processing status observed in the two cell lines is likely to reflect different expression levels of endogenous endoproteases involved in this process. By replacement of the Thr residue by a Lys at position -2, preceding the cleavage site in the rfX propeptide, a furin consensus cleavage sequence was created (Rudolph et al., 1997). Thus, complete propeptide removal was accomplished in HEK 293 cells expressing 1.5µg rfX/10^6 cells x day. Upon amplification, CHO cells were recently established that stably express up to 78µg rfX/10^6 cells x day (Himmelspach et al., submitted). At this high expression level, only 50% of the molecules were present as heterodimers. The remaining molecules were found to represent the single chain precursor. N-terminal amino acid sequence analysis revealed that propeptide removal had also become incomplete.

In order to achieve complete processing of precursors at this enormous yield, a downstream in vitro processing procedure must be established. Different C-terminally truncated, secreted furin derivatives were constructed and stably expressed. In order to facilitate purification, affinity epitopes had been added at the C-terminus. One rfurin derivative, termed rfurinΔCys-4xGly-10xHis, has retained the 577 N-terminal amino acids, followed by a short amino acid spacer, in order to ensure steric flexibility, and a 10 residues Histidine tag, and was proven to be particularly suitable. By a one step procedure, using a Ni^{2+}-NTA resin, the molecules were purified almost to homogeneity by an imidazole gradient (Preininger et al., 1998). rfurinΔCys-4xGly-10xHis was found to efficiently perform both rfX cleavage reactions, i.e. single chain precursor processing as well as propeptide removal, in vitro.

rfX expression in furin-deficient cells resulted in the abolition of single chain precursor cleavage, whereas propeptide removal was still accomplished. Thus, an endoprotease different from furin mediates rfX propeptide removal, while endogenous furin was demonstrated to be the physiological endoprotease responsible for single chain precursor cleavage (Himmelspach et al., submitted). The protease cleavage site Arg-Val-Thr-Arg[40] in the propeptide does not represent a typical furin site. Nonetheless, furin was able to successfully mediate propeptide removal in vitro (extracellularly). Thus, furin exhibits relaxed specificity in vitro and may use related, but atypical, sites.

5.4. USE OF FURIN IN TRANSGENIC ANIMALS

Protein C (PC) also is a vitamin K dependent plasma glycoprotein, which acts, in concert with its cofactor protein S, as a regulator of haemostasis by inactivating coagulation factors Va and VIIIa. After signal peptide removal, two additional endoproteolytic cleavages C-terminal to the amino acid sequences Ile-Arg-Arg-Lys[42] and His-Leu-Lys-Arg[199] must be performed for maturation of PC. The first reaction releases the 24 amino acid propeptide. Similar to fX, the second cleavage is required for the conversion of the single chain polypeptide into a heterodimer consisting of a 157 amino acids light chain and a 303 amino acids heavy chain, both of which are linked by a disulphide-bond. The two C-terminal amino acid residues Lys-Arg[199] subsequently are removed by exoproteolytic trimming (fig. 3). Recombinant protein C (rPC) represents a potential therapeutic agent for the treatment of patients suffering from congenital protein C deficiency, disseminated intravascular coagulation and other thrombotic complications.

Human protein C was the first vitamin K-dependent protein successfully produced at high yield in the milk of genetically engineered pigs (Velander et al., 1992; Lee et al., 1995) and mice (Velander et al., 1992A; Drohan et al., 1994; Drews et al., 1995). Microinjection of a PC encoding DNA sequence, under the control of the whey acidic protein promoter, allowed the specific expression of the foreign protein in the mammary glands of lactating animals. Maximal secretion rates of 1mg rPC/ml in swine and 1.65mg rPC/ml in mice were achieved. Similar to mammalian cell systems, limitations of the endogenous processing machinery also occured in the mammary gland tissue of these transgenic animals, resulting in the secretion of 20-30% propeptide-containing and of 40-60% single chain rPC forms in the milk.

In order to achieve complete processing, bigenic mice were subsequently established by co-injection of the PC gene and full length furin cDNA (Drews et al., 1995). rPC molecules present in the milk of these animals were correctly and almost completely converted into their mature forms. Further analysis revealed that a large amount of furin immuno-reactive material, approximately 80 kDa in size, was present in the milk (Paleyanda et al., 1997). These secreted furin molecules, presumably 'shed' furin, were found to mediate single chain precursor processing and propeptide removal extracellularly, as demonstrated by mixing the milk from transgenic mice expressing only rPC with milk derived from bigenic PC/furin animals.

6. Summary and perspectives

The discovery of the mammalian family of subtilisin-like pro-protein convertases has led to a better understanding of the processes involved in the maturation of precursor proteins. Recombinant forms of these proteases were found to improve maturation of recombinant target proteins requiring endoproteolytic cleavages. Coexpression of rvWF with full length rfurin and of rfIX with an experimentally truncated form of furin, which is rapidly released into the conditioned medium, both successfully mediated complete precursor processing in tissue culture. rPC processing in the milk of transgenic animals was similarly accomplished by full length furin coexpression. As a rule of thumb, up to 200ng target protein precursor/10^6 cells x day can be cleaved by endogenous furin present in CHO cells. Processing of target proteins at expression levels up to 2μg and 20μg/10^6 cells x day requires coexpression of full length rfurin and truncated rfurin, respectively (Preininger et al., 1998).

It may be possible to modulate processing efficiency, to some extent, by mutational optimisation of the cleavage site. In addition to Arg residues at amino acid positions -1 and -4 (preceding the cleavage site), basic amino acids at positions -2 and -6 were demonstrated to be beneficial for cleavage. Using prorenin cleavage site mutants, sensitivity to cleavage by rfurin was found to decrease in the following order: Arg-X-Arg-X-Lys-Arg (best cleavage) > X-X-Arg-X-Lys-Arg > Arg-X-Arg-X-X-Arg > Arg-X-X-X-Lys-Arg > X-X-Arg-X-X-Arg (worst cleavage; Takahashi et al., 1994). The nature of the amino acid residues at position +1 and -3 may also have some modulatory influence. Amino acids with a hydrophobic aliphatic side chain such as Val, Leu and Ile at position +1 strongly impair furin cleavage; similarly, acidic residues at position -3 are unsuitable (Jean et al., 1995). Substitution of Tyr by Ala at position +1 resulted in enhanced propeptide removal from fIX precursors by endogenous furin (Meulien et al., 1990).

In cases of recombinant protein expression exceeding yields of 20μg/10^6 cells x day, complete processing can be achieved by longer exposure of the target protein precursor to rfurin (when coexpressed) in the conditioned medium. Where prolonged exposure cannot be tolerated, due to lability of the target protein, an in vitro cleavage process needs to be established. Truncated and affinity epitope tagged rfurin derivatives were expressed and could be purified almost to homogeneity by a one step purification procedure. Purified protein precursor molecules were shown to be rapidly and specifically cleaved by purified rfurin in vitro.

Large amounts of rfurin derivatives may be required for industrial scale-up, e.g. for their immobilisation on a column matrix. Unfortunately, recombinantly expressed rfurin cannot be recovered in a functional form from prokaryotes. By engineering the catalytic domain of the prokaryotic convertase subtilisin BPN' an enzyme, called furilisin, with a substrate cleavage specificity resembling the one exhibited by furin was generated (Ballinger et al., 1996). Recombinantly engineered bacterial endoproteases may thus eventually prove to be very useful tools.

Significant overexpression of full length rfurin, compared to endogenous furin, did not improve the degree of rvWF propeptide removal intracellularly. Attempts to further increase full length rfurin expression by amplification also failed, possibly because of potential toxicity to the host cell, and/or uncontrolled intracellular dissemination and mislocalisation (Ayoubi et al., 1996). The inability of intracellular full length rfurin molecules to contribute to rvWF precursor processing, beyond the endogenous processing level, suggests the need for cofactors which may be crucial to rvWF precursor processing but which are present in only limiting amounts. Further support for the presence of intracellular cofactors stems from the relaxed specificity of rfurin in vitro observed at rfX maturation. Cofactors may direct the substrate molecules into the vicinity of furin and/or aid the target molecules to acquire a favourable conformation which renders the molecules more accessible to furin and more susceptible to cleavage. Smaller furin molecules, such as 'shed' rfurin or experimentally truncated forms, may not need cofactors as the result of reduced steric constraints due to their smaller size. Yet, it remains possible also that pro-proteins, found to be cleaved by furin in coexpression experiments or in vitro studies, are actually processed by proteases different from but related to furin under physiological conditions (i.e. in vivo). The maturation of a given pro-protein by endogenous PACE4, PC5B or PC7, which exhibit a cleavage specificity and a tissue distribution similar to that of furin, cannot be excluded on the basis of these experimental approaches. For a definitive answer, expression of the target protein precursor must be performed in furin-deficient cells, e.g. LoVo, RPE.40 or FD11 (Takahashi et al., 1995A; Spence et al., 1995; Gordon et al., 1997).

Recombinant proteins of pharmaceutical interest may exhibit enormous complexity, requiring a wide variety of post-translational modifications frequently crucial to the proper function of the protein. While amplification of heterologous genetic information in tissue culture cells, the most suitable hosts for complex protein production, has become quite a routine process, post-translational modifications often become incomplete. Thus, recombinant cell clones, ultimately used for production, favour optimal functional activity and complete post-translational modifications, at the expense of the yield potentially achievable. Employing recombinant furin, endoproteolytic maturation of pharmaproteins is the first post-translational modification whose insufficiency could be successfully overcome at high yield and industrial scale expression.

7. Acknowledgements

We thank A. Preininger, G. Mohr, A. Kyd-Rebenburg, V. Stichler and W. Wernhart for their contributions to rvWF and rfX precursor processing by furin. Also, we are grateful to M. O' Rourke for critical reading of the manuscript.

8. References

Anderson, E.D., VanSlyke, J.K., Thulin, C.D., Jean, F., and Thomas, G. (1997) Activation of the furin endoprotease is a multiple-step process: requirements for acidification and internal propeptide cleavage. *EMBO J.* **16,** 1508 - 1518.

Anson, D.S., Austen, D.E.G., and Brownlee, G.G. (1985) Expression of active human clotting factor IX from recombinant DNA clones in mammalian cells. *Nature* **315,** 683 - 685.

Ayoubi, T.A.Y., Meulemans, S.M.P., Roebroek, A.J.M., and Van de Ven, W.J.M. (1996) Production of recombinant proteins in chinese hamster ovary cells overexpressing the subtilisin-like proprotein converting enzyme furin. *Mol. Biol. Rep.* **23,** 87 - 95.

Barr, P.J., Mason, O.B., Landsberg, K.E., Wong, P.A., Kiefer, M.C., and Brake, A.J. (1991) cDNA and gene structure for a human subtilisin-like protease with cleavage specificity for paired basic amino acid residues. *DNA and Cell Biol.* **10,** 319-328.

Ballinger, M.D., Tom, J., and Wells, J.A. (1996) Furilisin: a variant of subtilisin BPN' engineered for cleaving tribasic substrates. *Biochemistry* **35,** 13579 - 13585.

Benjannet, S., Rondeau, N., Day, R., Chrétien, M., and Seidah, N.G. (1991) PC1 and PC2 are proprotein convertases capable of cleaving proopiomelanocortin at distinct pairs of basic residues. *Proc. Natl. Acad. Sci. USA* **88,** 3564 - 3568.

Bentley, A.K., Rees, D.J., Rizza, C., and Brownlee, G.G. (1986) Defective propeptide processing of blood clotting factor IX caused by mutation of arginine to glutamine at position -4. *Cell* **45,** 343 - 348.

Bonthron, D.T., Handin, R.I., Kaufman, R.J., Wasley, L.C., Orr, E.C., Mitsock, L.M., Ewenstein, B., Loscalzo, J., Ginsburg, D., and Orkin, S.H. (1986) Structure of pre-pro-von Willebrand factor and its expression in heterologous cells. *Nature* **324,** 270 - 273.

Bosshart, H., Humphrey, J., Deignan, E., Davidson, J., Drazba, J., Yuan, L., Oorschot, V., Peters, P.J., and Bonifacino, J.S. (1994) The cytoplasmic domain mediates localization of furin to the trans-Golgi network. En route to the endosomal/lysosomal system. *J. Cell Biol.* **126,** 1157 - 1172.

Bravo, D.A., Gleason, J.B., Sanchez, R.I., Roth, R.A., and Fuller, R.S. (1994) Accurate and efficient cleavage of the human insulin proreceptor by the human proprotein-processing protease furin. *J. Biol. Chem.* **269,** 25830 - 25837.

Brennan, S.O., and Nakayama, K. (1994) Cleavage of proalbumin peptides by furin reveals unexpected restrictions at the P_2 and P'_1 sites. *FEBS Lett.* **347,** 80 - 84

Bauerfeind, R., and Huttner, W.B. (1993) Biogenesis of constitutive secretory vesicles, secretory granules and synaptic vesicles. *Curr. Opin. Cell Biol.* **5,** 628 - 635.

Busby, S., Kumar, A., Joseph, M., Halfpap, L., Insley, M., Berkner, K., Kurachi, K., and Woodbury, R. (1985) Expression of active human factor IX in transfected cells. *Nature* **316,** 271 - 273.

Chanat, E., Weiss, U., Huttner, W.B., and Tooze, S.A. (1993) Reduction of the disulphide bond of chromogranin B (secretogranin I) in the trans-Golgi network causes its missorting to the constitutive secretory pathways. *EMBO J.* **12,** 2159 - 2168.

Cool, D.R., Fenger, M., Snell, C.R., and Loh, Y.P. (1995) Identification of the sorting signal motif within pro-opiomelanocortin for the regulated secretory pathway. *J. Biol. Chem.* **270,** 8723 - 8729.

Cool, D.R., Normant, E., Shen, F., Chen, H.C., Pannell, L., Zhang, Y., and Loh, Y.P. (1997) Carboxypeptidase E is a regulated secretory pathway sorting receptor: genetic obliteration leads to endocrine disorders in Cpe(fat) mice. *Cell* **88,** 73 - 83.

Creemers, J.W.M., Jackson, R.S., and Hutton, J.C. (1998) Molecular and cellular regulation of prohormone processing. *Sem. Cell and Dev. Biol.* **9,** 3 - 10.

Creemers, J.W.M., Siezen, R.J., Roebroek, A.J.M., Ayoubi, T.A.Y., Huylebroeck, D., and Van de Ven, W.J.M. (1993) Modulation of furin-mediated proprotein processing activity by site-directed mutagenesis. *J. Biol. Chem.* **268,** 21826 - 21834.

Creemers, J.W.M., Vey, M., Schäfer, W., Ayoubi, T.A.Y., Roebroek, A.J.M., Klenk, H.-D., Garten, W., and Van de Ven, W.J.M. (1995) Endoproteolytic cleavage of its propeptide is a prerequisite for efficient transport of furin out of the endoplasmic reticulum. *J. Biol. Chem.* **270,** 2695 - 2702.

De Bie, I., Marcinkiewicz, M., Malide, D., Lazure, C., Nakayama, K., Bendayan, M., and Seidah, N.G. (1996) The isoforms of proprotein convertase PC5 are sorted to different subcellular compartments. *J. Cell Biol.* **135,** 1261 - 1275.

De la Salle, H., Altenburger, W., Elkaim, R., Dott, K., Dieterlé, A., Drillien, R., Cazenave, J.-P., Tolstoshev, P., and Lecocq, J.-P. (1985) Active γ-carboxylated human factor IX expressed using recombinant DNA techniques. *Nature* **316,** 268 - 270.

Dittie, A.S., Thomas, L., Thomas, G., and Tooze, S.A. (1997) Interaction of furin in immature secretory granules from neuroendocrine cells with the AP-1 adaptor complex is modulated by casein kinase II phosphorylation. *EMBO J.* **16,** 4859 - 4870.

Diuguid, D.L., Rabiet, M.J., Furie, B.C., Liebman, H.A., and Furie, B. (1986) Molecular basis of haemophilia B: a defective enzyme due to an unprocessed propeptide is caused by a point mutation in the factor IX precursor. *Proc. Natl. Acad. Sci. USA* **83,** 5803 - 5807.

Drews, R., Paleyanda, R.K., Lee, T.K., Chang, R.R., Rehemtulla, A., Kaufman, R.J., Drohan, W.N., and Lubon, H. (1995) Proteolytic maturation of protein C upon engineering the mouse mammary gland to express furin. *Proc. Natl. Acad. Sci. USA* **92,** 10462 - 10466.

Foster, D.C., Rudinski, M.S., Schach, B.G., Berkner, K.L, Kumar, A.A., Hagen, F.S., Sprecher, C.A., Insley, M.Y., and Davie, E.W. (1987) Propeptide of human protein C is necessary for gamma-carboxylation. *Biochem.* **26,** 7003 - 7011.

Fischer, B., Mitterer, A., Schlokat, U., DenBouwmeester, R., and Dorner, F. (1994) Structural analysis of recombinant von Willebrand factor: identification of hetero- and homo-dimers. *FEBS Lett.* **351,** 345 - 348.

Fischer, B.E., Schlokat, U., Mitterer, A., Reiter, M., Mundt, W., Turecek, P.L., Schwarz, H.P., and Dorner, F. (1995) Structural analysis of recombinant von Willebrand factor produced at industrial scale fermentation of transformed CHO cells co-expressing recombinant furin. *FEBS Lett.* **375,** 259 - 262.

Fischer, B.E., Schlokat, U., Reiter, M., Mundt, W., and Dorner, F. (1997) Biochemical and functional characterization of recombinant von Willebrand Factor produced at large scale. *Cell. Mol. Life Sci.* **53,** 943 - 950.

Fuller, R.S., Brake, A., and Thorner, J. (1989) Yeast prohormone processing enzyme (KEX2 gene product) is a Ca^{2+}-dependent serine protease. *Proc. Natl. Acad. Sci. USA* **86,** 1434 - 1438.

Furlan, M. (1996) Von Willebrand factor: molecular size and functional activity. *Ann. Hematol.* **72,** 341 - 348.

Galeffi, P., and Brownlee, G.G. (1987) The propeptide region of clotting factor IX is a signal for a vitamin K dependent carboxylase: evidence from protein engeneering of amino acid -4. *Nucl. Acids Res.* **15,** 9505 - 9513.

Gensberg, K., Jan, S., and Matthews, G.M. (1998) Subtilisin-related serine proteases in the mammalian constitutive secretory pathway. *Sem. Cell and Dev. Biol.* **9,** 11 - 17.

Gordon, V.M., Rehemtulla, A., and Leppla, S.H. (1997) A role for PACE4 in the proteolytic activation of anthrax toxin protective antigen. *Infect. Immun.* **65,** 3370 - 3375.

Gut, A., Kappeler, F., Hyka, N., Balda, M.S., Hauri, H.P., and Matter, K. (1998) Carbohydrate-mediated Golgi to cell surface transport and apical targeting of membrane proteins. *EMBO J.* **17,** 1919 - 1929.

Hamilton, M.A., and Charlebois, T.S. (1997) Extracellular propeptide processing of recombinant human factor IX by a secreted form of the endoprotease, PACE, in M.J.T. Carrondo, B. Griffiths and J.L.P. Moreira (eds.), *Animal Cell Technology*, Kluwer Academic Publishers, pp. 495 - 501.

Hatsuzawa, K., Murakami, K., and Nakayama, K. (1992A) Molecular and enzymatic properties of furin, a Kex2-like endoprotease involved in precursor cleavage at Arg-X-Lys/Arg-Arg sites. *J. Biochem.* **111,** 296 - 301.

Hatsuzawa, K., Nagahama, M., Takahashi, S., Takada, K., Murakami, K., and Nakayama, K. (1992) Purification and characterization of furin, a kex2-like processing endoprotease, produced in chinese hamster ovary cells. *J. Biol. Chem.* **267,** 16094 - 16099.

Himmelspach, M., Schlokat, U., Pfleiderer, M., Fischer, B., Antoine, G., Falkner, F.G., and Dorner, F. Recombinant human factor X: high yield expression and maturation by furin mediated in vitro processing. Submitted.

Hosaka, M., Nagahama, M., Kim, W.S., Watanabe, T., Hatsuzawa, K., Ikemizu, J., Murakami, K., and Nakayama, K. (1991) Arg-X-Lys/Arg-Arg motif as a signal for precursor cleavage catalyzed by furin within the constitutive secretory pathway. *J. Biol. Chem.* **266,** 12127 - 12130.

Jean, F., Boudreault, A., Basak, A., Seidah, N.G., and Lazure, C. (1995) Fluorescent peptidyl substrates as an aid in studying the substrate specificity of human prohormone convertase PC1 and human furin and designing a potent irreversible inhibitor. *J. Biol. Chem.* **270,** 19225 - 19231.

Jones, B.G., Thomas, L., Molloy, S.S., Thulin, C.D., Fry, M.D., Walsh, K.A., and Thomas, G. (1995) Intracellular trafficking of furin is modulated by the phosphorylation state of a casein kinase II site in its cytoplasmic tail. *EMBO J.* **14,** 5869-5883.

Kaufman, R.J. (1998) Post-translational modifications required for coagulation factor secretion and function. *Thromb. Haemost.* **79,** 1068-1079.

Kaufman, R.J., Wasley, L.C., Furie, B.C., Furie, B., and Shoemaker, C.B. (1986) Expression, purification, and characterization of recombinant γ-carboxylated factor IX synthesized in chinese hamster ovary cells. *J. Biol. Chem.* **261,** 9622 - 9628.

Lankhof, H., Van Hoeij, M., Schiphorst, M.E., Bracke, M., Wu, Y.-P., Ijsseldijk, M.J.W., Vink, T., De Groot, P.G., and Sixma, J.J. (1996) A3 domain is essential for interaction of von Willebrand factor with collagen type III. *Thromb. Haemost.* **75,** 950 - 958.

Leduc, R., Molloy, S.S., Thorne, B.A., and Thomas, G. (1992) Activation of human furin precursor processing endoprotease occurs by an intramolecular autoproteolytic cleavage. *J. Biol. Chem.* **267,** 14304 - 14308.

Lee, T.K., Drohan, W.N., and Lubon, H. (1995) Proteolytic processing of human protein C in swine mammary gland. *J. Biochem.* **118,** 81 - 87.

Lethagen, S.R. (1995) Pathogenesis, clinical picture and treatment of von Willebrand's disease. *Annals of Medicine* **27,** 641 - 651.

Leyte, A., Voorberg, J., Van Schijndel, H.B., Duim, B., Pannekoek, H., and Van Mourik, J.A. (1991) The pro-polypeptide of von Willebrand factor is required for the formation of a functional factor VIII-binding site on mature von Willebrand factor. *Biochem. J.* **274,** 257 - 261.

Liu, G., Thomas, L., Warren, R.A., Enns, C.A., Cunningham, C.C., Hartwig, J.H., and Thomas, G. (1997) Cytoskeletal protein ABP-280 directs the intracellular trafficking of furin and modulates proprotein processing in the endocytic pathway. *J. Cell Biol.* **139,** 1719 - 1733.

Lubon, H., Paleyanda, R.K., Velander, W.H., and Drohan, W.N. (1996) Blood proteins from transgenic animal bioreactors. *Transfus. Med. Rev.* **10,** 131 - 143.

Lusson, J., Benjannet, S., Hamelin, J., Savaria, D., Chrétien, M., and Seidah, N.G. (1997) The integrity of the RRGDL sequence of the proprotein convertase PC1 is critical for its zymogen and C-terminal processing and for its cellular trafficking. *Biochem. J.* **326,** 737 - 744.

Malide, D., Seidah, N.G., Chrétien, M., and Bendayan, M. (1995) Electron microscopic immunocytochemical evidence for the involvement of the convertases PC1 and PC2 in the processing of proinsulin in pancreatic beta-cells. *J. Histochem. Cytochem.* **43,** 11 - 19.

Marks, M.S., Woodruff, L., Ohno, H., and Bonifacino, J.S. (1996) Protein targeting by tyrosine- and di-leucine-based signals: evidence for distinct saturable components. *J. Cell Biol.* **135,** 341 - 354.

Meulien, P., Nishino, M., Mazurier, C., Dott, K., Pietu, G., Jorieux, S., Pavirani, A., Girma, J.P., Oufkir, D., Courtney, M., and Meyer, D. (1992) Processing and characterization of recombinant von Willebrand factor expressed in different cell types using a vaccinia virus vector. *Thromb. Haemost.* **67,** 154 - 160.

Meulien, P., Balland, A., Lepage, P., Mischler, F., Dott, K., Hauss, C., Grandgeorge, M., and Lecocq, J.P. (1990) Increased biological activity of a recombinant factor IX variant carrying alanine at position +1. *Prot. Engineering* **3,** 629 - 633.

Molloy, S.S., Thomas, L., VanSlyke, J.K., Stenberg, P.E., and Thomas, G. (1994) Intracellular trafficking and activation of the furin proprotein convertase: localization to the TGN and recycling from the cell surface. *EMBO J.* **13,** 18 - 33.

Molloy, S.S., Bresnahan, P.A., Leppla, S.H., Klimpel, K.R., and Thomas, G. (1992) Human furin is a calcium-dependent serine endoprotease that recognizes the sequence Arg-X-X-Arg and efficiently cleaves anthrax toxin protective antigen. *J. Biol. Chem.* **267,** 16396 - 16402.

Mori, K., Kii, S., Tsuji, A., Nagahama, M., Imamaki, A., Hayashi, K., Akamatsu, T., Nagamune, H., and Matsuda, Y. (1997) A novel human PACE4 isoform PACE4E is an active processing protease containing a hydrophobic cluster at the carboxy terminus. *J. Biochem. (Tokyo)* **121,** 941 - 948.

Nagahama, M., Ikemizu, J., Misumi, Y., Ikehara, Y., Murakami, K., and Nakayama, K. (1991) Evidence that differenciates between precursor cleavages at dibasic and Arg-X-Lys/Arg-Arg sites. *J. Biochem.* **110,** 806 - 811.

Nakagawa, T., Murakami, K., and Nakayama, K. (1993) Identification of an isoform with an extremely large Cys-rich region of PC6, a Kex2-like processing endoprotease. *FEBS Lett.* **327,** 165 - 171.

Nakayama, K. (1997) Furin: a mammalian subtilisin/Kex2p-like endoprotease involved in processing of a wide variety of precursor proteins. *Biochem. J.* **327,** 625 - 635.

Nakayama, K., Watanabe, T., Nakagawa, T., Kim, W.-S., Nagahama, M., Hosaka, M., Hatsuzawa, K., Kondoh-Hashiba, K., and Murakami, K. (1992) Consensus sequence for precursor processing at mono-arginyl sites. *J. Biol. Chem.* **267,** 16335- 16340.

Normant, E., and Loh, Y.P. (1998) Depletion of carboxypeptidase E, a regulated secretory pathway sorting receptor, causes misrouting and constitutive secretion of proinsulin and proenkephalin, but not chromogranin A. *Endocrinology* **139,** 2137 - 2145.

Oda, K., Misumi, Y., Ikehara, Y., Brennan, S.O., Hatzuzawa, K., and Nakayama, K. (1992) Proteolytic cleavages of proalbumin and complement pro-C3 in vitro by a truncated soluble form of furin, a mammalian homologue of the yeast Kex2 protease. *Biochem. Biophys. Res. Comm.* **189,** 1353 - 1361.

Ohno, H., Fournier, M.C., Poy, G., and Bonifacino, J.S. (1996) Structural determinants of interaction of tyrosine-based sorting signals with the adaptor medium chains. *J. Biol. Chem.* **271,** 29009 - 29015.

Orci, L., Ravazzola, M., Amherdt, M., Madsen, O., Vassalli, J.D., and Perrelet, A. (1985) Direct identification of prohormone conversion site in insulin-secreting cells. *Cell* **42,** 671 - 681.

Paleyanda, R.K., Drews, R., Lee, T.K., and Lubon, H. (1997) Secretion of human furin into mouse milk. *J. Biol. Chem.* **272,** 15270 - 15274.

Preininger, A., Schlokat, U., Mohr, G., Himmelspach, M., Stichler, V., Kyd-Rebenburg, A., Plaimauer, B., Turecek, P.L., Schwarz, H.P., Falkner, F.G., Fischer, B.E., and Dorner, F. (1998) Strategies for recombinant furin employment in a biotechnological process: complete target protein precursor cleavage. *Cytotechnology* In press.

Rehemtulla, A., Dorner, A.J., and Kaufman, R.J. (1992) Regulation of PACE propeptide-processing activity: requirement for a post-endoplasmic reticulum compartment and autoproteolytic activation. *Proc. Natl. Acad. Sci. USA* **89,** 8235 - 8239.

Rehemtulla, A., and Kaufman, R.J. (1992A) Preferred sequence requirements for cleavage of pro-von Willebrand factor by propeptide-processing enzymes. *Blood* **79,** 2349 - 2355.

Roebroek, A.J.M., Creemers, J.W.M., Ayoubi, T.A.Y., and Van de Ven, W.J.M. (1994) Furin-mediated proprotein processing activity: involvement of negatively charged amino acid residues in the substrate binding region. *Biochimie* **76,** 210 - 216.

Roebroek, A.J.M., Schalken, J.A., Bussemakers, M.J.G., van Heerikhuizen, H., Onnekink, C., Debruyne, F.M.J., Bloemers, H.P.J., and Van de Ven, W.J.M. (1986) Characterization of human c-fes/fps reveals a new transcription unit (fur) in the immediately upstream region of the proto-oncogene. *Mol. Biol. Rep.* **11,** 117 - 125.

Rouillé, Y., Duguay, S. J., Lund, K., Furuta, M., Gong, Q., Lipkind, G., Oliva Jr., A.A., Chan, S.J., and Steiner, D.F. (1995) Proteolytic processing mechanisms in the biosynthesis of neuroendocrine peptides: the subtilisin-like proprotein convertases. *Front. Neuroendocrinol.* **16,** 322 - 361.

Rouillé, Y., Westermark, G., Martin, S.K., and Steiner, D.F. (1994) Proglucagon is processed to glucagon by prohormone convertase PC2 in αTC1-6 cells. *Proc. Natl. Acad. Sci. USA* **91,** 3242 - 3246.

Rudolph, A.E., Mullane, M.P., Porche-Sorbet, R., and Miletich, J.P. (1997) Expression, purification and characterization of recombinant human factor X. *Prot. Express. Purif.* **10,** 373 - 378.

Schäfer, W., Stroh, A., Berghofer, S., Seiler, J., Vey, M., Kruse, M.L., Kern, H.F., Klenk, H.D., and Garten, W. (1995) Two independent targeting signals in the cytoplasmic domain determine trans-Golgi network localization and endosomal trafficking of the proprotein convertase furin. *EMBO J.* **14,** 2424-2435.

Scheiffele, P., Peranen, J., and Simons, K. (1995) N-glycans as apical sorting signals in epithelial cells. *Nature* **378,** 96 - 98.

Schlokat, U., Fischer, B.E., Herlitschka, S., Antoine, G., Preininger, A., Mohr, G., Himmelspach, M., Kistner, O., Falkner, F.G., and Dorner, F. (1996) Production of highly homogeneous and structurally intact recombinant von Willebrand Factor multimers by furin-mediated propeptide removal in vitro. *Biotechnol. Appl. Biochem.* **24,** 257 - 267.

Schlokat, U., Himmelspach, M., Falkner, F.G., and Dorner, F. (1997) Permanent gene expression in mammalian cells: gene transfer and selection, in H. Hauser and R. Wagner (eds.), *Mammalian Cell Biotechnology in Protein Production,* Walter de Gruyter, Berlin, New York, pp. 33 - 52.

Seidah, N.G., Chrétien, M., and Day, R. (1994) The family of subtilisin/kexin like pro-protein and pro-hormone convertases: divergent or shared functions. *Biochimie* **76,** 197 - 209.

Seidah, N.G., Day, R., Hamelin, J., Gaspar, A., Collard, M.W., and Chrétien, M. (1992) Testicular expression of PC4 in the rat: Molecular diversity of a novel germ cell-specific Kex2/subtilisin-like proprotein convertase. *Mol. Endocrinol.* **6,** 1559 - 1570.

Seidah, N.G., Day, R., Marcinkiewicz, M., and Chrétien, M. (1998) Precursor convertases: An evolutionary ancient, cell-specific, combinatorial mechanism yielding diverse bioactive peptides and proteins. *Ann NY Acad. Sci.* **839,** 9 - 24.

Seidah, N.G., Hamelin, J., Mamarbachi, M., Dong, W., Tadros, H., Mbikay, M., Chrétien, M., and Day, R. (1996) cDNA structure, tissue distribution, and chromosomal localization of rat PC7, a novel mammalian proprotein convertase closest to yeast kexin-like proteinases. *Proc. Natl. Acad. Sci. USA* **93,** 3388 - 3393.

Siezen, R.J., Creemers, J.W.M., and Van de Ven, W.J.M. (1994) Homology modelling of the catalytic domain of human furin. A model for the eucaryotic subtilisin-like proprotein convertases. *Eur. J. Biochem.* **222,** 255 - 266.

Smeekens, S.P., Avruch, A.S., LaMendola, J., Chan, S.J., and Steiner, D.F. (1991) Identification of a cDNA encoding a second putative prohormone convertase related to PC2 in AtT20 cells and islets of Langerhans. *Proc. Natl. Acad. Sci. USA* **88,** 340 - 344.

Smeekens S.P., Montag, A.G., Thomas, G., Albiges-Rizo, C., Carroll, R., Benig, M., Phillips, L.A., Martin, S., Ohagi, S., Gardner, P. et al. (1992) Proinsulin processing by the subtilisin-related proprotein convertases furin, PC2, and PC3. *Proc. Natl. Acad. Sci. USA* **89,** 8822 - 8826.

Smeekens, S.P., and Steiner, D.F. (1990) Identification of a human insulinoma cDNA encoding a novel mammalian protein structurally related to the yeast dibasic processing protease Kex2. *J. Biol. Chem.* **265,** 2997 - 3000.

Spence, M.J., Sucic, J.F., Foley, B.T., and Moehring, T.J. (1995) Analysis of mutations in alleles of the fur gene from an endoprotease-deficient chinese hamster ovary cell strain. *Somat. Cell Mol. Gen.* **21,** 1-18.

Tanaka, S., Kurabuchi, S., Mochida, H., Kato, T., Takahashi, S., Watanabe, T., and Nakayama, K. (1996) Immunocytochemical localization of prohormone convertases PC1/PC3 and PC2 in rat pancreatic islets. *Arch. Histol. Cytol.* **59,** 261 - 271.

Takahashi, S., Hatsuzawa, K., Watanabe, T., Murakami, K., and Nakayama, K. (1994) Sequence requirements for endoproteolytic processing of precursor proteins by furin: transfection and in vitro experiments. *J. Biochem.* **116,** 47 - 52.

Takahashi, S., Nakagawa, T., Banno, T., Watanabe, T., Murakami, K., and Nakayama, K. (1995) Localization of furin to the trans-Golgi network and recycling from the cell surface involves Ser and Tyr residues within the cytoplasmic domain. *J. Biol. Chem.* **270,** 28397 - 28401.

Takahashi, S., Nakagawa, T., Kasai, K., Banno, T., Duguay, S.J., Van de Ven, W.J.M., Murakami, K., and Nakayama, K. (1995A) A second mutant allele of furin in the processing-incompetent cell line, LoVo. *J. Biol. Chem.* **270,** 26565 - 26569.

Tsuji, A., Hine, C., Tamai, Y., Yonemoto, K., Mori, K., Yoshida, S., Bando, M., Sakai, E., Mori, K., Akamatsu, T., and Matsuda, Y. (1997) Genomic organization and alternative splicing of human PACE4 (SPC4), kexin-like processing endoprotease. *J. Biochem.* **122,** 438 - 452.

Urbé, S., Tooze, S.A., and Barr, F.A. (1997) Formation of secretory vesicles in the biosynthetic pathway. *Biochem. Biophys. Acta* **1358,** 6-22.

Van de Loo, J.W., Creemers, J.W., Bright, N.A., Young, B.D., Roebroek, A.J., and Van de Ven, W.J. (1997) Biosynthesis, distinct post-translational modifications, and functional characterization of lymphoma proprotein convertase. *J. Biol. Chem.* **272,** 27116 - 27123.

Van den Ouweland, A.M.W., van Groningen, J.J.M., Roebroek, A.J.M., Onnekink, C., and Van de Ven, W.J.M. (1989) Nucleotide sequence analysis of the human *fur* gene. *Nucleic Acids Res.* **17,** 7101 - 7102.

Van de Ven, W.J.M., Voorberg, J., Fontijn, R., Pannekoek, H., van den Ouweland, A.M.W., van Duijnhoven, H.L.P., Roebroek, A.J.M., and Siezen, R.J. (1990) Furin is a subtilisin-like pro-protein processing enzyme in higher eucaryotes. *Mol. Biol. Rep.* **14,** 265 - 275.

Velander, W.H., Johnson, J.L., Page, R.L., Russell, C.G., Subramanian, A., Wilkins, T.D., Gwazdauskas, F.C., Pittius, C., and Drohan, W.N. (1992) High-level expression of a heterologous protein in the milk of transgenic swine using the cDNA encoding human protein C. *Proc. Natl. Acad. Sci. USA* **89,** 12003 - 12007.

Velander, W.H., Page, R.L., Morcol, T., Russell, C.G., Canseco, R., Young, J.M., Drohan, W.N., Gwazdauskas, F.C., Wilkins, T.D., and Johnson, J.L. (1992A) Production of biologically active protein C in the milk of transgenic mice. *Ann. NY Acad. Sci.* **665,** 391 - 403.

Vermeer, C. (1990) γ-Carboxyglutamate-containing proteins and the vitamin K-dependent carboxylase. *Biochem. J.* **266,** 625 - 636.

Verweij, C.L., Hart, M., and Pannekoek, H. (1987) Expression of variant von Willebrand factor (vWF) cDNA in heterologous cells: requirement of the pro-polypeptide in vWF multimer formation. *EMBO J.* **6,** 2885 - 2890.

Vey, M., Schäfer, W., Berghöfer, S., Klenk, H.D., and Garten, W. (1994) Maturation of the trans-Golgi network protease furin: compartmentalization of propeptide removal, substrate cleavage, and COOH-terminal truncation. *J. Cell Biol.* **127,** 1829 - 1842.

Voorhees, P., Deignan, E., van Donselaar, E., Humphrey, J., Marks, M.S., Peters, P.J., and Bonifacino, J.S. (1995) An acidic sequence within the cytoplasmic domain of furin functions as a determinant of trans-Golgi network localization and internalization from the cell surface. *EMBO J.* **14,** 4961 - 4975.

Vidricaire, G., Denault, J.B., and Leduc, R. (1993) Characterization of a secreted form of human furin endoprotease. *Biochem. Biophys. Res. Commun.* **195,** 1011 - 1018.

Watanabe, T., Murakami, K., and Nakayama, K. (1993) Positional and additive effects of basic amino acids on processing of precursor proteins within the constitutive secretory pathway. *FEBS Lett.* **320,** 215 - 218.

Walter, P., and Johnson, A.E. (1994) Signal sequence recognition and protein targeting to the endoplasmic reticulum membrane. *Annu. Rev. Cell Biol.* **10,** 87-119.

Ware, J., Diuguid, D.L., Liebman, H.A., Rabiet, M.-J., Kasper, C.K., Furie, B.C., Furie, B., and Stafford, D.W. (1989). Factor IX San Dimas. *J. Biol. Chem.* **264,** 11401 - 11406.

Wasley, L.C., Rehemtulla, A., Bristol, J.A., and Kaufman, R.J. (1993) PACE/Furin can process the vitamin K-dependent pro-factor IX precursor within the secretory pathway. *J. Biol. Chem.* **268,** 8458 - 8465.

Watzke, H.H., and High, K.A. (1995) Factor X, in K.A. High and H.R. Roberts (eds.), *Molecular Basis of Thrombosis and Hemostasis,* Marcel Dekker, Inc., pp 239 - 255.

Wise, R.J., Barr, P.J., Wong, P.A., Kiefer, M.C., Brake, A.J., and Kaufman, R.J. (1990) Expression of a human proprotein processing enzyme: correct cleavage of the von Willebrand factor precursor at a paired basic amino acid site. *Proc. Natl. Acad. Sci. USA* **87,** 9378 - 9382.

Wise, R.J., Dorner, A.J., Krane, M., Pittman, D.D., and Kaufman, R.J. (1991) The role of von Willebrand factor multimers and propeptide cleavage in binding and stabilization of factor VIII. *J. Biol. Chem.* **266,** 21948 - 21955.

Wolf, D.L., Sinha, U., Hancock, T.E., Lin, P.-H., Messier, T.L., Esmon, C.T., and Church, W.R. (1991) Design of constructs for the expression of biologically active recombinant human factors X and Xa. *J. Biol. Chem.* **266,** 13726 - 13730.

Xu, H., and Shields, D. (1994) Prosomatostatin processing in permeabilized cells. *J. Biol. Chem.* **269,** 22875 - 22881.

Zhong, M., Benjannet, S., Lazure, C., Munzer, S., and Seidah, N.G. (1996) Functional analysis of human PACE4-A and PACE4-C isoforms: identification of a new PACE4-CS isoform. *FEBS Lett* **396,** 31 - 36.

Zhou, A., Martin, S., Lipkind, G., LaMendola, J., and Steiner, D.F. (1998) Regulatory roles of the P domain of the subtilisin-like prohormone convertases. *J. Biol. Chem.* **273,** 11107 - 11114.

BACULOVIRUS EXPRESSION VECTORS: IMPROVING VIRUS-INFECTED CELL VIABILITY AND RECOMBINANT PROTEIN SECRETION

R.D. POSSEE[1]*, L. McCARROLL[2], C.J. THOMAS[2], S.G. MANN[2] and L.A. KING[2]

[1]NERC Institute of Virology and Environmental Microbiology, Mansfield Road, Oxford, OX1 3SR UK. [2]School of Biological and Molecular Sciences, Oxford Brookes University, Gipsy Lane Campus, OX3 OBP UK.

** Corresponding author, fax +44 1865 281696, e-mail rpossee@worf.molbiol.ox.ac.uk*

1. Introduction

Baculoviruses are arthropod-specific viruses with a genome of double-stranded DNA (90-160 kilobase pairs, kbp). No baculoviruses have been described that replicate in vertebrates or plants. The viruses have to be propagated in arthropod larvae or in arthropod cells. Baculoviruses have been developed as successful expression vectors for the following reasons: (a) baculoviruses are easy to work with and safe to handle, (b) recombinant baculoviruses are genetically stable, (c) baculoviruses can be used as helper-independent vectors of foreign gene products, (d) large genes and cistrons can be expressed and (e) high levels of the desired expression products are often obtained. Although the system was originally established to enable the synthesis of individual, native gene products, a range of other expression vectors are now available. Some permit the production of recombinant proteins fused with various heterologous sequences to facilitate easy purification. Other vectors can be used to produce multiple proteins in insect cells. Recombinant products can also be incorporated as structural proteins into the budded form of the baculovirus particle. It is also proposed to use baculoviruses as biosafe vectors for human somatic gene therapy. The preparation of recombinant baculoviruses is straightforward and has benefited from the development of a number of novel procedures that enable the non-insect virologist to use the system with ease. Recent reviews provide an historical perspective of the development of the baculovirus expression systems and describe various features which are not summarised in this article (King and Possee, 1992; O'Reilly *et al*, 1992; Possee, 1997; Jarvis, 1997)

The baculovirus replication cycle in its insect host is complicated, with some virus genes expressed early and in low amounts while others are made in great abundance very late in the infection course (Friesen, 1997; Lu and Miller, 1997). By substituting the coding sequence of a foreign gene for the coding sequence of one of the very late, non-essential baculovirus genes, or by placing the foreign gene sequence under a duplicated copy of a very late virus gene promoter, a foreign gene may be expressed to high level. Although most foreign gene products expressed from a very late baculovirus promoter are made in large quantities, this is not always the case. Conversely. some genes which are expressed poorly in bacteria or in other systems are expressed well using baculoviruses and insect cells.

The potential of a baculovirus as an expression system needs to be considered in terms of the natural course of virus infection in the host cell. Although there are baculoviruses that are specific for coleopteran, hymenopteran or other arthropod species (e.g. shrimps, crabs etc.), most of the expression work has utilised baculoviruses of certain Lepidoptera. The most widely used baculovirus expression vector derives from the *Autographa californica* (Alfalfa looper) multiple nucleopolyhedrovirus (AcMNPV). A closely related baculovirus, *Bombyx mori* (Bm) NPV, isolated from the silkworm has also been developed as an expression vector. This Chapter will concentrate mainly on AcMNPV with occasional reference to BmNPV where appropriate and describe recent developments that promise to improve further this already highly successful expression vector system.

2. Baculovirus replication

Several hundred baculoviruses have been described, although a paucity of information makes it difficult to establish genetic relationships. By convention, baculoviruses are named in accordance with their host of origin, although (at least for some of the insect-specific baculoviruses) members cross-infect several host species. This system makes it likely that some viruses have been isolated on more than one occasion and given different names. For example, viruses very similar to AcMNPV have also been isolated from *Trichoplusia ni* (cabbage looper), *Galleria mellonella* (Great wax moth) and *Rachiplusia ou* (Soybean looper).

The classification system of baculoviruses has undergone several revisions in recent years, but currently is based on the existence and type of occlusion structures or bodies that baculoviruses make during an infection (Volkman *et al.*, 1995). These occlusion bodies are termed polyhedra or granules (Funk *et al.*, 1997). Polyhedra contain virus nucleocapsids that are encapsulated by a membrane to form virus particles. These virus particles are surrounded by a large crystalline matrix of protein composed of multiple copies of a single species termed the polyhedrin protein (29 kDa for AcMNPV). The entire structure is called a polyhedron and is bounded by another membrane, reported to be composed of carbohydrate with a phosphorylated 34 kDa protein covalently associated via a thiol linkage. Polyhedra are formed within the nucleus of an infected cell and are released from it when lysis occurs. The virus particles within polyhedra may contain single or multiple nucleocapsids. This trait appears to be genetically determined, although some viruses exhibit considerable variation in the number of nucleocapsids packaged within each virus particle. Hence, the nucleopolyhedovirus genus of the *Baculoviridae* comprises those viruses that have several virus particles with one or more nucleocapsids in each polyhedron (e.g. AcMNPV). A second genus, the granuloviruses, contains members with one nucleocapsid per virus particle and single virus particles in each occlusion body or granule. The granules are also formed within the nuclei of infected cells; although sometimes the nuclear membrane breaks down to form a nucleus-cytoplasm milieu. The primary function of the polyhedron and granule is to protect the occluded virus particles in the environment until they encounter an insect host. In the high pH environment of the insect larval midgut the polyhedra are solubilised and liberate the virus particles to infect susceptible cells.

In addition to the occluded virus species, the nucleopolyhedroviruses and granuloviruses make an alternative infectious form of the virus during the replication process. This is a virion that buds from the cell plasma membrane. It comprises a single viral nucleocapsid enveloped by lipid derived from the infected cell surface as it buds through the plasma membrane. The viral envelope also contains a virus-encoded glycoprotein (GP67; Blissard and Rohrmann, 1989; Whitford *et al.*, 1989) which is not found in the virus particles occluded by polyhedra. These budded viruses (BV) serve to spread infection from cell to cell within the host. After several days, the virus-infected insect dies and liquefies to liberate polyhedra from the cadaver that disseminate infection to other susceptible insect hosts.

Most studies on the replication of AcMNPV have utilised insect cell cultures, which are broadly analogous to the spread of virus within an insect host, after initial replication in the midgut epithelium. Tissue-specific differences in virus replication are lost using this approach, but much valuable information has been accumulated from these studies conducted *in vitro*. The infection begins with adsorption and endocytosis of the virus particle followed by uncoating as the viral membrane probably fuses with endosomes; GP67 is believed to have a pivotal role in these events (Williams and Faulkner, 1997). Thereafter, virus DNA is introduced into the cell nucleus. At least four phases of viral mRNA transcription and translation are recognised, although sometimes the first two phases, immediate early and delayed early are grouped together as early. Transcription of early genes is mediated by the action of a cellular DNA dependent RNA polymerase II; the virus DNA is infectious (Friesen, 1997). A subset of these genes is then responsible for initiating virus DNA replication, which is required for transcription of the late and very late baculoviruses genes (Lu *et al.*, 1997; Lu and Miller, 1997). The late gene products are required for the formation of infectious virus particles that bud out of the cell from about 12 h p.i. The very late virus genes are transcribed from about 12 h p.i., but reach maximum levels of expression at about 24-48 h p.i. Both late and very late virus genes are transcribed by an alpha-amanitin-resistant RNA polymerase that is probably encoded by the virus genome, although host cell proteins may have a role (Guarino *et al.*, 1998). Baculovirus gene expression can be considered as a cascade in which the products of each phase of transcription are required for activation of the next set of virus genes.

In the very late phase of baculovirus gene expression two major virus proteins are synthesised; polyhedrin and p10. Together, these proteins can account for about 25-50% of the total cellular protein as assessed by staining of extracts fractionated in polyacrylamide gels. The production of polyhedrin is required for the formation of occlusion bodies containing infectious virus particles. The polyhedrin gene can be inactivated without affecting budded virus production, but polyhedron formation is prevented. The role of the p10 protein is ill defined, but it can be deleted from the virus genome without affecting the synthesis of polyhedra or their infectivity *in vivo*. However, polyhedra formed in the absence of P10 appear more fragile than their wild type counterparts and are easily fragmented by physical stress. They appear to lack the polysaccharide-containing membrane that normally surrounds each polyhedron. The appearance of virus-infected cells after ultrathin sectioning suggests that the p10 protein aids in the application of this membrane to the surface of the polyhedron.

3. Baculovirus expression vectors

The two viruses most commonly used as expression vectors are AcMNPV and BmNPV. The use of BmNPV is attractive because silkworm larvae can be used as an *in vivo* bioreactor for the production of recombinant proteins. Although it is possible to insert genes directly into the virus genome via ligation of DNA molecules *in vitro*, most recombinant baculoviruses are made by an indirect route, involving the use of a plasmid transfer vector. Briefly, a transfer vector is constructed by inserting a portion of the baculovirus genome spanning the polyhedrin gene in a bacterial plasmid. The polyhedrin gene coding region is replaced by the sequence to be expressed. Many different transfer vectors are now available that permit the insertion of foreign coding regions with end's compatible with a wide range of restriction endonucleases, so the requirement for tailing of fragments is minimal. This plasmid is mixed with infectious virus DNA and used to co-transfect insect cells. Homologous recombination between the two molecules effects transfer of the foreign gene into the virus genome while remaining under the control of the polyhedrin gene promoter. The recombinant viruses, lacking the polyhedrin gene, are identified in a simple plaque assay by the absence of polyhedron production. The appropriate virus plaques are isolated, amplified to working virus stocks and used to infect cells to monitor recombinant protein production.

Although the derivation of recombinant virus is based on simple technologies, for the inexperienced user the procedure can be time consuming and frustrating. The rate of recombination between virus DNA and plasmid transfer vector are about 1-2% and make the identification of polyhedrin-negative plaques difficult. Such plaques are often contaminated with wild type virus that produces polyhedra. This problem has been largely solved by a number of ingenious methods. These include: linearisation of parental virus DNA prior to cotransfection of insect cells (Kitts and Possee, 1993); efficient direct ligation of foreign DNA into the virus genome before transfection Ernst *et al.*, 1994); transposon mediated production of recombinant virus genomes in bacterial cells followed by transfection of insect cells to derive infectious virus (Luckow *et al.*, 1993). Essentially, it is now possible to isolate recombinant baculoviruses without difficulty.

4. Improvements to baculovirus expression vectors

A comprehensive set of reagents is now available to produce recombinant proteins in baculovirus-infected cells. A survey of the literature reveals that many hundreds of different proteins have now been synthesised using the system. Many of them were complex proteins requiring extensive post-translational modification (Jarvis, 1997). These have been used as tools to study protein structure and function, as candidate vaccines or diagnostic reagents and as substrates in enzyme assays or targets for drugs. Almost without exception, the recombinant proteins were produced by replacing either the coding sequence of the polyhedrin or p10 genes and expressing the foreign genes in the very late phase of virus replication. In some examples, the levels of plasma membrane-bound or secreted foreign proteins produced in baculovirus-infected insect cells, although high in comparison with other expression systems, do not match native polyhedrin and p10 protein synthesis. This may not be surprising, given that the polyhedrin protein (29 kDa) and p10 proteins (10 kDa) are relatively simple proteins that do not appear to require much post-translational processing. Therefore, although baculoviruses have been used very successfully as expression vectors, there is still scope for improving the system.

A further complication is that the production of recombinant protein is accomplished as a batch process in which virus is used to infect insect cells in vessels of varying size. After several days, the cells die and are harvested to derive the protein products. The identity of most of the virus genes responsible for DNA replication and very late gene expression are known. However, it is not yet possible to construct a virus-free system in a transgenic insect cell for the continuous production of recombinant protein via the use of the polyhedrin gene promoter. Such a system would be advantageous, since it would remove the requirement for producing virus inoculum prior to infecting insect cells for foreign protein synthesis and also avoid the need to remove infectious virus particles from the final protein preparation. Immediate early AcMNPV gene promoters have been used to derive insect cell lines that express foreign protein continuously, but the yields are low and do not rival the intact virus genome as an expression system (Jarvis, 1993). In the future, baculovirus-based expression systems will remain dependent on the use of infectious viruses to induce foreign gene expression in insect cells. Any improvements to the system will be to modulate native virus gene expression to enhance recombinant protein production.

The recombinant baculovirus, although devoid of the polyhedrin gene, is essentially an unmodified, wild type virus. It is the product of natural selection in insects to produce a parasite that can maintain itself in host populations. The AcMNPV-infected *Trichoplusia ni* shown in Figure 1 illustrates the effect of the virus

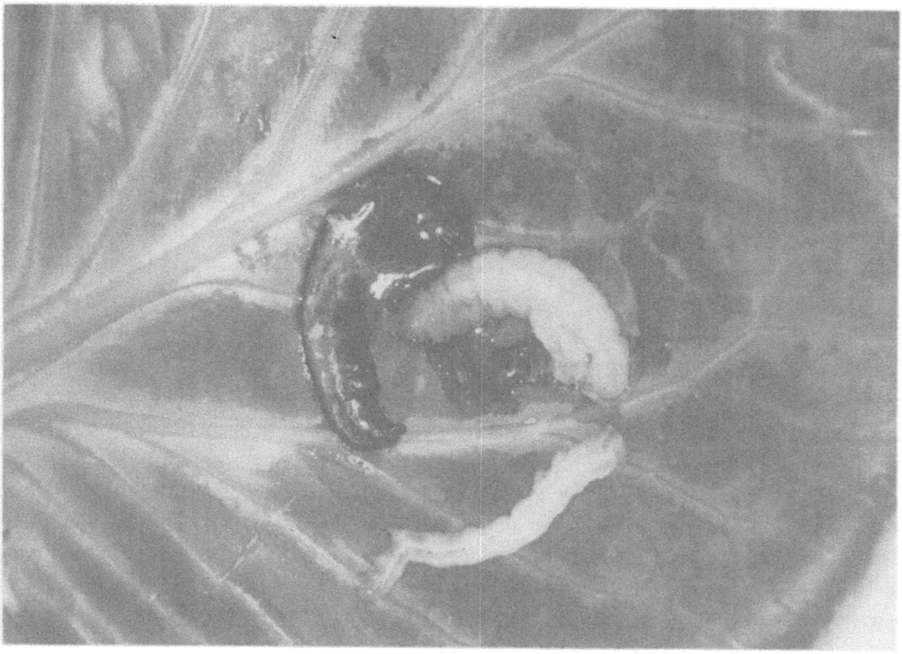

Figure 1. Liquefaction of virus-infected larvae. Fourth instar *T. ni* larvae were fed AcMNPV (left), Ac*chiA*⁻/*cath*⁻.*lac*Z (bottom) or mock-infected. Larvae were reared at 24°C and monitored daily. The insects were photographed at 8 days p.i.

on the insect in the terminal stages of infection. The same effect is seen in insects infected with polyhedrin-negative, non-occluded virus particles. The insect liquefies via the production of the virus-encoded chitinase and cysteine proteinase (cathepsin) in the late stages of virus replication. When the chitinase (*chiA*) and cathepsin (*cath*) genes are deleted from the AcMNPV genome the virus can still cause infection and death in the insect host, but it fails to liquefy (Fig. 1; Ohkawa *et al.*, 1994; Slack *et al.*, 1995; Hawtin *et al.* (1997). The insect cuticle is a mixture of chitin and protein, which are thought to be digested by the chitinase and cathepsin. Sequential removal of protein and chitin exposes more sites susceptible to both enzymes. Both of these gene products are also produced in virus-infected insect cell cultures, so it is pertinent to consider their effect on recombinant protein production.

4.1 CATHEPSIN

Both AcMNPV and BmNPV contain *cath*, a proteinase gene that belongs to the papain superfamily (Rawlings *et al.*, 1992; Ohkawa *et al.*, 1994). A number of observations suggested that expression of the baculovirus *cath* might be detrimental to recombinant protein production in virus-infected cells. Inactivation of *cath* in either AcMNPV or BmNPV showed that insects infected with these mutants failed to liquefy (Ohkawa *et al.*, 1994; Slack *et al.*, 1995). In BmNPV-infected *B. mori* larvae, the tracheal epithelia were largely destroyed, the haemolymph became milky white due to the release of polyhedra into the haemocoel and lipid transport proteins (lipophorins) were degraded. Insects infected with a *cath*-deficient virus (CPd) retained an intact tracheal epithelium with cells containing many polyhedra in hypertrophied nuclei. The haemolymph remained largely clear, contained few polyhedra and showed little degradation of lipophorins. These results were consistent with high levels of protease activity in the haemolymph of silkworm larvae infected with BmNPV and its absence in the haemolymph of insects infected with CPd (Suzuki *et al.*, 1997).

Foreign coding regions were introduced into CPd at the polyhedrin gene locus under the control of the polyhedrin gene promoter to test expression. A recombinant virus, CPdPL was constructed that contained the luciferase gene from the firefly, *Photinus pyalis*. Luciferase activity in the haemolymph of insects infected with CPdPL was compared with that induced in an insect infected with a control virus (WTPL) containing both intact *cath* and the luciferase coding region. In the late stages of virus infection, luciferase activity decreased in insects infected with WTPL. An immunoblotting procedure also showed that luciferase was subject to degradation in WTPL-infected insects as evidenced by the appearance of smaller products, whereas in the absence of protease it remained largely intact. Similar results were obtained when human growth hormone (hGH) was produced by recombinant baculoviruses lacking *cath*. The hGH remained intact in the absence of protease, but was cleaved to smaller peptides if *cath* was expressed in insect cells. These data suggest that the removal of *cath* from recombinant BmNPV, and probably AcMNPV, would be beneficial for foreign protein production in insect cells. To date, these studies have not been replicated in insect cell culture with either virus, so it is difficult to conclude that cathepsin production *in vitro* is detrimental to foreign protein production. Confocal immunofluorescence microscopy studies have shown that cathepsin is located in the lysosome of virus-infected cells (C.J Thomas, unpublished data). If these cells remain intact for the major period of recombinant production, the cathepsin may never come into contact with the desired protein.

4.2 CHITINASE

A chitinase gene is encoded by the AcMNPV and BmNPV genomes. The predicted protein sequence of the AcMNPV chitinase shares 57% identity and 88% similarity with a chitinase from *Serratia marcescens* (Ayres *et al.*, 1994). Both proteins have a similar size; 551 amino acids for AcMNPV and 563 amino acids for *S. marcescens*. The conservation between the two sequences is maintained throughout most of the coding regions, except for the N- and C-termini. The AcMNPV chitinase has a predicted eukaryotic signal peptide at the N-terminus, whereas the *S. marcescens* chitinase has a prokaryotic signal sequence. At the C-terminus, the AcMNPV chitinase has a KDEL endoplasmic reticulum retention signal that is absent from the bacterial protein (Hawtin *et al.*, 1995). The sequence similarities between the virus and prokaryotic genes are suggestive of a common origin.

The AcMNPV *chiA* is expressed as a late gene from about 8 h p.i. in virus-infected cells. Transcription begins from a TAAG motif 12 nucleotides upstream from the translational initiation codon (Hawtin *et al.*, 1995). High levels of both endo- and exochitinase activities reach a peak in virus-infected cells by 12 h p.i. and remain stable throughout the remainder of the infection. Western blot analysis of virus-infected cell extracts with a chitinase-specific antibody showed that a product of about 58 kDa was present from 11 until at least 18 h p.i. Interestingly, the intracellular level of chitinase, as judged by immunoblotting, continues to accumulate beyond 12 h p.i., when the increase in enzyme activity has ceased. It is not known whether or not this is due to inactivation of the enzyme after 12 h p.i., resulting in a steady state level of activity, or due to retention of the chitinase in an insoluble form in virus-infected cells. A freeze-thaw method is used to extract chitinase from virus-infected cells. It is possible that after 12 h p.i. the enzyme is only released inefficiently from cells.

Deletion of *chiA* from the AcMNPV genome (Ac*chiA*$^-$) has no effect on the production of infectious budded virus in cell culture or on the infectivity of polyhedra *in vivo*. However, striking differences are observed in the condition of the insect larvae after death; instead of a rapid liquefaction, the cadaver remains intact (Hawtin *et al.*, 1997). This reproduces the effect observed when *cath* is deleted from the virus genome (Slack *et al.*, 1995). The yield of polyhedra from an Ac*chiA*$^-$-infected insect is slightly higher than from a host infected with the wild type virus (C.J. Thomas, unpublished data). The reasons for this are not known, but the increase is probably a consequence of the lack of liquefaction enabling virus-infected cells to continue polyhedrin production for longer.

At least 90% of the chitinase activity induced in AcMNPV-infected cells in culture remains intracellular (Hawtin *et al.*, 1995). Immunofluorescence staining with a polyclonal chitinase-specific antibody revealed that the protein was in the cytoplasm of the virus-infected cells. These data were inconsistent with the presence of the eukaryotic signal peptide at the N terminus of the chitinase protein. It was anticipated that the signal peptide should have attached the nascent chitinase to the endoplasmic reticulum, enabling entry to the secretory pathway and subsequent release from the cell. However, the results are consistent with the presence of a KDEL ER-localisation signal at the C terminus of the protein. This was confirmed by dual staining of AcMNPV-infected cells with the chitinase-specific antibody and a monoclonal antibody specific for the HDEL ER retention motif (Thomas *et al.*, 1998). Both antibodies highlighted an irregular reticulate staining pattern around the nucleus. The nuclear membrane was also stained, but no fluorescence was observed at the plasma membrane. In virus-infected cells, the ER becomes dilated due to the effects of infection. Immunogold labelling and electron microscopy were used to confirm the localisation of chitinase in AcMNPV-infected cells. Gold particles were observed around the nuclei and also within vacuolar-like structures apparent throughout the cytoplasm. When insect cells were infected with Ac*chiA*$^-$, these cytoplasmic vacuoles were absent. Immunoflurorescent and immunogold labelling of the same cells proved negative.

The aberrant appearance of the endoplasmic reticulum in AcMNPV in the very late stages of infection, presumably a consequence of chitinase localisation, suggested that the passage of proteins through the secretory pathway might be compromised. This would not be a problem for the virus in the normal course of infection, because the production of most of the membrane-bound and other glycoproteins (e.g. GP67) probably occurs in the late phase of gene expression, when chitinase accumulation in the cells has yet to peak. This hypothesis was tested by constructing recombinant viruses containing the urokinase plasminogen activator coding region (UK) under the control of the polyhedrin gene promoter that either retained (AcUK) or lacked (AcUK/*chiA*$^-$) the complete *chiA* coding region (Fig. 2). Insect cell cultures were infected with either virus and the extracellular and intracellular levels of urokinase activity determined between 0 and 72 hours post-infection using a fibrin well assay.

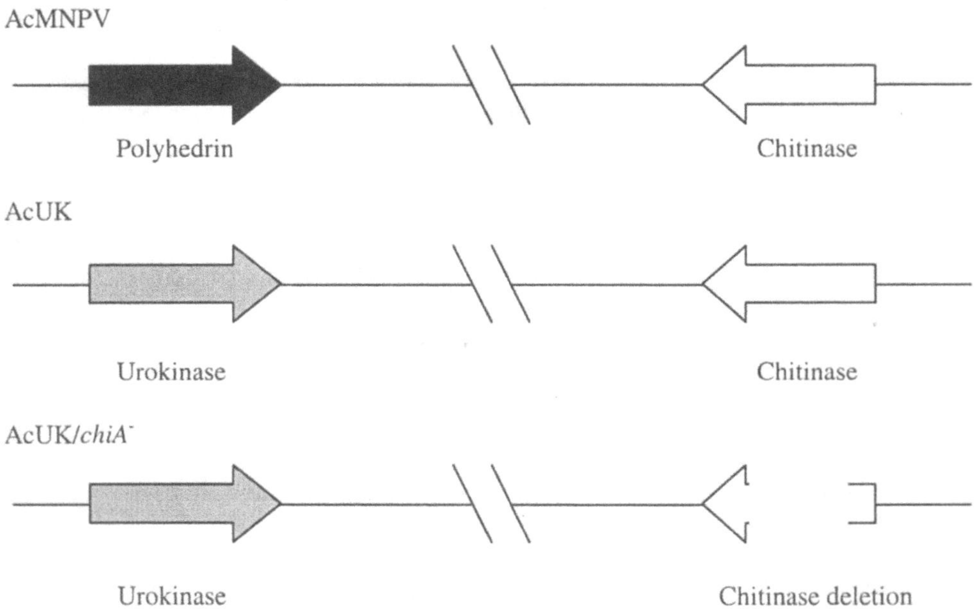

Figure 2. Genome organisation of recombinant baculoviruses. The intact polyhedrin and chitinase genes are shown in AcMNPV. The AcUK contains the urokinase plasminogen activator in place of the polyhedrin gene coding region, but under the control of the polyhedrin gene promoter. The chitinase gene remains intact in AcUK. In AcUK/*chiA*$^-$, 638 base pairs were removed from *chiA*.

Figure 3a shows that between 24 and 72 h p.i., there was approximately two-fold greater urokinase activity in the medium supporting cells infected with AcUK/*chiA*$^-$. In contrast, intracellular levels of urokinase were greater in cultures infected with AcUK (Fig. 3b). Another interesting observation was made when the viability of virus-infected cells maintained in suspension was determined using trypan blue exclusion. Up to 24 h p.i., cells infected with AcUK or AcUK/*chiA*$^-$ were almost

A

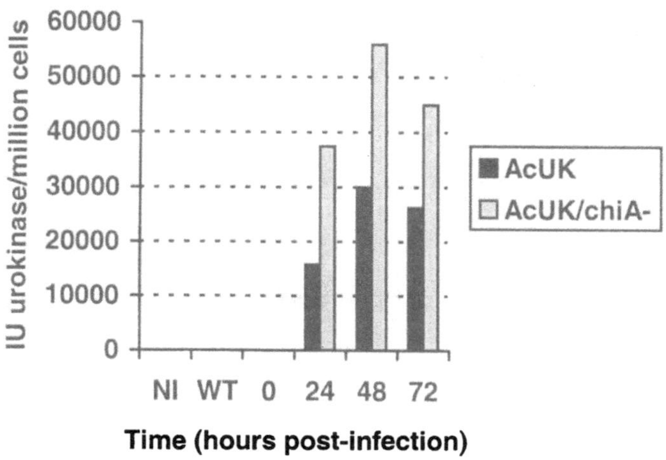

B

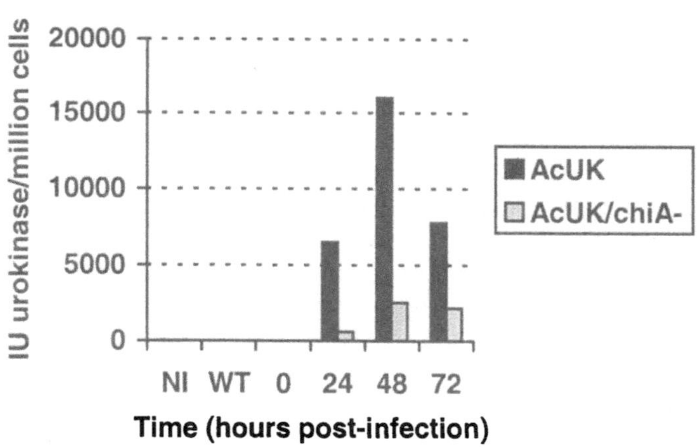

Figure 3. Urokinase production by recombinant baculovirus-infected cells. Suspension cultures of *Spodoptera frugiperda* cells were inoculated with AcMNPV, AcUK, AcUK/*chiA*⁻ (10 pfu/cell) or mock-infected. Samples were withdrawn from each culture at the indicated times and separated into fractions containing cells or medium. (A) The media samples were assayed for urokinase using a fibrin well assay without further treatment. (B) The cells were frozen and thawed three times and the lysates assayed for urokinase activity. The results are expressed in international units of urokinase per million cells.

100% viable (Fig. 4). Thereafter, at 48 h p.i., less than 20% of AcUK-infected cells were alive, whereas nearly 80% of AcUK/*chiA*⁻ -infected cells were viable. By 72 h p.i., negligible numbers of AcUK-infected cells were viable, but over 20% of AcUK/*chiA*⁻ -infected cells excluded trypan blue.

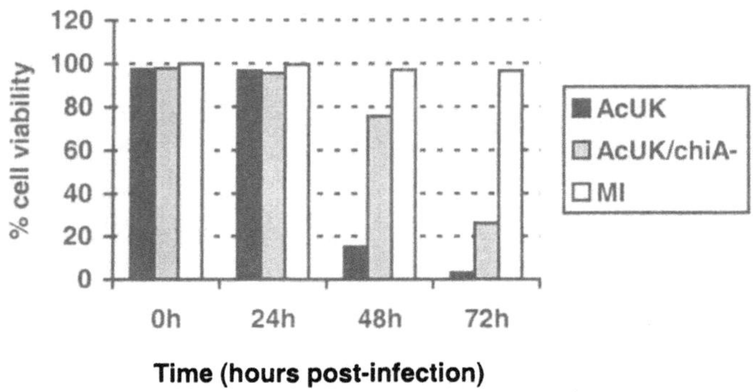

Figure 4. Viability of AcUK, AcUK/*chiA*⁻ or mock-infected *S. frugiperda* cells. The viability of the cells was determined using trypan blue exclusion.

The improved secretion of urokinase in recombinant baculovirus-infected cells may be attributed to two factors. The prevention of virus-encoded chitinase production and its accumulation in the ER of the insect cells may simply remove a logjam in protein secretion. Alternatively, the increased longevity of the virus-infected cells in the absence of chitinase may preserve the protein secretory pathway for longer. Both factors are probably important, but the observation that urokinase secretion was greater at 24 h p.i. in the absence of chitinase when there was no difference in the viability of AcUK and AcUK/*chiA*⁻ -infected cells suggests that chitinase is normally a significant factor in blocking protein secretion. Any protein that passes through the secretory pathway would be better produced using a baculovirus expression vector that lacks *chiA*.

4.3 REMOVAL OF OTHER BACULOVIRUS GENES FROM EXPRESSION VECTORS

The removal of p10 from the AcMNPV genome also has an effect on the stability and viability of virus-infected cells. Insect cells infected with unmodified AcMNPV were reported to begin to lyse by two days p.i., whereas cells infected with a virus lacking functional p10 were more stable and remained intact for at least two weeks (Williams *et al.*, 1989). We repeated these experiments to show that cells infected with AcMNPV did lyse more readily than those infected with a p10-deficient virus (Fig. 5).

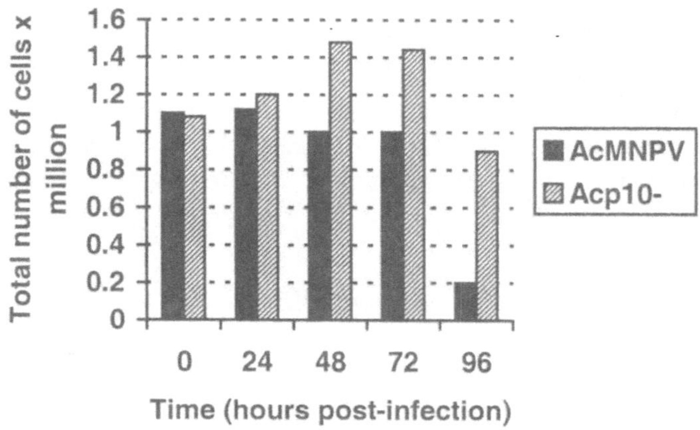

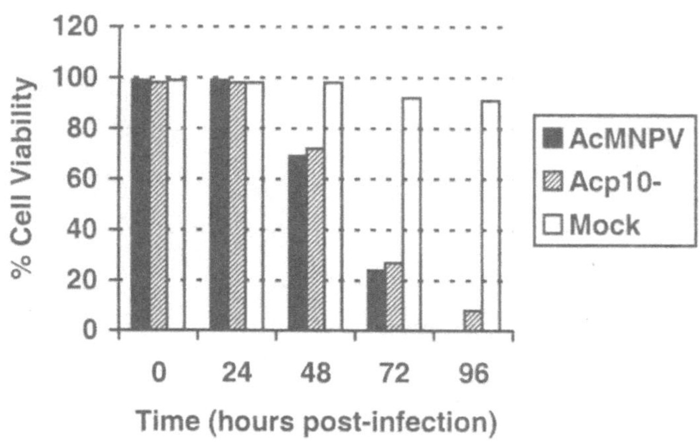

Figure 5. Stability and viability of AcMNPV and Acp10⁻-infected *S. frugiperda* cells. Suspension cell cultures were infected with AcMNPV, Acp10⁻ or mock-infected. At the indicated time, the concentration (A) and viability of the cells were determined.

We noted, however, that the total number of cells in AcMNPV-infected cultures remained static until 72 h p.i., but declined to about 20% of the original number by 96 h p.i. Cells infected with the virus lacking p10 increased in number between 24 and 72 h p.i. and were only reduced to about 90 % of the original number by 96 h p.i. Companion studies to assess the viability of these cells showed that there was little difference in cultures infected with either virus up to 72 h p.i. By 96h p.i., however, all cells infected with AcMNPV were dead. A small proportion of the cells infected with the *p10* mutant was viable. It is unclear whether or not the removal of p10 from AcMNPV could have a significant effect on foreign protein production in this system. However, the fact that the p10 promoter is highly active in the very late phase of virus gene expression in insect cells suggests that its removal can only be beneficial for foreign gene expression via the polyhedrin gene promoter by reducing the competition for transcription factors.

5. Concluding remarks

The removal of the AcMNPV *chiA* was shown to be beneficial for the improved secretion of a secreted recombinant protein. The AcMNPV genome contains several other baculovirus genes are known to be non-essential for replication of the virus in cell culture (O'Reilly, 1997). The function of most of these genes remains unknown, but some may have deleterious effects on foreign protein production in the insect in the very late stage of virus infection. It will be interesting to test the effect of removing each gene on recombinant protein production. Another reason for removing non-essential baculovirus genes from the expression vector is to reduce the competition for transcription and translation factors in the virus genome. There is little point in retaining the expression of gene products for which there is no useful purpose in the recombinant virus-infected cell.

6. References

Ayres, M.D., Howard, S.C., Kuzio, J., Lopez-Ferber, M. and Possee, R.D. (1994) The complete DNA sequence of *Autographa californica* nuclear polyhedrosis virus. *Virology* **202,** 586-605.

Blissard, G.W. and Rohrmann, G.F. (1989) Location, sequence, transcriptional mapping and temporal expression of the gp64 envelope glycoprotein gene of the *Orgyia pseudotsugata* multicapsid nuclear polyhedrosis virus. *Virology* **170**, 537-542

Ernst, W.J., Grabherr, R.M. and Katinger, H.W. (1994) Direct cloning into the *Autographa californica* nuclear polyhedrosis virus for generation of recombinant baculoviruses. *Nucleic Acids Res.* **22,** 2855-2857.

Friesen, P.D. (1997) Regulation of baculovirus early gene expression, in L.K. Miller (ed.), *The Baculoviruses*, Plenum Press, New York, pp. 141-170.

Funk, C.J., Braunagel, S.C. and Rohrmann, G.F. (1997) *Baculovirus structure*, in L.K. Miller (ed.), *The Baculoviruses*, Plenum Press, New York, pp. 7-27.

Guarino, L.A., Xu B., Jin, J and Dong, W. (1998) A virus-encoded RNA polymerase purified from baculovirus-infected cells. *J. Virol.* **72,** 7985-7991.

Hawtin, R.E., Arnold, K., Ayres, M.D., Zanotto, P.M de A., Howard, S.C., Gooday, G.A., Chappell, L.H.,

Kitts, P.A., King, L.A., and Possee, R.D. (1995) Identification and preliminary characterisation of a chitinase gene in the *Autographa californica* nuclear polyhedrosis virus genome. *Virology* **212,** 673-685

Hawtin, R.E., Zarkowska, T., Arnold, K., Thomas, C.A., Gooday, G.W., King, L.A. and Possee, R.D. (1997) Liquefaction of *Autographa californica* nucleopolyhedrovirus-infected insects is dependent on the integrity of virus-encoded chitinase and cathepsin genes. *Virology* **238**, 243-253.

Jarvis, D.L. (1993) Effects of baculovirus infection on IE1-mediated foreign gene expression in stably transformed insect cells. *J. Virol.* **67,** 2583-2588.

Jarvis, D.L. (1997) Baculovirus expression vectors, in L.K. Miller (ed.), *The Baculoviruses*, Plenum Press, New York, pp. 389-431.

King, L.A. and Possee, R.D. (1992) *The Baculovirus Expression System: A Laboratory Guide*, Chapman and Hall, London.

Kitts, P.A. and Possee, R.D. (1993) A method for producing recombinant baculovirus expression vectors at high frequency. *Biotechniques* **14,** 810-817.

Lu, A. and Miller, L.K. (1997) Regulation of baculovirus late and very late gene expression, in L.K. Miller (ed.), *The Baculoviruses*, Plenum Press, New York, pp. 193-216.

Lu, A., Krell, P.J., Vlak, J.M. and Rohrmann, G.F. (1997) Baculovirus DNA replication, in L.K. Miller (ed.), *The Baculoviruses*, Plenum Press, New York, pp. 171-191.

Luckow, V.A., Lee, S.C., Barry, G.F. and Olins, P.O. (1993) Efficient generation of infectious recombinant baculoviruses by site-specific transposon-mediated insertion of foreign genes into a baculovirus genome propagated in *Escherichia coli*. *J. Virol.* **67,** 4566-4571.

O'Reilly, D.R. (1997) Auxiliary Genes of baculoviruses, in L.K. Miller (ed.), *The Baculoviruses*, Plenum Press, New York, pp. 267-300.

Ohkawa, T., Majima, K. and Maeda, S. (1994) A cysteine protease encoded by the baculovirus *Bombyx mori* nuclear polyhedrosis virus. *J. Virol.* **68,** 6619-6625.

O'Reilly, D.R., Miller, L.K. and Luckow, V.A. (1992) *Baculovirus Expression Vectors*, W.H. Freeman, New York.

Possee, R.D. (1997) Baculoviruses as expression vectors. *Current Opinion In Biotechnology* **8**, 569-572.

Rawlings, N.D., Pearl, L.H. and Buttle, D.F. (1992) The baculovirus *Autographa californica* multiple nuclear polyhedrosis virus genome includes a papain-like sequence. *Biol. Chem. Hoppe-Seyler* **373,** 1211-1215.

Slack, J.M., Kuzio, J. and Faulkner, P. (1995). Characterisation of *v-cath*, a cathepsin L-like proteinase expressed by the baculovirus *Autographa californica* multiple nuclear polyhedrosis virus. *J. Gen. Virol.* **76,** 1091-1098.

Suzuki, T., Kanaya, T., Okazaki, H., Ogawa, K., Usami, A., Watanabe, H., Kadono-Okuda, K., Yamakawa, M., Sato, H., Mori, H., Takahashi, S. and Oda, K. (1997) Efficient protein production using a *Bombyx mori* nuclear polyhedrosis virus lacking the cysteine proteinase gene. *J. Gen. Virol.* **78,** 3073-3080.

Thomas, C.J., Brown, H.L., Hawes, C.R., Lee, B. Y., Min, M.-K., Linda A. King, L.A. and Possee, R.D. (1998) Localisation of a baculovirus-induced chitinase in the insect cell endoplasmic reticulum. *J. Virol.* **72,** 10207-10212.

Volkman, L.E., Blissard, G.W., Friesen, P., Keddie, B.A., Possee, R.D. and Thielmann, D.A. (1995) Family Baculoviridae: In: *Virus Taxonomy: Sixth Report of the International Committee on Taxonomy of Viruses*, pp. 104-113. Edited by F.A. Murphy, C.M. Fauquet, D.H.L. Bishop, S.A. Ghabrial, A.W. Jarvis, G.P. Martelli, M.A. Mayo and M.D. Summers. Springer-Verlag Wien, New York.

Whitford, M., Stewart, S., Kuzio, J. and Faulkner, P. (1989) Identification and sequence analysis of a gene encoding gp67 an abundant envelope glycoprotein of the baculovirus, *Autographa californica* nuclear polyhedrosis virus. *J. Virol.* **63,** 1393-1398.

Williams, G.V., Rohel, D.Z., Kuzio, J. and Faulkner, P. (1989) A cytopathological investigation of *Autographa californica* nuclear polyhedrosis virus p10 gene function using insertion/deletion mutants. *J. Gen. Virol.* **70,** 187-202.

Williams, G.V. and Faulkner, P. (1997) Cytological changes and viral morphogenesis during baculovirus infection, in L.K. Miller (ed.), *The Baculoviruses*, Plenum Press, New York, pp. 61-107.

THE REGULATION OF APOPTOSIS IN ANIMAL CELLS

S.L. MCKENNA*, R.J. CARMODY* and T.G. COTTER[†]

Tumour Biology Laboratory, Department of Biochemistry,
University College Cork, Cork, Republic of Ireland.
Fax: 353-21-904068, email: t.g.cotter@ucc.ie
**These authors have contributed equally to this work.*
† Corresponding author.

1.0 Introduction

It has now been established that physiologically controlled cell death is an integral and indispensable component of metazoan life. Normal development necessitates a measured amount of cell deletion as does homeostatic maintenance and protection of the organism from pathogens. The process whereby a cell orchestrates its own destruction is now referred to as *apoptosis* or *programmed cell death* (PCD). Apoptosis is a phrase originally employed to denote a collective set of features observed in dying cells (Kerr *et al*, 1972). These include cell shrinkage, chromatin condensation and membrane blebbing. These visible features reflect an orderly internal process of cellular dismantling, packaging and surface alterations which make the cell recognisable and safely digestible for neighbouring or phagocytic cells. In contrast to necrosis (or pathological cell death), where cell contents are released, apoptosis does not induce an inflammatory response and it is a relatively unobtrusive exit from life.

Accumulating evidence suggests that every cell harbours a suicidal tendency and that its survival depends upon continuous provocation to live. Both internal and external factors exert an influence on cell survival. Maintaining the correct balance between life and death requires a considerable degree of regulation. Those mechanisms which invoke cell death and conversely those regulatory mechanisms which help to maintain the desire for life are of fundamental significance to cell biology.

The consequences of deregulated apoptosis are severe. Insufficient cell death can lead to the development of neoplasia whereas inappropriate cell death is a feature of auto-immune and degenerative diseases. The exploration of this relatively new frontier in cell science has raised expectations for novel therapeutic options for presently incurable diseases. Not surprisingly therefore, recent years have witnessed a surge of interest and information regarding the incidence and regulation of apoptotic cell death. In this Chapter we will discuss evidence regarding the role of apoptosis in multicellular existence, and review recent advances in the identification of its regulatory mediators. An initial section describing the characteristic features of apoptotic cells including some methodology for recognition may prove useful to unfamiliar readers. The importance of apoptosis will be evidenced in a section which outlines several biological regulatory mechanisms which depend upon it. Major families of apoptotic regulators and control points will then be reviewed, followed by some general discussion on the wider impact of current apoptosis research.

2.0 Biological Features of Apoptotic Cells

The major unifying feature of apoptotic cell death in all cell types is the prolonged retention of plasma membrane integrity despite extensive proteolytic and degradative events within its confines. Other cumulative morphological and biochemical events signify apoptosis in animal cells. Morphological features indicative of apoptosis in animal cells include nuclear alterations, chromatin condensation and fragmentation of the nucleus; cell shrinkage; surface alterations such as membrane blebbing; and the formation of membrane bound vesicles termed apoptotic bodies (see Fig 1.I). Recent studies have helped to elucidate some of the biochemical events underlying these features and their importance for safe elimination of the dying cell.

2.1 NUCLEAR EVENTS

The nucleus undergoes dramatic alterations during apoptosis where nuclear shrinkage, chromatin condensation and/or the formation of crescent shaped deposits along the nuclear envelope are commonly observed. The fragmentation of DNA into nucleosomal sized fragments of 180-200 base pairs is a further nuclear associated event during apoptosis (Cohen and Duke, 1984). Indeed the formation of a ladder pattern, following the electrophoresis of apoptotic DNA, has long been considered to be a biochemical hallmark of apoptosis and is frequently employed as a diagnostic tool in apoptosis research (see Fig 1.II). However it should be noted that such low molecular weight DNA fragmentation is not a universal feature of apoptosis (Oberhammer *et al.*, 1993) and in some cell types only high molecular weight fragments of 300-500 kilo base pairs may be detected (Walker *et al.*, 1995). Both types of DNA fragmentation can be detected by enzymatic fluorescent labelling of DNA strand breaks followed by analysis using either fluorescence microscopy or flow cytometry (see Fig 1.II). The presence or absence of low/high molecular weight DNA fragmentation may be dependent on the cell type or death-inducing stimulus and may be also be a consequence of differential endonuclease activation.

The search for the endonuclease(s) responsible for DNA fragmentation during apoptosis has resulted in the proposal of several previously described, as well as novel enzymes, as candidates for the apoptotic endonuclease (Hughes and Cidlowski, 1994). However the most convincing candidate to date is the recently identified caspase activated DNase (CAD) (Enari *et al.*, 1998). This molecule is usually found in the cytoplasm in a latent, inactive form, bound to an inhibitor (ICAD) which stabilises and inactivates its nuclease activity. It appears that specific proteolytic digestion (see later section on caspases) of ICAD during apoptosis results in CAD activation and its translocation to the nucleus where it effects DNA fragmentation (Sakahira *et al.*, 1998). The proteolytic activation of CAD represents the only apoptosis specific pathway for the fragmentation of nuclear DNA to date, although its existence does not preclude the involvement of other candidate enzymes during the apoptotic process.

2.1 CELL SHRINKAGE

Alterations in cell volume and granularity appear to be universal characteristics of apoptosis. Cell shrinkage during apoptosis has been suggested to occur as a result of a net movement of fluid out of the cell, although the exact mechanism by which this takes place is currently unclear. It has been proposed that the fusion of golgi and endoplasmic recticulum membranes with the plasma membrane facilitates this movement of fluid from the cell (Morris *et al.*, 1984), while more recent data suggests a role for the active efflux of sodium and potassium ions through the Na^+, K^+-ATPase pump and Ca^{2+}-dependent channel (McCarthy and Cotter, 1997). Inhibitor based studies indicate that this active ion efflux appears to be necessary for cell shrinkage and subsequent apoptotic body formation (McCarthy and Cotter, 1997). The proteolytic cleavage of several cytoskeletal components such as actin, fodrin, spectrin, Gas2, gelsolin, focal adhesion kinase (FAK) and p21-activated kinase 2 (PAK2) is important for the cytoskeletal restructuring required for cell shrinkage and apoptotic body formation (see later section on caspases).

2.2 APOPTOTIC CELL RECOGNITION AND PHAGOCYTOSIS

Plasma membrane alterations appear to be required for cell shrinkage, membrane vesicle formation and the eventual phagocytosis of apoptotic bodies. Membrane integrity is essential to prevent the leakage of the intracellular contents and to permit cell shrinkage. The activation of tissue type transglutaminase, an enzyme which catalyses the crosslinking of e(g-glutamyl) lysine groups between protein substrates, appears to play an important role in the maintenance of membrane integrity (Fesus *et al.*, 1987) until phagocytosis is complete.

The translocation of phosphatidylserine (PS) from the inner leaflet to the outer leaflet of the plasma membrane is an early event during apoptosis and is believed to serve as an "eat me" signal for the phagocytosis of the dying cell (Martin *et al.*, 1995). Exposure of PS on the surface of the plasma membrane of apoptotic cells appears to be a universal feature of this form of cell death and has allowed the development of diagnostic flow cytometric death assays employing fluorochrome labelled annexin-V, a PS binding protein (see Fig 1.IV). The mechanism of PS exposure is unclear but it has been reported to depend on the activity of a scramblase enzyme (Verhoven *et al.*, 1995), with the level of PS exposure modulated by aminophospholipid translocase activity (Bratton *et al.*, 1997). Evidence exists for the recognition of PS by macrophages in a dose-dependent manner (Fadok *et al.*, 1992), which has led to the suggestion that a receptor for PS must exist which facilitates recognition and subsequent phagocytosis of apoptotic cells. Several candidate receptors for apoptotic cell recognition and clearance have been proposed and include the vitronectin receptor a_vb_3 (Savill *et al.*, 1990), CD14 (Devitt *et al.*, 1998), scavenger receptors (Fadok *et al.*, 1998) and the ABC1 transporter (Luciani and Chimini, 1996) which bears homology to Ced-7 (known to mediate corpse engulfment in *C. elegans*) (Wu and Horvitz, 1998a).

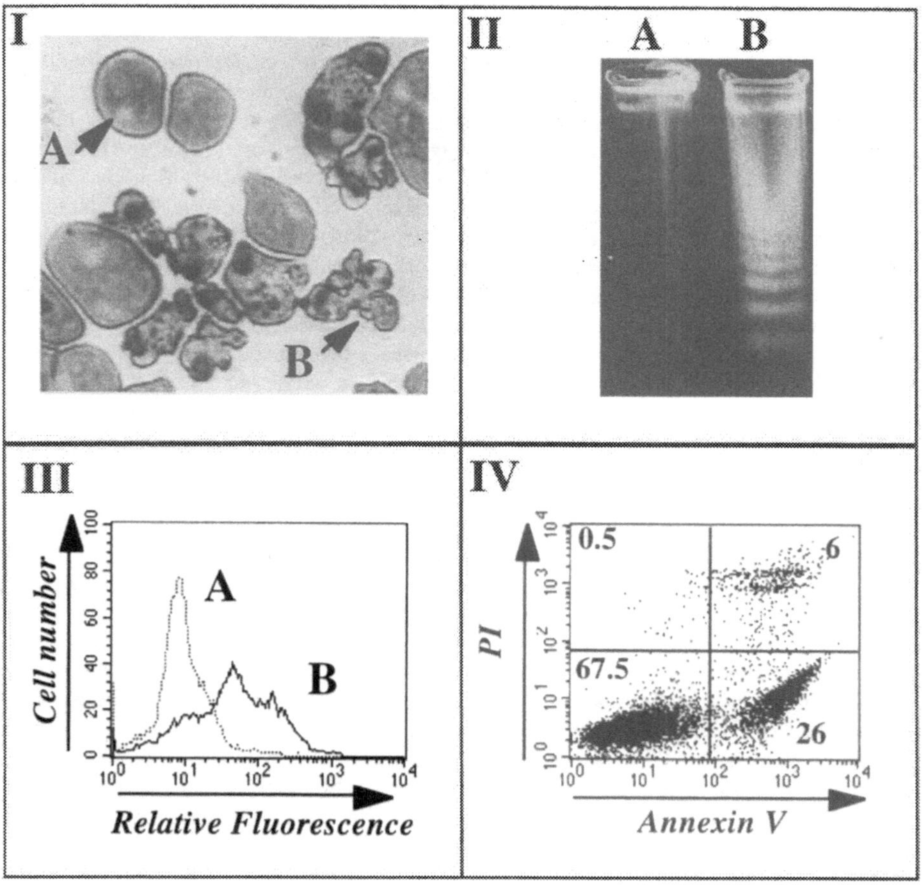

Figure 1. Identification of apoptosis in a variety of cell types.
I. *Morphology.* (A human T-cell line (Jurkats) subjected to ultraviolet *(u.v).* irradiation for three minutes and then cultured for three hours). Cells were stained as described in (Cotter, 1997) and analysed by light microscopy. **A** is an example of a normal cell in which the nuclear material is diffuse and occupies the majority of the cell volume. **B** is a typical apoptotic cell. The nucleus is condensed and fragmented and the cell confines are irregular due to membrane blebbing.
II. *Detection of DNA fragmentation by agarose gel electrophoresis.* Human myeloid leukaemia cells (HL60) were treated with a cytotoxic drug (etoposide, 40 μM for 6 hours). Cells were then lysed in DNA extraction buffer and electrophoresed through a 1.2% agarose gel as described in (Cotter *et al.*, 1997,). Untreated cells (**A**) possess only high molecular weight DNA which does not migrate into the gel, whereas treated cells (**B**) show a typical nucleosomal ladder pattern due to endonucleolytic cleavage of DNA into multiples of nucleosome sized units (180-200bp).
III *Detection of DNA fragmentation by TUNEL labelling and flow cytometry.* HL60 cells were *u.v.* irradiated as described above. Cells were then fixed and DNA strand breaks were enzymatically tagged using the TUNEL assay as described in Carmody *et al.*, 1998). Peak **A** shows the fluorescence exhibited by normal untreated cells. Apoptotic cells incorporate the fluorescent tag at DNA strand breaks and thus show increased fluorescence (peak **B**).
IV. Detection of phospholipid asymmetry by Annexin staining. Mouse myeloma cells (NSO) were treated for 18 hours with 25mM ammonia and then stained with both annexin and propidium iodide as described in Carmody *et al.*, 1998. Annexin stains both necrotic (primary and secondary) and apoptotic cells. Apoptotic cells (lower right quadrant) can be distinguished from necrotic cells (upper right quadrant) as their membranes are still intact and therefore they do not take up the DNA stain - propidium idodide (PI).

The class B scavenger receptor CD36 (also a thrombospondin receptor) has been implicated in the uptake of neutrophils by macrophages (Savill *et al.*, 1992) and has also been proposed to mediate the phagocytosis of photoreceptor outer segments by retinal pigment epithelium (Ryeom *et al.*, 1996). The glycophosphatidylinositol (GPI) anchored CD14 receptor on the surface of human macrophages is thought to interact with bacterial lipopolysaccharide (LPS) to trigger an inflammatory response. However it appears that CD14 may also be involved in apoptotic body recognition by binding certain lipids without triggering macrophage activation (Devitt *et al.*, 1998), including anionic phospholipids similar to phosphatidylserine (Yu *et al.*, 1997); although direct binding of PS to CD14 has yet to be demonstrated. CD14's involvement in cell recognition is supported by the finding that cells lacking CD14 exhibit a decreased ability to recognise and phagocytose apoptotic cells (Callahan *et al.*, 1997). How CD14 regulates different macrophage responses to non-self LPS and self PS on apoptotic cells, however, is currently unknown. Whether any of these candidate receptors are specific for PS remains to be determined and the possibility exists that multiple receptors for PS are present on phagocytes.

The ability of monosaccharides to prevent macrophage binding of apoptotic thymocytes (Duvall *et al.*, 1985) and the altered expression of carbohydrates on the surface of apoptotic cells (Falasca *et al.*, 1996) has indicated that sugar changes on the surface of apoptotic cells may be important in their recognition and subsequent phagocytosis. These findings have led some to speculate that altered membrane asymmetry may also make carbohydrate ligands available on the cell surface.

The *Caenorhabditis elegans*. gene *ced-5* has also been demonstrated to play a role in the engulfment of cell corpses in the nematode. It encodes a protein bearing homology to the human DOCK180, a cytoplasmic SH3 domain-containing protein (Wu and Horvitz, 1998b). Mutational studies with Ced-5 have led to suggestions that it is involved in the cytoskeletal reorganisation necessary for an engulfing cell to complete phagocytosis of an apoptotic cell. The ability of CD36 and vitronectin receptors to signal through tyrosine kinases suggests a possible link between these receptors and the SH3 containing DOCK180 and Ced-5 molecules. Such a link would integrate cell recognition and its engulfment into a single signal pathway, however further evidence is required to establish this hypothesis.

3.0 Apoptosis in Multicellular Survival

Following the first morphological description of apoptosis (Kerr *et al.*, 1972) the pace of apoptosis research was initially painfully slow. As reports of its incidence trickled into the literature it became apparent that features of this process have been highly conserved throughout metazoan evolution. More than a decade however elapsed before the first genetic regulators of cell death were identified (Ellis and Horvitz, 1986). Subsequent research demonstrated that all cells harbour a genetically encoded program for self destruction and that the process is of fundamental significance in cell biology. This acknowledgement, coupled with the recent advances in the recognition of apoptosis, has led to its widespread consideration in biological studies. Consequently, instances and proposed reasons for harbouring a program for cell death have been accumulating. Reported incidences of apoptosis induction which are imperative for the survival of multicellular organisms include: Developmental cell death, tissue homeostasis in the adult, defence against pathogens, DNA damage and oncogene deregulation.

3.1 DEVELOPMENT

Controlled cell deletion was first identified in developmental studies where it was referred to as programmed cell death (PCD). PCD plays a crucial role in the removal of unnecessary tissue in the sculpting of limbs and organs. For example the sculpting of digits by the programmed cell death of inter-digital tissue (Kerr *et al.*, 1987) and the formation of lumina following the death of internal cells in solid structures (Coucouvanis and Martin, 1995). PCD has also been reported in palatal fusion (Glucksmann, 1951), remodelling of cartilage and bones (Lewinson and Silberman, 1992; Furtwangler *et al.*, 1985), formation of the retina (Penfold and Provis, 1986), and in the development of the nervous system (Barde, 1989; Oppenheim, 1991; and see following section). Although the phrase 'programmed cell death' was originally associated with developmental cell death it is now a term which is frequently used to refer to any incidence of cell death that is directed by an internal programme (Jacobson *et al.*, 1997).

While a simple invertebrate such as the nematode, *C. elegans,* can tolerate the absence of apoptosis in development (Ellis *et al.*, 1991), dysfunction of central mediators in higher animals such as specific caspases (pro-apoptotic effectors; see later section) in mice is lethal in embryonic stages, or shortly after birth, due to excessive accumulation of cells. (Hakem *et al.*, 1998; Kuida *et a*l., 1996). Conversely a deficiency in an important anti-apoptotic (or survival) function is also lethal early in life due to intolerable cell depletion (e.g. loss of IGF-1, see following section).

3.2 SURVIVAL FACTORS AND THE MAINTENANCE OF HOMEOSTASIS

In all normal proliferating adult tissues mitosis must be counterbalanced by apoptosis in order to maintain cell numbers and functional integrity. Evidence suggests that the major mechanism restraining population size is the inherent requirement of cells for specific survival factors. This concept was first introduced by Martin Raff in 1992, who proposed that all cells express a default program for cell death and depend upon survival signals from other cells for their viability (Raff 1992). While this may have seemed a rather radical concept at the time, the weight of evidence supporting it has now led to its general acceptance.

Survival factors may be soluble mediators or they may be derived from cell adhesion molecules at cell-cell/matrix contacts (reviewed by O'Connor, 1998; Ruoslahti and Reed, 1994). The limited availability of survival signals will restrain specific cell types to the tissue producing the required set of survival factors and restrict population expansion. The specific factor (or combination of factors) which any given cell requires will depend upon the type of cell and its differentiation status. Many mediators previously known as growth factors, cytokines, or hormones have now also been demonstrated to provide crucial survival signals to their target cells. The importance of survival factors can be exemplified by considering two well studied systems; (a) the nervous system and (b) the haematopoietic system.

(a) Excessive production and apoptosis of neurons is a feature of normal development. The viability of sensory neurons is dependant upon neurotrophic factors, such as nerve growth factor (NGF), brain derived growth factor (BDGF) and neurotrophins-3 and -4, which are secreted by the cells which the neurones will innervate. Thus although many neurones may reach a target tissue, competition for a limited amount of survival factor will ensure the correct numerical balance between the innervating neurones and target cells (Ockel *et al.*, 1996; Lindsay, 1996; Lewin, 1996). Neurotrophic survival factors also maintain the continued viability of nerve cells, and their production may be elevated during injury to improve viability. Motor neurones similarly rely on survival factors from the tissue which they innervate. Factors shown to promote survival and/or prevent injury induced death include fibroblast growth factors, transforming growth factor beta, IGF-1, and neurotrophins-4 and -5. (Gouin *et al.*, 1993; Hughes *et al.*, 1993; Koliatsos *et al.*, 1994).

(b) Specialised functions of the haematopoietic system require a continuous turnover of its effector cells. Proliferation, differentiation, apoptosis and various other functional activities are largely regulated by a diverse group of protein factors collectively referred to as cytokines. Accumulating evidence suggests that many differentiation pathways may be pre-programmed in haematopoietic cells and that the major role of cytokines may be to promote the survival of the various sub-populations. Thus, the concentration and combination of cytokines in the haematopoietic environment would dictate which lineages of cells would survive through to terminal differentiation, and also regulate the numbers from each lineage that are produced (reviewed in McKenna and Cotter, 1997; Smithgall, 1998). Some cytokines promote the survival of more than one lineage of cells, whereas others are cell type and differentiation status specific. The survival of multi-lineage progenitor cells is dependent upon the availability of colony stimulating factors such as interleukin-3 (IL-3), stem cell factor (SCF), granulocyte-macrophage colony stimulating factor (GM-CSF) and also upon stromal interactions in the bone marrow. As cells differentiate along their respective lineages they rely upon a specific subset of cytokines for their viability. For example, erythroid progenitor cells are produced in excess and compete for a limited amount of erythropoietin (Koury and Bondurant, 1990), and myeloid cells compete for limited amounts of colony stimulating factor-1 (CSF-1) and granulocyte colony-stimulating factor (G-CSF) (Lowery, 1995). Thus, availability of these cytokines will be the major arbiter of the final concentration of cellular subsets in the peripheral blood. Similarly cells of the lymphoid lineages are dependant on the above mentioned multi-lineage survival factors and also various lineage specific survival factors to maintain their population boundaries.

Some factors can promote the survival of a range of cell types and protect against a variety of apoptosis inducing stimuli. For example the insulin like growth factors (IGFs) are potent survival factors (reviewed in O'Connor 1998). IGF-1 can provide a survival function for motor neurones (Hughes *et al.*, 1993) and haematopoietic cells (Rodriguez-Tarduchy *et al.*, 1992). It can also protect cells from oncogene de-regulation (Harrington *et al.*, 1994), cytotoxic drugs (Sell *et al.*, 1995) and ultra violet (*u.v.)* irradiation (Kulik *et al.*, 1997). The importance of this particular survival factor is also evident from the consequences of its absence. Mice deficient in IGF-1 or its cognate receptor (IGF-1R) are underweight, demonstrate underdeveloped bone, muscle and lung tissue, and die shortly after birth (Liu *et al.*, 1993). It is perhaps unsurprising therefore that IGF-1 is also considered to be an essential media component for the *in vitro* propagation of mammalian cells (Goldring and Goldring, 1991).

All survival factors bind to highly specific receptors on target cells. Conformational changes following receptor ligation result in phosphorylation and activation of an intrinsic tyrosine kinase in the cytoplasmic domain of the receptor (e.g.

SCF, CSF-1, IGF-1) or, recruitment and phosphorylation of cytoplasmic non-receptor tyrosine kinases (e.g. GM-CSF, IL-3). The subsequent activation of downstream mediators delivers a survival signal which inhibits apoptosis in the respective cell. The components of survival pathways are currently the subject of intensive investigation. Activation of the GTP-binding protein Ras is a common event in tyrosine kinase signal transduction. Several investigators have shown that over-expression of Ras can rescue cells from apoptosis following growth factor withdrawal or detachment of adherent cells from the extra-cellular matrix (Kinoshita *et al.*, 1995; Khwaja *et al.*, 1997). Signal transduction mediators downstream from Ras include the Raf/ERK/MAP kinase cascade, the phosphatidylinositol 3-OH kinase (PI(3)K) pathway and the Ral GDS pathway. (Downward, 1998; and See Fig 2). PI(3)K can also be activated independently of Ras by binding to other docking proteins at the activated receptor (e.g. IGF-1; Sun *et al.*, 1995). As activation of many of these receptors can also deliver a mitogenic signal, investigators have employed specific kinase inhibitors and constructed sophisticated mutant proteins in attempts to define those pathways exclusively responsible for survival. PI(3)K signalling to its downstream target - protein kinase B (PKB/Akt) - has been shown to be the major anti-apoptotic pathway mediating serum survival signals in Rat-1 fibroblasts (Kennedy *et al.*, 1997; Kauffmann-Zeh *et al.*, 1997), IGF-1 mediated survival of neuronal and haematopoietic cells (Dudek *et al.*, 1997; Minshall *et al.*, 1996) and survival signals from adhesion molecules (Khwaja *et al.*, 1997). IL-3 mediated survival is however mediated by at least two independent pathways one of which is PI(3)K mediated and the other which is reported to involve the Raf/MAPK cascade (Minshall *et al.*, 1996; Kinoshita *et al.*, 1997). The mechanism by which these signalling molecules influence life and death decisions is unclear, however it is possible that they may alter specific Bcl-2 family interactions (see later section on Bcl-2 family members and roles as apoptotic mediators). Bcl-2 family members are functionally influenced by phosphorylation, but little is known at present about the regulation of their phosphorylation status. Recent studies have alluded to the presence of such a mechanism. Phosphorylation of Bad following IL-3 survival signalling leads to its binding to the cytosolic protein 14-3-3 and sequestering of Bad away from Bcl-xL. This releases Bcl-xL's anti-apoptotic function and may be responsible for tipping the balance away from death and towards survival (Zha *et al.*, 1996) and see Fig. 2. Other protein kinases have also been implicated in the regulation of apoptosis but their major effector targets remain to be elucidated (Bokoch, 1998).

3.3 PROTECTION FROM ONCOGENE DEREGULATION

Several studies have shown that attempts to force cells into cycle by deregulating genes associated with proliferation, will invariably result in apoptosis. For example, inappropriate activation of the *c-myc* gene results in concomitant proliferation and apoptosis and thus no net gain in population size. Cells expressing de-regulated *myc* can be rescued from apoptosis by co-expression of anti-apoptotic genes such as *bcl-2*, or mutant *p53*, or by provision of survival factors such as PDGF or IGF-1 (Fanidi *et al.*, 1992; Lotem and Sachs, 1995; Harrington *et al.*, 1994). Similarly the E1A viral oncoprotein can only drive proliferation in the presence of an inhibitor of apoptosis (Rao *et al*, 1992). The *ras* oncogene is capable of stimulating both mitogenic and survival pathways, yet its oncogenic variants also initiate apoptosis and require a secondary co-operating anti-apoptotic event (often the loss of p53) to achieve cellular transformation (reviewed by Evan and Littlewood, 1998).

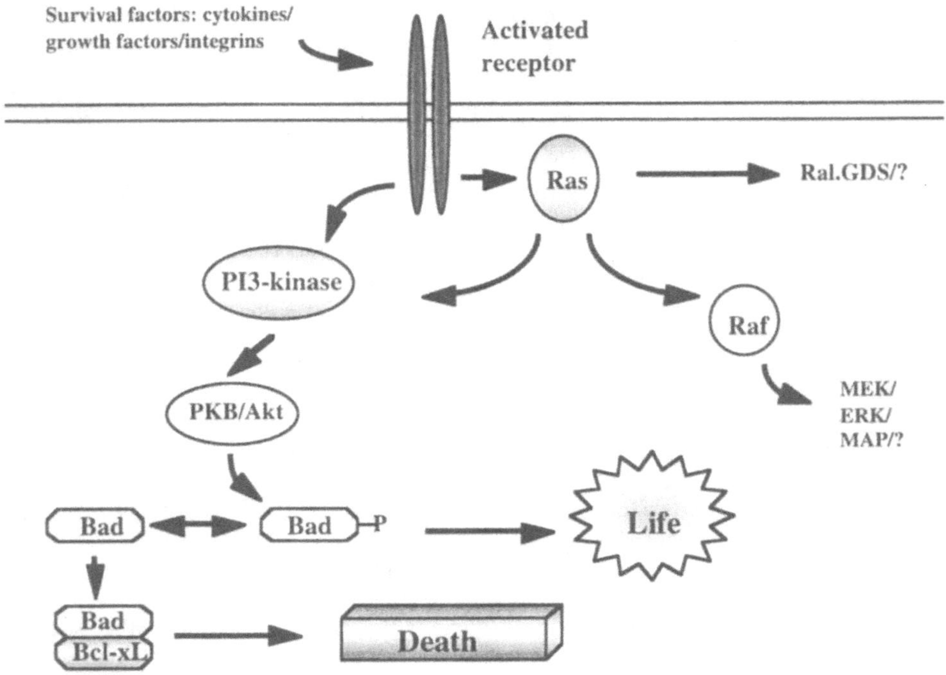

Figure 2. Survival pathways and apoptosis
Binding of survival factors to specific receptors delivers an anti-apoptotic signal to the recipient cell. PI-3 kinase phosphorylation of Akt has been shown to mediate a survival signal for many cell types. Phosphorylation of Bad by Akt results in its binding to the cytosolic protein 14-3-3, and its sequestering away from Bcl-xL. This allows Bcl-xL to exert its anti-apoptotic activity and thereby retain cell viability. Ras has also been reported to deliver a survival signal through the MAP kinase pathway.

In vivo evidence has also demonstrated the requirement for gene co-operation. Bcl-2 (see subsequent section) is over-expressed as a result of the t(14:18) chromosome translocation associated with follicular lymphoma. Transgenic mice expressing *bcl-2* developed high grade lymphomas following a long latent period during which time they acquired additional genetic changes. Half of these were alterations in the *c-myc* proto-oncogene. Mice transgenic for *myc* alone did not show hyperplasia. The generation of double *myc/bcl-2* transgenic mice further demonstrated the ability of these genes to co-operate as they rapidly generated a highly malignant disease (reviewed by Cory *et al.*, 1994).

These studies suggest that the cellular proliferative machinery is inextricably linked to apoptosis. Mitogenic signals can only be effective if they are activated in conjunction with survival signals. In addition to restraining population size (as discussed previously) this requirement for disruption of at least two signalling pathways will undoubtedly reduce the incidence of neoplastic transformation due to gene mutation, and also afford protection against oncogenic viruses. It is notable that loss of the tumour suppressor gene *p53* is often a co-operating lesion in transformation. The mechanism by which p53 'senses' oncogene de-regulation and induces apoptosis is unknown.

3.4 DNA DAMAGE AND THE STRESS RESPONSE

It is ultimately preferable to for an organism to delete a cell with damaged DNA rather than risk oncogene deregulation and possible malignancy. Thus internal sensors of DNA damage are linked to initiators of apoptotic pathways. Known sensors of DNA damage include poly(ADP-ribose) polymerase (PARP), DNA dependant protein kinase (DNAPK) and the gene product mutated in the disorder ataxia telangiectasia, (ATM). All of these molecular sensors have been placed upstream of the tumour suppressor gene *p53* which appears to be the ultimate decision maker regarding the appropriate course of action (reviewed in Agarwal *et al.*, 1998).

p53 can direct two different responses depending upon the extent of DNA damage. In the case of low levels of damage p53 transcriptionally activates several genes whose collective function is to arrest the cell cycle. Cell cycle arrest allows time for damage assessment and possible DNA repair. In the case of extensive and unsustainable damage p53 directs the cell to undergo apoptotic cell death (see reviews by Agarwal *et al.*, 1998; Morgan and Kastan, 1997) and see Fig 3. The mechanisms by which p53 induces apoptosis are poorly understood. Both transcriptionally dependent (Yonish-Rouach *et al.*, 1995; Sabbatini *et al.*, 1995) and independent induction has been reported (Caelles *et al.*, 1994; Haupt *et al.*, 1995). Activation of *bax* (Miyashita and Reed, 1995) and *Fas* (Owen-Schaub *et al.*, 1995) genes, and/or repression of the *bcl-2* gene (Miyashita *et al.*, 19941), may play a role in transcription dependent p53 mediated apoptosis.

p53 has also been reported to mediate the response to survival factor withdrawal in some cell types (Lin and Benchimol, 1995; Blandino *et al.*, 1995). In addition, although p53 deficiency can be non-lethal in development, a significant proportion of knockout mice have developmental abnormalities associated with neural tissue (Armstrong *et al*; 1995; Sah *et al.*, 1995), and specifically fail to undergo tissue modelling in the developing eye (Reichel *et al.*, 1998).

The importance of p53 in maintaining genome integrity and suppressing oncogene de-regulation in the adult is evidenced by the fact that families harbouring a p53 mutation have a high probability (50%) of developing cancer at an early age, and p53 deficient mice are highly susceptible to spontaneous tumour formation. In addition, p53 mutations occur in more than half of all human tumours, and represent the most common genetic alteration in human cancer.

Other cellular responses following exposure to stressful (or pro-apoptotic) stimuli include the activation of stress activated protein kinases (SAPKs). Such kinases include the c-jun amino terminal kinase (JNK), p38, ERK/MAP kinases and possibly ceramide activated kinase (Cosulich and Clark, 1996; Mathias *et al.*, 1991). JNK activity is elevated following exposure to a variety of DNA damaging agents including cis-platinum, alkylating agents, Ara-C and irradiation (Kharbanda *et al.*, 1995 a and b). JNK and ERK were both found to be critical for the induction of rat PC12 nerve cell apoptosis following withdrawal of nerve growth factor (Xia *et al.*, 1995) and expression of dominant negative JNK in T-cells (or a Fas interacting protein Daxx- see later section) blocked stress kinase activation and conferred resistance to apoptosis induced by Fas and other death inducing stimuli (Yang *et al.*, 1997; Verheij *et al.*, 1996). Evidence suggests that SAPK's may also influence life and death decisions by phosphorylating critical members of the Bcl-2 family (Mandrell *et al.*, 1997; and see later section on Bcl-2 family).

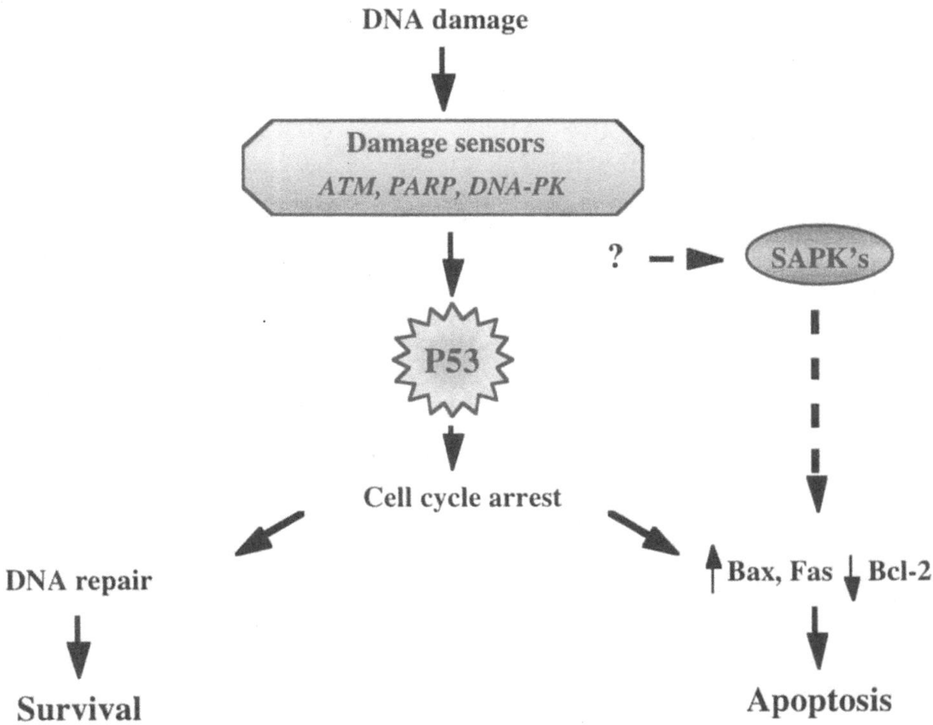

Figure 3. DNA damage and the stress response
Various types of DNA damage are detected by specific sensors in the nucleus. Several of these sensors have been directly linked to p53. The p53 mediated response depends upon the extent of the DNA damage. p53 can arrest the cell cycle, possibly allowing time for damage assessment and repair. In the case of extensive or unsustainable damage p53 directs apoptosis by a mechanism which may involve upregulation of inducers of apoptosis or downregulation of inhibitors. DNA damage can also activate stress activated protein kinases (SAPKs) which have been shown to influence cell susceptibility to apoptosis.

3.5 APOPTOSIS AS A DEFENCE AGAINST PATHOGEN INVASION

Reported incidences of apoptotic cell death in unicellular eukaryotes and even in prokaryotes suggests that a suicidal death program may have evolved as a mechanism of colony protection following viral infection (Haecker and Vaux, 1994). An anti-viral suicide response mechanism exists in certain strains of *E. coli* (Yu and Snyder, 1994), and in the hypersensitivity response of plants; the detection of a pathogen by a single cell initiates a suicide response in the infected cell, and in cells in its immediate vicinity (Greenberg, 1996). The fact that a large range of potent viruses have now been found to carry anti-apoptotic genes largely substantiates this theory (Granville *et al.*, 1998).

Defence against disease is undoubtedly more complicated in higher animals, but it still relies heavily on specialised methods of inducing apoptosis in excessive white blood cells, as well as in infected or abnormal cells. Auto-reactive immature T-cells are continuously being negatively selected in the thymus and undergo apoptosis. The retraction of lymphocyte populations following the resolution of an infection requires both the removal of survival factors and the interaction of death-inducing ligands with specialist death receptors (see later section on death receptors). Cytotoxic T-cells have developed elaborate independent pathways for inducing apoptosis in host cells harbouring intracellular pathogens, or in abnormal cells such as cancerous cells. A number of strategies (death receptor ligation and activation of downstream caspases; see subsequent sections) are employed to induce death in the target cell. Evidently the evolution of anti-apoptotic viral strategies has necessitated the parallel evolution of the cytotoxic T-cells which must eliminate them.

Loss of function of specialised inducers of apoptosis in the haematopoietic system predisposes an individual to disease, for example; auto-immunity, inflammation or haematological malignancies (reviewed in McKenna and Cotter 1997).

4.0 Major Mediators of Apoptotic Pathways

The first evidence that cell death was genetically regulated came from developmental studies of the nematode worm *C. elegans*. Of the 1090 somatic cells formed during the development of this hermaphrodite, 131 normally undergo programmed cell death. Mutants defective in this developmental cell death provided powerful tools for genetic analysis. Two genes *ced-3* and *ced-4,* were found to be required for normal developmental cell death. Another gene, *ced-9,* was identified as a negative regulator of cell death and acts antagonistically to *ced-3* and *ced-4* to suppress cell death. (Ellis and Horvitz 1986, Ellis *et al.*, 1991, Hengartner *et al.*, 1992). The discovery of human homologues (Ced-3/caspase-3; Ced-9/Bcl-2; Ced-4/Apaf-1) for these genes intensified apoptosis research. Indeed recent research has shown that higher organisms have evolved large gene families homologous to the pivotal regulators in *C. elegans*. The requirement for many apoptotic regulators is probably a reflection of the extent of cellular diversity and range of death inducing stimuli in higher organisms. In this section we will overview major families of apoptotic mediators and where possible orientate their hierarchical position in pathways to cell death.

4.1 THE CASPASE FAMILY

The first mammalian homologue of Ced-3 to be identified was the interleukin-1β converting enzyme (ICE), which is a cysteine protease responsible for the cleavage of the pro-inflammatory cytokine pro-interleukin-1β (Yuan *et al.*, 1993). Subsequent studies have revealed a large family of related proteolytic enzymes which have distinct roles in inflammation and apoptosis. Included in this family is CPP32/caspase-3 which is the actual (or closest) human homologue of Ced-3 and is a major effector protease in apoptotic pathways (Fernandes-Alnemri *et al.*, 1994). As all of these enzymes have the absolute requirement for an aspartate residue adjacent to their cleavage site they were named caspases. The '**c**' denotes the fact that they are cysteine proteases whereas the '**asp**ase' reflects their specificity for an aspartate (Almeri *et al.*, 1996). Each caspase is numbered based on the chronological order of its discovery. Fourteen mammalian caspases have been identified to date, two of which are murine (11 and 12) and have no known human counterparts. They can be further subgrouped into three families based on their sequence homology and substrate specificity (reviewed in Nicholson and

Thornberry 1997; Thornberry and Lazebnik, 1998). Caspase-13 (Humke *et al.*, 1998) and -14 (Van de Craen *et al.*, 1998) have been described most recently.

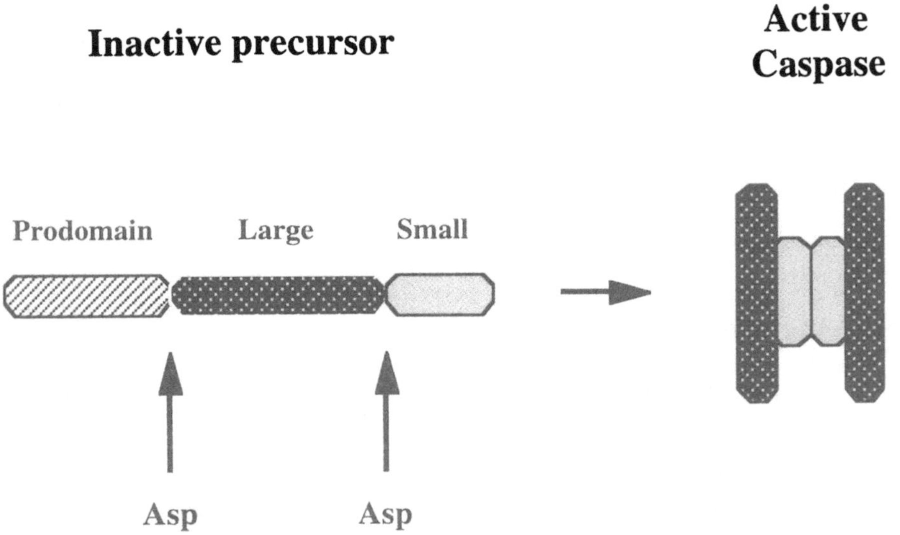

Figure 4. Schematic representation of caspase processing.
Caspases are constitutively produced as inactive precursors. Precursors consist of a prodomain, a large domain and a small domain. Proteolytic processing at aspartate residues releases the large and small domains which heterodimerise and associate to form a tetramer. Each heterodimer contributes an independent catalytic site to the active enzyme.

Caspases are synthesised as pro-enzymes which are composed of three domains: an NH_2- terminal domain, a large subunit and a small subunit (see Fig 4). NH_2 terminal domains are highly variable in sequence and length and are involved in the regulation of activation. Proteolytic processing at aspartate residues generates an active enzyme. Crystal structures of caspase-1 and -3 suggest that heterodimers of large and small subunits associate to form a (a_2b_2) tetramer which constitutes an active enzyme (Walker *et al.*, 1994; Wilson *et al.*, 1994; Rotonda *et al.*, 1996). The fact that these domains are cleaved at caspase consensus sites suggests that they may be able to activate their own precursors and participate in a proteolytic amplification cascade.

Caspases are constitutively expressed in most cells and can be rapidly activated following apoptosis-inducing stimuli. Evidence suggests that multiple members do participate in proteolytic cascades and that their activation represents a point of no return for the cell (Enari *et al.*, 1996; Orth *et al.*, 1996; Li *et al.*, 1997). Their destructive capability therefore necessitates stringent control of upstream mediators. Caspases at the apex of proteolytic cascades are referred to as initiator caspases and they are activated by the binding of other proteins or co-factors to their regulatory pro-domains. Examples of initiator caspases are caspase-8 (FLICE) and caspase-9 (ICE-LAP6, Mch6). Recent studies suggest that these caspases activate proteolytic cascades in response to different types of stimuli. Caspase-8 is activated by interaction of its 'death effector domain' (DD) with the DD of an adapter molecule (FADD), which has itself been activated by the ligation of specific death receptors (e.g. Fas or TNF; see later section). Caspase-9 has been associated with death induced by cytotoxic agents (Hakem *et al.*, 1998; Kuida *et al.*, 1998). Caspase-9 is activated by interaction of its CARD (caspase recruitment domain) with the CARD of apoptotic protease activating factor 1 (Apaf-1). Apaf-1 has regions of homology to Ced-4 and is considered to be its functional equivalent (Zou *et al.*, 1997). Apaf-1 requires cytochrome-*c* release from the mitochondria and dATP in order to activate caspase-9 (see subsequent section on mitochondria and Apafs). Other caspases with large pro-domains also have CARD's (e.g. Caspase-4) and may also associate with Apaf-1 in mammalian cells (Hu *et al.*, 1998).

Availability of activating protein domains or co-factors may be only one of a range of sophisticated mechanisms controlling caspase activation. An inert caspase (cFLIP), which is similar to caspase-8 but lacks catalytic residues is thought to act like a decoy, probably by competing for the co-factor FADD (see section on Death Receptors). An inhibitory protein with a CARD domain (ARC-apoptosis repressor with caspase recruitment domain) has also been recently identified (Koseki *et al.*, 1998). Several viral proteins are also known to be specific inhibitors of caspases. Viral inhibitors of caspase-8 (v-FLIPS) may also compete for the cofactor FADD (see following section on death receptors). Crm A is a caspase inhibitor encoded by cowpox virus which preferentially inhibits caspase-1 (ICE) (Miura *et al.*, 1993), and p35 is a Baculovirus gene which inhibits various members of the caspase family (Bump *et al.*, 1995; Xue and Horvitz *et al.*, 1995). A family of IAP proteins from baculoviruses also inhibit a broad range of caspases (Clem *et al*, 1996), and is known to have 5 mammalian homologues. (Roy *et al.*, 1995; Rothe *et al.*, 1995; Duckett *et al.*, 1996; Liston *et al.*, 1996; Uren *et al.*, 1996; Ambrosini *et al.*, 1997). The precise targets of these proteins and their role in the regulation of caspase activity remains to be established.

Initiator caspases activate a proteolytic cascade leading to the downstream activation of 'effector' caspases. These proteases have small pro-domains and are responsible for the dismantling of the cellular apparatus, for example caspase-3, -7 and -6 (see Fig 5).

Proteolytic digestion is however not indiscriminate, caspases display a high degree of target specificity. The substrate specificity of caspases is determined by their recognition of at least four amino acids NH_2-terminal to their cleavage site, and it also relies to some extent on 3-dimensional structure as not all proteins containing recognition sequences are targets for cleavage. As the list of known targets continues to grow it appears that the path to cellular shutdown encompasses a considerable degree of deliberation. Some substrates are inactivated, presumably because their normal activity conflicts with apoptosis, whereas other proteins are activated and participate in the process of apoptosis. Enzymes which play a critical role in DNA replication (replication factor C), mRNA splicing (U1-70K), and DNA repair (PARP and DNAPK) are among those targets which are cleaved and inactivated by caspases (Casciola-Rosen *et al.*, 1994; Lazebnik *et al.*, 1994; Song *et al.*, 1996). This may abolish unnecessary energy consumption by proteins which could be considered functionally redundant in apoptosis. Also inactivated by caspases are mediators of important survival pathways, for example PKB/Akt and Raf-1 (Widmann *et al.*, 1998), Bcl-2 family members (Cheng *et al.*, 1997) and retinoblastoma protein - an important regulator of the cell cycle (An and Ping Dou, 1996). Other inactivated proteins include (ICAD) the inhibitor of an endonuclease responsible for DNA fragmentation (CAD) (Enari *et al.*, 1998), and MDM-2 a negative regulator of the tumour suppressor p53 (Erhardt *et al.*, 1997).

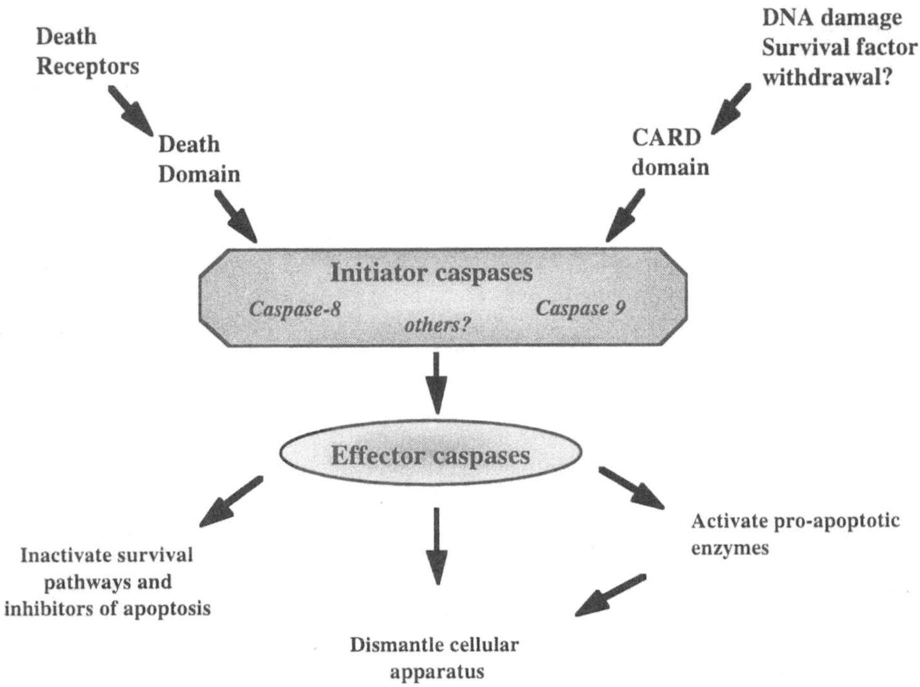

Figure 5. Caspase activation and consequences for the cell
Initiator caspases are activated by their interaction with death domain (DD) or caspase recruitment domain (CARD) containing proteins. The initiator caspase then activates a proteolytic cascade leading to the effector caspases. Effector caspases inactivate pathways which conflict with apoptosis, activate proteins which participate in apoptosis, and dismantle the cellular apparatus.

Substrates which are activated following cleavage by a caspase include p21 activated kinase 2 (PAK2) which is thought to play a role in cytoskeketal re-organisation and possibly in signal transduction during apoptosis. The cleavage and activation of a negative regulator of Ras activity (RasGAP) has been reported early in apoptosis and may serve to inhibit Ras mediated survival pathways (Wen *et al.*, 1998). PKCδ, PKCθ and MEKK1 are kinases which are also activated by the removal of their regulatory domains (reviewed by Bokoch, 1998). Other targets of caspases are structural components of the cell, (e.g. nuclear lamins, actin and Gas2) the cleavage of which is likely to facilitate chromatin condensation and cellular disassembly (reviewed by Porter *et al.*, 1997 and Bokoch, 1998).

As the list of caspases is rapidly expanding, so questions have been asked regarding their functional specificity or redundancy. Much of the evidence regarding caspase targets has been generated from *in vitro* studies. The fact that a caspase (e.g. caspase-3) is expressed in a wide range of cells and is capable of cleaving several targets does not signify a universal requirement in all cell types, moreover, the *in vitro* targets of caspases identified to date may not represent their critical *in vivo* targets. Studies with knockout mice have helped to assign the functions of some individual caspases *in vivo*. Mice have been generated which are deficient in caspases-1, -2, -3, -9 and -11 (Kuida *et al.*, 1995, 1996; Li *et al.*, 1995; Bergeron *et al.*, 1998; Wang *et al.*, 1998; Woo *et al.*, 1998; Kuida *et al.*, 1998; Hakem *et al.*, 1998). These studies suggest that there is not a global requirement for specific caspases as the knockout mice exhibit tissue specific and stimulus-specific deficiencies in cell death. Interestingly these studies have provided further evidence for the sequential activation of caspases, and also indicate the presence of several apoptotic pathways. Mice deficient in caspase-2 develop normally and do not have severe phenotypic abnormalities. Caspase-2 was however shown to be an essential inducer of apoptosis in female germ cells and in mediating apoptosis induced by granzyme B (a protease released by cytotoxic T-cells). Surprisingly, caspase-2 was also found to be an inhibitor of cell death in certain neuronal populations (Bergeron *et al.*, 1998). Caspase-1 and -11 deficient mice also develop normally. Their thymocytes are however defective in Fas-mediated cell death, but are not resistant to apoptosis induced by other stimuli (Kuida *et al.*, 1995, Wang *et al.*, 1998). Caspase-11 was also shown to interact with caspase -1 and to be essential for its activation (Wang *et al.*, 1998).

Caspase-3 and -9 knockouts are affected more severely in development. The majority of mice die perinatally (caspase-9 $^{-/-}$) or postnatally (caspase-3$^{-/-}$) with a vast excess of cells in the brain. (Kuida *et al.*, 1998; Hakem *et al.*, 1998 Kuida *et al.*, 1996; Woo *et al.*, 1998). Studies with various embryonic cells from these mice demonstrated stimulus and cell specific resistance to apoptosis. Notably, caspase-9 deficient brain cells and thymocytes are unable to activate caspase-3. Caspase-3 activity can however be restored by the addition of *in vitro* translated caspase 9, affirming that caspase-9 is upstream in a proteolytic cascade leading to caspase-3 (Kuida *et al.*, 1998). Certain types of apoptosis-inducing stimuli do not appear to involve these caspases (e.g. *u.v.* irradiation) and they are not always mutually activated suggesting the presence of several apoptosis pathways (Hakem *et al.*, 1998).

Thus, in contrast to a simple organism such *C. elegans* which employs only one caspase (Ced-3), mammals have evolved a family of highly specific proteases. This evidently reflects a more complex process of cellular destruction, possibly requiring the inactivation of several tissue specific survival pathways. In addition, the use of stimulus specific pathways may enable a tighter regulation of the cellular response to apoptosis inducing stimuli.

4.2 DEATH RECEPTORS

The death receptors are members of the tumour necrosis factor (TNF) receptor superfamily and are so called because they contain a homologous cytoplasmic sequence termed the death domain (Nagata, 1997). The death domain allows the death receptors to engage the cellular apoptotic program but it may also mediate separate, including anti-apoptotic, signals.

Of the death receptors identified to date the best characterised are Fas (CD95/Apo-1) and TNFR1 (p55/CD120a) (Gruss and Dower, 1995). Others include DR3 (TRAMP/Apo3/LARD/WSL-1) (Chinnaiyan *et al.*, 1996), DR4 (Pan *et al.*, 1997) and DR5 (Apo2/TRAIL-2/TRICK-2/KILLER) (Walczak *et al.*, 1997). The ligands of the death receptors are all members of the tumour necrosis factor (TNF) gene family and are structurally related. Fas ligand (FasL) binds Fas, TNF binds TNFR1, Apo3 ligand

(Apo3L) binds DR3 (Marsters *et al.*, 1998) and Apo2 ligand (Apo2L/TRAIL) binds DR4 (Pan *et al.*, 1997) and DR5 (Ling *et al.*, 1998).

4.2.1 Fas

Fas receptor-mediated signalling has been implicated in the deletion of autoreactive peripheral lymphocytes, the killing of target cells by cytotoxic T-cells and natural killer cells and the maintenance of immuno-privilege status in the eye and testes (Nagata, 1997). An additional role for the Fas signalling pathway appears to be the elimination of activated immune cells by tumour cells which express FasL (Hahne *et al.*, 1996).

The Fas receptor signalling complex represents the archetype of the death receptors and involves the trimerisation of receptor proteins upon ligand binding (Huang *et al.*, 1996). Receptor crosslinking promotes the binding of the Fas associated death domain (FADD/Mort-1) protein (Boldin *et al.*, 1995). FADD also contains a death effector domain (DED), an example of a CARD, which enables its interaction with procaspase-8 (Boldin *et al.*, 1996). Caspase-8 oligomerisation following recruitment by FADD allows its activation by autocatalyic proteolysis (Muzio *et al.*, 1998). Active caspase-8 is then liberated from the receptor complex and may then proteolytically activate other effector caspases, commiting the cell to apoptosis (Fig. 6). Mutational studies employing a dominant negative form of FADD (FADD-DN) have demonstrated that FADD is essential for apoptosis signaling through Fas (Yeh *et al.*, 1998; Zhang *et al.*, 1998). However, the finding that FADD knockout in mice results in abnormal T cell response and prenatal fatality indicates that FADD is required for signaling pathways other than Fas (Yeh *et al.*, 1998; Zhang *et al.*, 1998).

A recently described family of proteins, vFLIPs (Thome *et al.*, 1997), and cellular homolgue cFLIP (CASH/CASPER/I-FLICE/FLAME) (Hu *et al.*, 1997), also contain death domains allowing their interaction with FADD. These appear to act as inhibitors of Fas signalling by competing with caspase-8 for FADD, although their exact role is unclear due to conflicting reports that their overexpression may result in either an inhibition or activation of apoptosis (Goltsev *et al.*, 1997; Srinivasula *et al.*, 1997). Another cytoplasmic protein, Daxx, also binds Fas and activates a FADD-independent apoptotic pathway involving c-Jun NH_2-terminal kinase (JNK) (Yang *et al.*, 1997). However, since some FADD depleted cells are completely resistant to Fas induced apoptosis (Zhang *et al.*, 1998), Daxx does not mediate Fas signalling in all cell types.

4.2.2 TNFR1

TNF ligation to TNFR1 results in receptor trimerisation and recruitment of the death domain containing TRADD protein (Hsu *et al.*, 1995). TRADD can subsequently recruit a number of proteins to the receptor complex, including TRAF2 (Hsu *et al.*, 1996) and RIP (Ting *et al.*, 1996), both of which lead to the activation of the transcription factors NF-κB and JNK/AP1; and FADD, which results in apoptosis through the activation of caspase-8 (Hsu *et al.*, 1996) (fig. 6). TRADD may also bind the death domain-containing adaptor protein RAIDD which binds both RIP and caspase-2, consequently inducing apoptosis (Duan and Dixit, 1997).

4.2.3 DR3

DR3 shows high sequence homolgy to TNFR1 and similarly may either activate NF-κB or induce apoptosis by interaction with TRADD, TRAF2, RIP, and FADD (Ashkenazi and Dixit, 1998). DR3 activity is regulated by the binding of ApoL3, a molecule which is very similar to TNF (Ashkenazi and Dixit, 1998). Differential expression patterns of TNF, Apo3L, TNFR1 and DR3 (Tartaglia and Goeddel, 1992) indicates that the DR3 and TNFR1 signaling pathways are likely to play separate physiological roles.

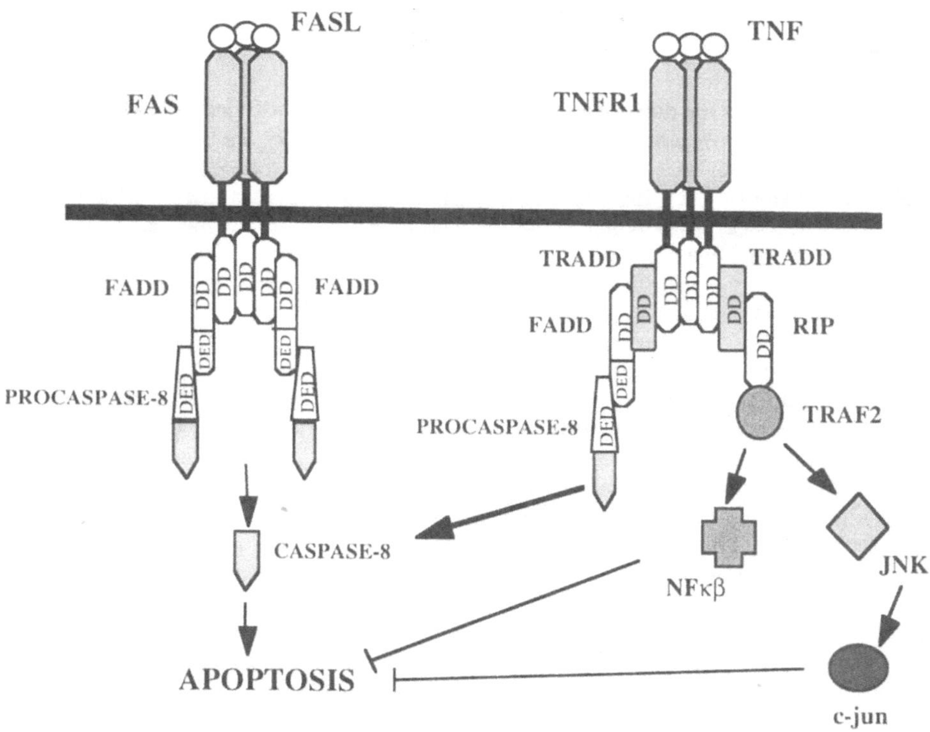

Figure 6. Death receptors
Signalling through the Fas and TNF Death Receptors. Receptor death domains (DD) allow the recruitment of adaptor proteins such as FADD, TRADD and RIP following ligand mediated receptor crosslinking. FADD may activate caspase-8 through its CARD death effector domain (DED) initiating apoptosis, while RIP, through recruitment of TRAF2, may induce NF-κB and jun N-terminal kinase signaling, thereby preventing apoptosis.

4.2.4 DR4 and DR5

Apo2L (TRAIL) exhibits the highest sequence similarity with FasL in the TNF gene family and appears to function primarily in the induction of apoptosis (Wiley *et al.*, 1995). Recent data implicates Apo2L in peripheral T-cell deletion and destruction of virally infected cells (Ashkenazi and Dixit, 1998). Apo2L induction of apoptosis appears to require caspase activity (Mariani *et al.*, 1997), although experiments with FADD-DN indicate that Apo2L caspase activation occurs through a FADD-independent pathway (Yeh *et al.*, 1998). Both DR4 and DR5 bind Apo2L, although there are conflicting findings as to whether signaling through these pathways involves FADD, TRADD, TRAF2 and RIP (Pan *et al.*, 1997; Walczak *et al.*, 1997).

A recently described group of truncated death receptor (DcRs) suggests a possible regulatory mechanism for the induction of apoptosis by Apo2L through DR4 and DR5 (Goldstein, 1997; Sheridan *et al.*, 1997). DcR1 and DcR2 both resemble DR4 and DR5, however, DcR1 is a GPI anchored cell surface protein (Mongkolsapaya *et al.*, 1998) while DcR2 lacks a functional death domain (Pan *et al.*, 1998). Emerging evidence suggsts that these receptors compete for Apo2L thus preventing its apoptotic signaling through either DR4 or DR5 (Marsters *et al.*, 1997).

4.3 BCL-2 FAMILY

The *bcl-2* gene has the dubious distinction of being the first oncogene shown to mediate its effects by an inhibition of apoptosis as opposed to the promotion of cell proliferation. It has been shown to be homologous to the *ced-9* gene in *C. elegans* (Hengartner and Horvitz, 1994) and is now but one member of a growing family of at least fifteen related genes which encode both positive (for example Bax, Bok, Bak, Bad, Bik, Bid) and negative regulators of apoptosis (for example Bcl-2, Ced-9, Mcl-1, Bcl-xL) (Chao and Korsmeyer, 1998). Bcl-2 contains four conserved motifs termed Bcl-2 homology (BH) domains, at least one of which is found in all Bcl-2 family members. The anti-apoptotic members of the Bcl-2 family are the most closely related to Bcl-2 and contain BH1 and BH2 domains, with those sharing most homology to Bcl-2 containing all four BH domains. The pro-apoptotic members of the family may be sub-divided into two groups; those containing the BH1, BH2 and BH3 domains (Bax, Bak and Bok) and those containing only the BH3 domain (for example, Bid, Bik and Bad). The ability of BH3 only containing members to exert pro-apoptotic effects suggests that the BH3 domain may be essential for this activity (Kelekar and Thompson, 1998).

Both the pro- and anti-apoptotic Bcl-2 family members may interact to form hetero- and homodimers on the mitochondrial, endoplasmic and nuclear membranes (Oltvai *et al.*, 1993). It is now believed that the relative ratios of these pro- and anti-apoptotic members is more important to the survival status of a cell than the expression of any one member in particular (Oltvai and Korsmeyer, 1994) (fig. 7). Heterodimerisation with other family members does not appear to be required for the function of the anti-apoptotic Bcl-2 proteins (Kelekar *et al.*, 1997), but is essential for the activity of BH3 only proteins (Kelekar and Thompson, 1998). Further evidence indicates that heterodimerisation may not be as significant for the other pro-apoptotic members function; emphasised by the ability of Bax homodimers to induce apoptosis (see *mitochondria and apoptosis* below). Heterodimerisation may also be preferential among Bcl-2 family members with interactions limited to a certain number of pro-and anti-apoptotic members for any particular protein (O'Connor *et al.*, 1998; Inohara *et al.*, 1998; Inohara *et al.*, 1997;Wang *et al.*, 1996) (fig. 7).

The biochemical basis for the action of the Bcl-2 family members has so far eluded researchers, partly due to the lack of significant sequence homology with other known proteins. However, co-immunoprecipitation experiments indicate that Bcl-2 and Bcl-xL may bind to at least eleven different proteins, including Raf-1, calcineurin, R-Ras, H-Ras, the p53 binding protein p53-bp2, the prion protein Pr-1, caspase-8, caspase-1, Ced-4 (Apaf-1) (see later section on Apaf-1) and proteins with unknown functions such as Bag-1, Nip-1, -2, and -3 (reviewed by Hacker and Vaux, 1995 and Reed, 1997). Significantly, Bax does not appear to interact with these proteins, suggesting that these interactions may be important for the death suppressing action of Bcl-2 and Bcl-xL. It has been suggested that Bcl-2 and other anti-apoptotic members of

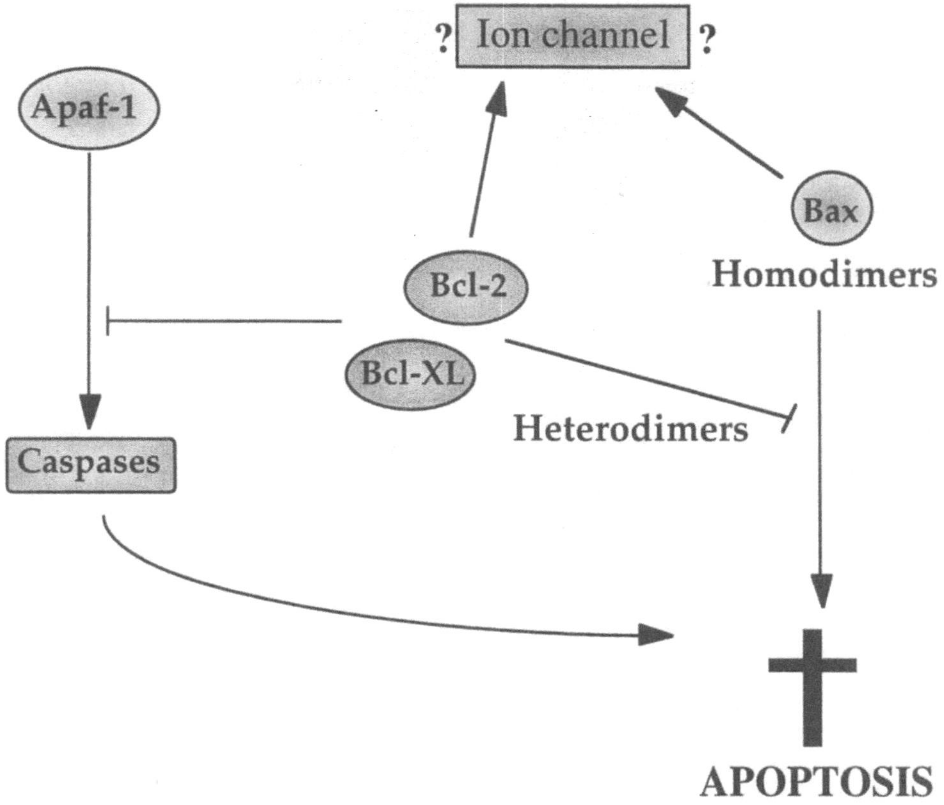

Figure 7. Regulatory mechanisms of Bcl-2 family proteins
Bcl-2/Bcl-xL may inhibit Bax homodimer driven apoptosis through heterodimerisation. Bcl-2/Bcl-xL may also interact with Apaf-1, preventing caspase activation and subsequent apoptosis. The significance of Bcl-2 family member ion channels on the regulation of apoptosis is presently unclear.

the family act by sequestering target proteins to their membrane localisation, thereby inactivating them or allowing them to interact with other membrane-associated proteins. In this scenario, the formation of heterodimers with the pro-apoptotic members of the Bcl-2 family, such as Bax, may prevent the docking/adapter function of Bcl-2 or Bcl-xL, thus frustrating their anti-apoptotic effects.

Recent structural studies carried out on Bcl-xL suggest homology with the pore forming domain of the diphtheria toxin (Muchmore *et al.*, 1996). This together with circumstantial evidence from studies in which Bcl-2 appeared to affect calcium fluxes across the mitochondrial, endoplasmic and nuclear membranes, indicates that Bcl-2 proteins might also have a membrane transport function (Baffy *et al.*, 1993; Lam *et al.*, 1994; Ryan *et al.*, 1994). It has been reported that, *in vitro*, Bcl-2 and Bcl-xL can form discrete ion channels, which suggests a second possible mechanism by which these proteins may effect their function (Minn *et al.*, 1997; Schendel *et al.*, 1997). The ability of Bax to heterodimerise with Bcl-2 and Bcl-xL raises the possibility that Bax may prevent Bcl-2 or Bcl-xL from forming ion channels which may be important for their anti-apoptotic effects or possibly that heterodimeric channels have different ion specificities and thereby different biochemical effects. The importance of ion channel formation by the Bcl-2 family remains to be fully understood but evidence suggests that ion channel formation is either not necessary for the induction of apoptosis or that it may be just one of a number of mechanisms through which the Bcl-2 proteins exert their effects (Borner *et al.*, 1994; Hockenbery *et al.*, 1993). The pro-apoptotic Bik does not possess the putative membrane integration domain yet retains the ability to form heterodimers, indicating that it can induce apoptosis without forming membrane channels.

Evidence exists for the regulation of Bcl-2 family members by cytokines and other survival signals (Chao *et al.*, 1998; Lin *et al.*, 1993; Kozopas *et al.*, 1993; Boise *et al.*, 1995; Maraskovsky *et al.*, 1997; Akashi *et al.*, 1997). Bax expression can be upregulated in some cells in response to p53 activation (Miyashita and Reed, 1995) while Bad can be phosphorylated by interleukin-3 signalling through the PKB/Akt and phosphoinositide-3 (PI(3)K) kinases (Zha *et al.*, 1996; Datta *et al.*, 1997; del Peso *et al.*, 1997, see also earlier section on survival signals). Modulation of Bcl-2 activity may occur as a result of phosphorylation of Ser70 (Ito *et al.*, 1997; Haldar *et al.*, 1998) as well as other sites (Chang *et al.*, 1997), possibly by the stress activated JNK (Maundrell *et al.*, 1997; Ling *et al.*, 1998).

4.4 MITOCHONDRIA AND APOPTOSIS

Emerging evidence now suggests that mitochondria play an important role in the execution of the apoptotic programme and may even represent a central regulator of the death process (Kroemer *et al.*, 1998). Cell free studies on apoptosis have demonstrated the requirement of a mitochondrial fraction in order for caspase activation and DNA fragmentation to occur, events which can be inhibited by Bcl-2/Bcl-xL (Mancini *et al.*, 1998). Further investigations revealed the release of several factors from mitochondria during apoptosis including cytochrome-*c* (Kluck *et al.*, 1997), AIF (apoptosis inducing factor) (Susin *et al.*, 1996) and in some cells, procaspase-3 (Mancini *et al.*, 1998). Cytochrome-*c* in the cytosol forms a complex with Apaf-1, the mammalian homologue of the *C. elegans* death gene *ced-4*, and procaspase-9 resulting in the activation of caspase-9 and initiatation of a caspase activation cascade (Li *et al.*, 1997) and (see *Apaf-1* below) (Fig. 8). The failure of caspase inhibitors to prevent cytochrome-*c* release from mitochondria in several cell systems indicates that this event is upstream of caspase activity (Bossey-Wetzel, *et al.*, 1998; Jurgenmeister *et al.*, 1998; Vander Heiden *et al.*, 1997).

Interaction between Ced-4, cytochrome-*c* and Ced-9 (the nematode homologue of Bcl-2) has also recently been demonstrated (Chinnaiyan *et al.*, 1997), as has interaction between Apaf-1 and Bcl-xL (Pan *et al.*, 1998). These interactions suggest a potential mechanism for caspase regulation by Bcl-2 through association with Apaf-1 (see Apaf-1 section). AIF has been reported to process procaspase-3 *in vitro* and disruption of its activity by a caspase inhibitor has led to suggestions that AIF may be a caspase (Susin *et al.*, 1996; Susin *et al.*, 1997).

Alterations in mitochondrial transmembrane potential ($\Delta\psi m$) have been reported as early events during apoptosis induced by a diverse range of stimuli (Kroemer *et al.*, 1998). It is believed that loss of $\Delta\psi m$ occurs via the permeability transition (PT) pore which allows the movement of solutes 1.5kD and smaller out of the mitochondria (Kroemer *et al.*, 1998). This opening of the PT pore can cause gross mitochondrial membrane disruption due to the expansion of the matrix space following osmolarity changes. The swelling of the matrix space can cause rupture of the outer membrane, since the inner membrane possesses a much larger surface area. Rupture of the outer membrane can then be envisaged to release cytochrome-*c*, AIF and procaspase-9 into the cytosol where they could inititate the apoptotic caspase cascade. Inhibitors of PT, which include Bcl-2 and Bcl-xL, have also been demonstrated to be efficient inhibitors of apoptosis (Zamzami *et al.*, 1996). In addition, Bax and other stimuli which directly affect PT such as oxidants and increases in intracellular Ca^{2+} have also been shown to induce membrane disruption and release of caspase-activating molecules (Zamzami *et al.*, 1996; Susin *et al.*, 1997).

However, recent studies have reported the release of cytochrome-*c* from the mitochondria preceeding loss of $\Delta\psi m$, indicating that the interaction of cytochrome-*c* with Apaf-1 and procaspase-9 may occur upstream of PT pore opening (Bosy-Wetzel *et al.*, 1998). This is further supported by the finding that in Apaf-1 deficient cells loss of $\Delta\psi m$ is prevented but cytochrome-*c* release still occurs (Yoshida *et al.*, 1998). This suggests that cytochrome-*c* acts upstream of both Apaf-1 and loss of $\Delta\psi m$ and that disruption of $\Delta\psi m$ is regulated by the interaction of Apaf-1 and cytochrome-*c*. It has been suggested that repeated, rapid opening and closing of the PT pore may allow the release of cytochrome-*c* into the cytosol in the absence of a detectable loss of $\Delta\psi m$ (Ichas and Mazat, 1998). This implies that the subsequent rupture of the outer mitochondrial membrane may serve as an amplification step for the release of caspase-activating proteins from the mitochondria (fig. 8). It should be noted, however, that mitochondrial membrane rupture is not observed in apoptotic cells *in vivo* (Kerr *et al.*, 1972), leading some to speculate on the possible existance of a large channel for the release of cytochrome-*c* from the mitochondria.

4.5 APAF-1

The recently identified human protein Apaf-1 has been demonstrated to be homologous to the pro-apoptotic *C.elegans* Ced-4 protein (Zou *et al.*, 1997). Similar to Ced-4, Apaf-1 has been shown to act upstream of caspase activity but downstream of Bcl-2 (Zou *et al.*, 1997). Sequence analysis suggests that Apaf-1 is not a caspase and that it carries out its regulatory function through protein-protein interaction. Recent data indicates that procaspase-9 binds to Apaf-1 in a reaction triggered by cytochrome-*c* and dATP (Li *et al.*, 1997). In this scenario Apaf-1 serves as a docking protein for both cytochrome-*c* and procaspase-9 allowing the subsequent activation of caspase-9 (by an unknown mechanism, possibly autocatalysis) and initiation of a cascasde of caspase activity (fig. 8). Cytochrome-*c* binds to Apaf-1 in the absence of dATP, but intracellular dATP levels are required to allow the binding of procaspase-9 (Li *et al.*, 1997). Interaction between Apaf-1 and procaspase-9 is mediated by a CARD on both proteins and it is possible that the binding of dATP to Apaf-1 induces a conformational change which exposes its CARD to procaspase-9.

The formation of an apoptosis-inducing complex by Apaf-1, cytochrome-*c* and procaspase-9 in a dATP-dependent manner is supported by the finding that cells depleted of dATP undergo necrosis in response to apoptosis-inducing stimuli, while restoration of intracellular dATP levels restores the apoptotic response (Leist *et al.*, 1997). The absolute requirement of cytochrome-*c* for Apaf-1-mediated procaspase-9 activation (Li *et*

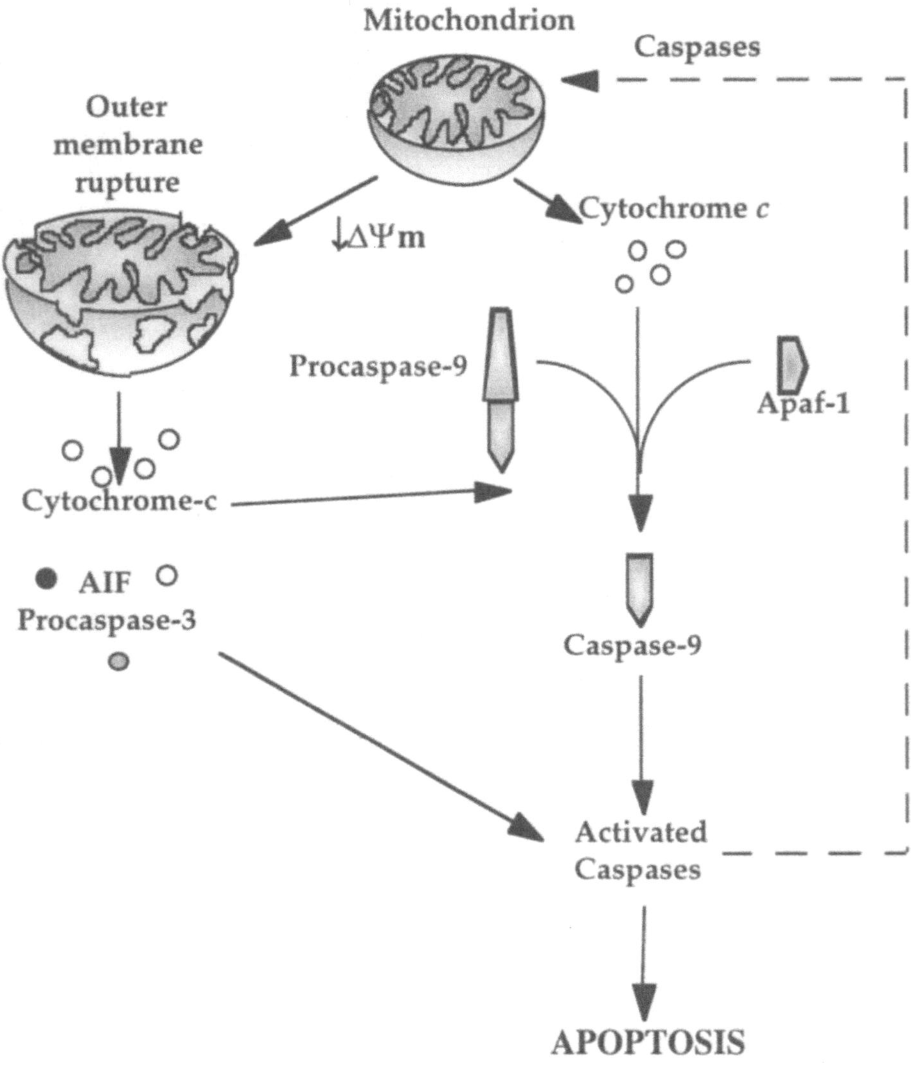

Figure 8. Central mediators of apoptosis: Mitochondria and APAF-1.
Caspase activation through mitochondrial signalling. Release of cytochrome-*c* from the mitochondria by a presently unidentified mechanism(s) allows the formation of a caspase-9 activating complex. In the presence of ATP cytochrome-*c*, Apaf-1 and procaspase-9 associate to induce the autocatalytic activation of caspase-9. Activated caspase-9 may then activate other caspases, commiting the cell to apoptosis. It has been proposed that activated caspases may then trigger further mitochondrial associated events such as depolarisation, outer membrane disruption and release of additional caspase activating molecules such as AIF, cytochrome-*c* and caspase-9. Thus mitochondrial depolarisation and outer membrane rupture may represent an amplification loop for caspase activation downstream of Apaf-1.

al., 1997) also correlates with the placing of Bcl-2 upstream of Apaf-1 activity, as Bcl-2 has been demonstrated to inhibit cytochrome-*c* release from the mitochondria (Kluck *et al.*, 1997). It is also worth noting that high levels of Bcl-2 can delay cell death even when cytochrome-*c* is already present in the cytosol (Zhivotosky *et al.*, 1998; Rosse *et al.*, 1998). This suggests that Bcl-2 may have two sites of action; inhibiting cytochrome-*c* release from the mitochondria as well as interfering with Apaf-1 binding of cytochrome-*c*, as has been demonstrated for Bcl-xL, thereby preventing caspase activation.

The involvment of procaspase-9 in the Apaf-1 and cytochrome-*c* complex has allowed caspase-9 to be placed highest of the caspases in the 'non-receptor initiated' apoptotic pathway. However, caspases-4, and -8, have also been demonstrated to interact *in vitro*
with Apaf-1 (Hu *et al.*, 1998; Li *et al.*, 1997), albeit with lower affinities, suggesting that Apaf-1 may regulate the activity of more than one iniatiator caspase.

The generation of Apaf-1 knockout mice has provided much support for Apaf-1's hypothesised role in apoptotic pathways. In these Apaf-1-deficient mutants all tissues whose development requires cell death (loss of interdigital web, palate formation, control of neural cell number, lens and retinal development) display abnormalities during the embryonic phase (Yoshida *et al.*, 1998). The absence of some of these defects in the known caspase knockout mice suggests that there are other, presently unknown, initiator caspases which can bind to Apaf-1 and mediate the activation of the protease cascade.

Some forms of apoptosis remain intact in Apaf-1 deficient cells, including Fas receptor mediated cell death; providing further evidence for the ability of this apoptotic pathway to access caspase activity directly without mitochondrial involvement. However, apoptosis induced by staurosporine and glucocorticoids was unaffected in Apaf-1 null cells, implying the existance of other, as of yet unidentified, Apaf-1-like molecules, or possibly other independent apoptotic pathways.

5.0 Conclusions and Future Perspectives

Apoptosis research has revealed a previously unforeseen intimacy between life and death. Living cells can be relatively easily coerced into death as they are entirely prepared for the event. Apoptotic programs co-exist with - and are intrinsically linked to - programs for mitogenesis, differentiation and possibly other cellular functions. Life and death decisions are subject to extensive regulation, as is the actual dismantling process of apoptosis itself. Much progress has been made regarding the identification of important regulatory mediators, and a renewed interest has been generated in several cytoplasmic constituents - in particular the role of cytochrome *c* and the mitochondria in apoptosis. Significant gaps in our knowledge of apoptotic pathways however remain.

Studies with several cell types have demonstrated that tissue-specific survival factors mediate viability. The important viability factors (or combination of) for many cell types still remain to be determined. In addition, the signalling components of survival pathways and also stress pathways need to be further elucidated. These pathways influence downstream regulators of apoptosis, but reported links between these components are tenuous at present. It is possible that the tissue-specific use of Bcl-2 family members (e.g. A1 and Mcl-1 are restricted to differentiating myeloid cells) may be a reflection of the cellular dependence upon anti-apoptotic signals originating from specific survival pathways.

The generation of mice deficient in certain apoptotic mediators has been particularly informative in delinating apoptotic pathways. Knockout mice for several of the caspases have now been generated. These studies suggest that while some caspases may be functionally redundant in certain cell types, there are tissues or specific apoptotic pathways which rely upon individual caspases. For example targeted disruption of caspases-1 and -11 had little or no effect upon developmental apoptosis, although Fas-mediated apoptosis and cytokine processing was disrupted. Deletion of caspases-3 and -9 produced more profound effects particularly in the development of the brain. Cells from these animals were also found to be resistant to apoptosis induced by cytotoxic agents. These studies also demonstrated that death inducing stimuli may activate one (or more) of a number of apoptotic pathways, and that cells utilise apoptotic pathways in a tissue specific and stimulus-specific manner.

The identification of key molecular components of apoptotic pathways such as cytochrome-*c*, Apaf-1, Bcl-2 and the caspases has encouraged much speculation about which step represents the point of no return for a cell on the verge of apoptosis. Fuel for this debate has been provided by both Apaf-1 knockout mutants and *in vitro* experiments on caspase-independent cell death. While the interdigital webs in Apaf-1 knockouts do not die on schedule they do appear to die at a later time (1-2 days later) (Yoshida *et al.*, 1998). Since Apaf-1 has been demonstrated to act downstream of cytochrome-*c*, this suggests that the cells are commited to die once cytochrome-*c* has been released from the mitochondria and that cells unable to activate caspases through Apaf-1 may simply take longer to die. This is further supported by several inhibitor based studies of cell death in the absence of caspase activity (Green and Reed, 1998). These cells die despite a block in caspase activation, albeit by a mechanism which may not be readily identifiable as apoptosis. It may also be argued that the release of cytochrome-*c* will inevitably result in disruption of the electron transport chain, failure of cellular respiration and the generation of reactive oxygen species - events which would be expected to render a cell non-viable, even in the absence of downstream caspase activity. However, evidence also exists for the regulation of cell death downstream of cytochrome-*c* release. Bcl-2 has been demonstrated to block cell death following cytochrome-c release into the cytosol. This, coupled with the reported inhibition of Apaf-1-dependent caspase-9 activation by Bcl-xL (Hu *et al.*, 1998), suggests that regulatory controls over cell death may exist downstream of cytochrome-*c*. Thus it may be that some cells can survive apoptosis iniatiation events upstream of the caspases possibly by *de novo* synthesis of mitochondria following a block in the caspase dependent apoptotic program (Mancini *et al.*, 1997).

While it is unclear at present whether or not cell viability is maintained following cytochrome-c release, it appears likely that the signal or mechanism for cytochrome-*c* release from the mitochondria could represent a target for not only the inhibition of apoptosis but for the maintenance of cell viability as well (i.e. the mode by which anti-apoptotic proteins such as Bcl-2 and Bcl-xL appear to act). Identification of the factors regulating cytochrome-*c* release may at present pose the next greatest challenge in apoptosis research.

Emerging evidence therefore suggests there may be a number of pathways and stimulus-specific control points mediating life and death decisions. Understanding the pivotal decision makers and the factors which influence them may enable future manipulation of apoptotic thresholds. This should facilitate the more rational design of therapeutic regimes for diseases which are now known to be largely a consequence of aberrant apoptosis. Degenerative conditions could benefit from replenishment of specific survival factors (or downstream activation of a survival pathway), or possibly by inhibition of the appropriate apoptotic pathway(s). Interestingly, expression of the caspase inhibitor p35 (see caspase section) prevented retinal degeneration in Drosophila (Davidson and Steller, 1998), and administration of a synthetic peptide inhibitor of caspases could reduce ischemia-reperfusion injury (Hara *et al.*, 1997) and FasL- or TNF-mediated liver damage in mice (Kunstle *et al.*, 1997).

Malignant cells undoubtedly avail of survival or anti-apoptotic mechanisms which are unavailable to their normal counterparts. Despite this however, other apoptotic pathways must be available as the cells are still capable of undergoing cell death, albeit in response to non-physiological cytotoxic stimuli. The main challenge in chemotherapeutic regimes is to induce cytotoxicity of the cancer cell with minimal damage to normal cells. Present cytotoxic regimes are relatively non-specific. It may be possible to re-open blocked apoptosis pathways further downstream (e.g. by disrupting Bcl-2 family member interactions or activating specific caspases), or to design combination therapy such that constituents induce a more cell-specific response and possibly activate non-overlapping pathways for maximum effect. It is also apparent that cells have different thresholds regarding their responsiveness to ligand binding at death receptors. It may therefore be possible to combine cytotoxic drugs with immunotherapy for a more tissue specific effect (Costa-Pereira and Cotter, 1998).

In addition to the obvious therapeutic implications, inhibition of apoptosis may also be useful to those who employ cells as 'factories' for the production of biological products such as vaccines or antibodies. The extension of cell viability at high cell densities frequently correlates with increased productivity in many commercially important cell lines or hybridomas. The addition of cell-specific survival factors to synthetic media and/or genetic manipulation of cell lines with anti-apoptotic genes has already proven to be useful for improvement of the viability of several important cell lines (reviewed in Carmody *et al.*, 1998 and see later Chapters this volume).

Overcoming default apoptosis programs is also a hurdle in the generation of new cell lines. Inadequate provision of survival factors may be responsible for the limited ability to establish cell lines from certain tissues. More research into tissue-specific survival factors may aid in the development of new cell lines for medical research and possibly also improve the survival of tissues for grafts or transplant.

In conclusion, recent research has unearthed a minefield in cell biology. The capacity for self destruction has evolved out of necessity and it is a integral component of life, particularly for multi-cellular organisms. It is anticipated that further research into this field will broaden our understanding of the evolution of the cell and its regulatory signalling networks, and hopefully yield therapeutic improvements for the future treatment of many presently incurable diseases.

Acknowledgements

We would like to acknowledge the EU Biotech programme, Irish Cancer Society, the Irish Health Research Board and RP Ireland Fighting Blindness for their generous financial support.

References

Agarwal, M.L., Taylor, W.R., Chernov, M.V., Chernova, O.B. and Stark, G.R. (1998) The p53 network, *J. Biol. Chem.* **273**, 1-4.

Akashi, K., Kondo, M., von Freeden-Jeffry, U., Murray, R. and Weissman, I.L. (1997) Bcl-2 rescues T lymphopoiesis in interleukin-7 receptor-deficient mice, *Cell* **89**, 1033-1041.

Alnemri, E.S., Livingston, D.J., Nicholson, D.W., Salvesen, G., Thornberry, N.A., Wong, W.W. and Yuan, J. (1996) Human ICE/CED-3 protease nomenclature, *Cell* **87**, 171.

Ambrosini, G., Adida, C. and Altieri, D.C. (1997) A novel anti-apoptosis gene, survivin, expressed in cancer and lymphoma, *Nat.Med.* **3**, 917-921.

An, B. and Ping Dou, Q. (1996) Cleavage of retinoblastoma protein during apoptosis: an interleukin 1 beta converting enzyme like protease, *Cancer Res.* **56**, 438-442.

Armstrong, J.F., Kaufman, M.H., Harrison, D.J. and Clark, A.R. (1995) High-frequency developmental abnormalities in p53 deficient mice, *Curr. Biol.* **5**, 931-936.

Ashkenazi, A. and Dixit, V.M. (1998) Death receptors: signaling and modulation, *Science* **281**, 1305-1308.

Baffy, G., Miyashita, T., Wiliamson, T.R. and Reed, J.C. (1993) Apoptosis induced by withdrawl of interleukin-3 from an IL-3 dependent hematopoetic cell line is associated with a repartitioning of intracellular calcium and is blocked by enforced bcl-2 oncoprotein production, *J. Biol. Chem.* **268**, 6511-6519.

Barde, Y.A. (1989) Trophic factors and neuronal survival, *Neuron* **2**, 1525-1534.

Bergeron, L., Perez, G.I., Macdonald, G., Shi, L., Sun, Y., Jurisicova, A., Varmuza, S., Latham, K.E., Flaws, J.A., Salter, J.C.M., Hara, H., Moskowitz, M.A., Li, E., Greenberg, A., Tilly, J.L. and Yuan, J. (1998) Defects in regulation of apoptosis in caspase-2 deficient mice, *Genes Dev.* **12**, 1304-1314.

Blandino, G., Scardigli, R., Rizzo, M.G., Crescenzi, M., Soddu, S. and Sacchi, A. (1995) Wild type p53 modulates apoptosis of normal, IL-3 deprived, hematopoitic cells, *Oncogene* **10**, 731-737.

Boise, L.H., Minn, A.J., Noel, P.J., June, C.H., Accavitti, M.A., Lindsten, T. and Thompson, C.B. (1995) CD28 costimulation can promote T cell survival by enhancing the expression of Bcl-XL, *Immunity* **3**, 87-98.

Bokoch, G.M. (1998) Caspase mediated activation of PAK2 during apoptosis: proteolytic kinase activation as a general mechanism of apoptotic signal transduction? *Cell Death Differ.* **5**, 637-645.

Boldin, M.P., Goncharov, T.M., Goltsev, Y.V. and Wallach, D. (1996) Involvement of MACH, a novel MORT1/FADD-interacting protease, in Fas/APO-1- and TNF receptor-induced cell death, *Cell* **85**, 803-815.

Boldin, M.P., Varfolomeev, E.E., Pancer, Z., Mett, I.L., Camonis, J.H. and Wallach, D. (1995) A novel protein that interacts with the death domain ofFas/APO1 contains a sequence motif related to the deathdomain, *J. Biol. Chem.* **270**, 7795-7798.

Borner, C., Martinou, I., Mattmann, C., Irmler, M., Schaerer, E., Martinou, J.C. and Tschopp, J.J. (1994) The protein bcl-2 alpha does not require membrane attachment, but two conserved domains to suppress apoptosis, *J. Cell Biol* **126**, 1059-1068.

Bossy-Wetzel, E., Newmeyer, D.D. and Green, D.R. (1998) Mitochondrial cytochrome c release in apoptosis occurs upstream of DEVD-specific caspase activation and independently of mitochondrial transmembrane depolarization, *EMBO J.* **17**, 37-49.

Bratton, D.L., Fadok, V.A., Richter, D.A., Kailey, J.M., Guthrie, L.A. and Henson, P.M. (1997) Appearance of phosphatidyltserine on apoptotic cells requires calcium mediated non-specific flip-flop and is enhanced by loss of the aminophspholipid translocase, *J. Biol. Chem.* **272**, 26159-26165.

Bump, N.J., Hackett, M., Hugunin, M., Seshagiri, S., Brady, K., Chen, P., Ferenz, C., Franklin, S., Ghayur, T., Li, P., Licari, P., Mankovich, J., Shi, L., Greenberg, A., Miller, L.K. and Wong, W.W. (1995) Inhibition of ICE family proteases by baculovirus anti-apoptotic protein P35, *Science* **269**, 1885-1888.

Caelles, C., Helmberg, A. and Karin, M. (1994) p53-dependant apoptosis in the absence of transcriptional activation of p53-target genes, *Nature* **370**, 220-223.

Callahan, M.K., Krahling, S.A., Williamson, P. and Schengel, R.A. (1997) The role of CD14 in the recognition and phagocytosis of apoptotic thymocytes by macrophages, *Mol. Cell. Biol.* **8** (Suppl.), 147a.

Carmody, R.J., Cotter, T.G. and McKenna, S.L. (1998) Programmed Cell Death (Apoptosis), In: *Encyclopedia of Animal and Plant Cell Technology* Ed: Miller, W.M. In press. Wiley, USA.

Carmody, R.J., Costa-Pereira, A.P., McKenna, S.L., and Cotter, T.G. (1998) Detection of molecular events during apoptosis by flow cytometry, In: *Aging: Methods and protocols,* Strehler, B. (ed), In press. The Humana Press Inc., USA.

Casciola-Rosen, L.A., Miller, D.K., Anhalt, G.J. and Rosen, A. (1994) Specific cleavage of the 70-kDa protein component of the U1 small nuclear ribonucleoprotein is a characteristic biochemical feature of apoptotic cell death, *J. Biol. Chem.* **269**, 30757-30760.

Chang, B.S., Minn, A.J., Muchmore, S.W., Fesik, S.W. and Thompson, C.B. (1997) Identification of a novel regulatory domain in Bcl-X(L) and Bcl-2, *EMBO J.* **16**, 968-977.

Chao, D.T. and Korsmeyer, S.J. (1998) BCL-2 family: regulators of cell death, *Annu. Rev. Immunol.* **16**, 395-419.

Chao, J.R., Wang, J.M., Lee, S.F., Peng, H.W., Lin, Y.H., Chou, C.H., Li, J.C., Huang, H.M., Chou, C.K., Kuo, M.L., Yen, J.J. and Yang-Yen, H.F. (1998) mcl-1 is an immediate-early gene activated by the granulocyte-macrophage colony-stimulating factor (GM-CSF) signaling pathway and is one component of the GM-CSF viability response, *Mol. Cell. Biol.* **8**, 4883-4898.

Cheng, E.H.-.Y., Kirsch, D.G., Clem, R.J., Ravi, R., Kastan, M.B., Bedi, A., Ueno, K. and Hardwick, J.M. (1997) Conversion of Bcl-2 to a Bax like death effector by caspases, *Science* **278**, 1966-1968.

Chinnaiyan, A.M., O'Rourke, K., Lane, B.R. and Dixit, V.M. (1997) Interaction of CED-4 with CED-3 and CED-9: a molecular framework for cell death, *Science* **275**, 1122-1126.

Chinnaiyan, A.M., O'Rourke, K., Yu, G.L., Lyons, R.H., Garg, M., Duan, D.R., Xing, L., Gentz, R., Ni, J. and Dixit, V.M. (1996) Signal transduction by DR3, a death domain-containing receptor related to TNFR-1 and CD95, *Science* **274**, 990-992.

Clem, R.J., Hardwick, J.M. and Miller, L.K. (1996) Anti-apoptotic genes of baculoviruses, *Cell Death Differ.* **3**, 9-16.

Cohen, J.J. and Duke, R.C. (1984) Glucocorticoid activation of a calcium-dependant endonuclease in thymocyte nuclei leads to cell death, *J. Immunol.* **132,** 38-42.

Cory, S., Harris, A.W. and Strasser, A. (1994) Insights from transgenic mice regarding the role of bcl-2 in normal and neoplastic lymphoid cells, *Philos. Trans. R. Soc.*

Lond. B. Biol. Sci. **345**, 289-295.

Costa-Pereira, A.P. and Cotter, T.G. (1998) Camptothecin sensitises androgen independant prostate cancer cells to anti-Fas induced apoptosis, *Br. J. Cancer* **In press**.

Cosulich, S. and Clark, P. (1996) Apoptosis: Does stress kill? *Curr. Biol.* **6**, 1586-1588.

Cotter, T.G. and Martin, S.J: *Techniques in apoptosis: A users guide*, Portland Press, London, (1996).

Coucouvanis, E. and Martin, G.R. (1995) Signals for death and survival: a two step mechanism for cavitation in the verterbrate embryo, *Cell* **83**, 279-287.

Datta, S.R., Dudek, H., Tao, X., Masters, S., Fu, H., Gotoh, Y. and Greenberg, M.E. (1997) Akt phosphorylation of BAD couples survival signals to the cell-intrinsic death machinery, *Cell* **91**, 231-241.

del Peso, L., Gonzalez-Garcia, M., Page, C., Herrera, R. and Nunez, G. (1997) Interleukin-3-induced phosphorylation of BAD through the protein kinase Akt, *Science* **278**, 687-689.

Devitt, A., Moffatt, O.D., Raykundalia, C., Capra, J.D., Simmons, D.L. and Gregory, C.D. (1998) Human CD14 mediates recognition and phagocytosis of apoptotic cells, *Nature* **392**, 505-509.

Downward, J. (1998) Ras signalling and apoptosis, *Curr. Opin. Genet. Dev* **8**, 49-54.

Duan, H. and Dixit, V.M. (1997) RAIDD is a new 'death' adaptor molecule, *Nature* **385**, 86-89.

Duckett, C.S., Nava, V.E., Gedrich, R.W., Clem, R.J., Vandongen, I.L., Gilfillan, M.C., Shiels, H., Hardwick, J.M. and Thomspon, C.B. (1996) A conserved family of cellular genes related to the baculovirus IAP gene and encoding apoptosis inhibitors, *EMBO J.* **15**, 2685-2694.

Dudek, H., Robert Datta, S., Franke, T.F., Birnbaum, M.J., Yao, R., Cooper, G.M., Segal, R.A., Kaplan, D.R. and Greenberg, M.E. (1997) Regulation of neuronal survival by the serine-threonine protein kinase Akt, *Science* **275**, 661-664.

Duvall, E., Wyllie, A.H. and Morris, R.G. (1985) Macrophage recognition of cells undergoing programmed cell death (apoptosis), *Immunology* **56**, 351-358.

Ellis, H.M. and Horvitz, H.R. (1986) Genetic control of programmed cell death in the nematode C.elegans, *Cell* **44**, 817-829.

Ellis, R.E., Yuan, J.Y. and Horvitz, H.R. (1991) Mechanisms and functions of cell death, *Ann. Rev. Cell Biol.* **7**, 663-698.

Enari, M., Sakahira, H., Yokoyama, H., Okawa, K., Iwamatsu, A. and Nagata, S. (1998) A caspase-activated Dnase that degrades DNA during apoptosis, and its inhibitor ICAD, *Nature* **391**, 43-50.

Enari, M., Talanian, R.V., Wong, W.W. and Nagata, S. (1996) Sequential activation of ICE-like proteases during Fas-mediated apoptosis, *Nature* **380**, 723-726.

Erhardt, P., Tomaselli, K.J. and Cooper, G.M. (1997) Identification of the MDM2 oncoprotein as a substrate for CPP32-like apoptotic proteases, *J. Biol. Chem.* **272**, 15049-15052.

Evan, G. and Littlewood, T. (1998) A matter of life and death, *Science* **281**, 1317-1322.

Fadok, V.A., Bratton, D.L., Frasch, S.C., Warner, M.L. and Henson, P.M. (1998) The role of phosphatidylserine in recognition of apoptotic cells by phagocytes, *Cell Death Differ.* **5**, 551-562.

Fadok, V.A., Voelker, D.R., Campbell, P.A., Cohen, J.J., Bratton, D.L. and Henson, P.M. (1992) Exposure of phosphatidylserine on the surface of apoptotic lymphocytes triggers specific recognition and removal by macrophages, *J. Immunol.* **148**, 2207-2216.

Falasca, L., Bergamini, A., Serafino, A., Balabaud, C. and Dini, L. (1996) Human Kupffer cell recognition and phagocytosis of apoptotic peripheral blood lymphocytes, *Exp. Cell Res.* **224**, 152-162.

Fanidi, A., Harrington, E.A. and Evan, G.I. (1992) Cooperative interaction between c-myc and bcl-2 proto-oncogenes, *Nature* **359**, 554-556.

Fernandes-Alnemri, T., Litwack, G. and Alnemri, E.S. (1994) CPP32 a novel human apoptotic protein with homology to *C.elegans* cell death ced-3 and mammalian interleukin-1B converting enzyme, *J. Biol. Chem.* **269**, 30761-30764.

Fesus, L., Thomazy, V. and Falus, A. (1987) Induction and activation of tissue transglutaminase during programmed cell death, *FEBS Lett.* **224**, 104-108.

Furtwangler, J.A., Hall, S.H. and Koskinen, M.L. (1985) Sutural morphogenesis in the mouse calvaria: the role of apoptosis, *Acta. Anat.* **124**, 74-80.

Glucksman, A. (1951) Cell deaths in normal vertebrate ontogeny, *Biol. Rev. Camb. Philos. Soc.* **26**, 59-86.

Goldring, M.B. and Goldring, S.R. (1991) Cytokines and cell growth control, *Crit. Rev. Eukaryot. Gene. Expr.* **1**, 301-26.

Golstein, P. (1997) Cell death: TRAIL and its receptors, *Curr. Biol.* **7**, R750-R753.

Goltsev, Y.V., Kovalenko, A.V., Arnold, E., Varfolomeev, E.E., Brodianskii, V.M. and Wallach, D. (1997) CASH, a novel caspase homologue with death effector domains, *J. Biol. Chem.* **272**, 19641-19644.

Gouin, A., Camu, W., Bloch-Gallego, E., Mettling, C. and Henderson, C.E. (1993) Growth and survival factors of spinal motorneurones, *C. R. Seances. Soc. Biol. Fil.* **187**, 47-61.

Granville, D.J., Carthy, C.M., Yang, D., Hunt, D.W.C. and McManus (1998) Interaction of viral proteins with host cell death machinery, *Cell Death Differ.* **5**, 653-659.

Green, D.R. and Reed, J.C. (1998) Mitochondria and Apoptosis, *Science*, **281**, 1309-1312.

Greenberg, J.T. (1996) Programmed cell death: A way of life for plants, *Proc. Natl. Acad. Sci. USA* **93**, 12094-12097.

Gruss, H.J. and Dower, S.K. (1995) Tumor necrosis factor ligand superfamily: involvement in the pathology of malignant lymphomas, *Blood.* **85**, 3378-3404.

Hacker G and Vaux DL. (1995) A sticky buisness, *Curr. Biol.* **5**, 622-624.

Hacker, G. and Vaux, D.L. (1994) Viral, worm and radical implications for apoptosis, *Trends Biochem. Sci.* **19**, 99-100.

Hahne, M., Rimoldi, D., Schroter, M., Romero, P., Schreier, M., French, L.E., Schneider, P., Bornand, T., Fontana, A., Lienard, D., Cerottini, J. and Tschopp, J. (1996) Melanoma cell expression of Fas(Apo-1/CD95) ligand: implications for tumor immune escape, *Science* **274**, 1363-1366.

Hakem, R., Hakem, A., Duncan, G.S., Henderson, J.T., Woo, M., Soengas, M.S., Elia, A., de la Pompa, J.L., Kagi, D., Khoo, W., Potter, J., Yoshida, R., Kaufman, S.A., Lowe, S.W., Penninger, J.M. and Mak, T.W. (1998) Differential requirement for Caspase 9 in apoptotic pathways in vivo, *Cell* **94**, 339-352.

Haldar, S., Basu, A. and Croce, C.M. (1998) Serine-70 is one of the critical sites for drug-induced Bcl2 phosphorylation in cancer cells, *Cancer Res.* **58**, 1609-1615.

Hara, H., Friedlander, R.M., Gagliardini, V., Ayata, C., Fink, K., Huang, Z., Shimizu-Sasamata, M., Yuan, J. and Moskowitz, M.A. (1997) Inhibition of interleukin 1 beta converting enzyme family proteases reduces ischemic and excitotoxic neuronal damage, *Proc. Natl. Acad. Sci. USA* **94**, 2007-2012.

Harrington, E.A., Bennett, M.R., Fanidi, A. and Evan, G.I. (1994) c-Myc-induced apoptosis in fibroblasts is inhibited by specific cytokines, *EMBO J.* **13**, 3286-3295.

Haupt, Y., Rowan, S., Shaulian, E., Vousden, K.H. and Oren, M. (1995) Induction of apoptosis in HeLa cells by transactivation deficient p53, *Genes Dev.* **9**, 2170-2183.

Hengartner, M.O. and Horvitz, H.R. (1994) C. elegans cell survival gene ced-9 encodes a functional homolog of the mammalian proto-oncogene bcl-2, *Cell* **76**, 665-676.

Hockenbery, D.M., Oltvai, Z.N., Yin, X.M., Milliman, C.L. and Korsmeyer, S.J. (1993) Bcl-2 functions in an antioxidant pathway to prevent apoptosis, *Cell* 75, 241-251.

Hsu, H., Shu, H.B., Pan, M.G. and Goeddel, D.V. (1996) TRADD-TRAF2 and TRADD-FADD interactions define two distinct TNF receptor 1 signal transduction pathways, *Cell* **84**, 299-308.

Hsu, H., Xiong, J. and Goeddel, D.V. (1995) The TNF receptor 1-associated protein TRADD signals cell death and NF-kappa B activation, *Cell* **81**, 495-504.

Hu, S., Vincenz, C., Ni, J., Gentz, R. and Dixit, V.M. (1997) I-FLICE, a novel inhibitor of tumor necrosis factor receptor-1- and CD-95-induced apoptosis, *J. Biol. Chem.* **272**, 17255-17257.

Hu, Y., Benedict, M.A., Wu, D., Inohara, N. and Nunez, G. (1998) Bcl-XL interacts with Apaf-1 and inhibits Apaf-1 dependent caspase-9 activation, *Proc. Natl. Acad. Sci USA.* **95**, 4386-4391.

Huang, B., Eberstadt, M., Olejniczak, E.T., Meadows, R.P. and Fesik, S.W. (1996) NMR structure and mutagenesis of the Fas (APO-1/CD95) death domain, *Nature* **384**,638-641.

Hughes, R.A., Sendtner, M. and Thoenen, H. (1993) Members of several gene families influence survival of rat motoneurones in vitro and in vivo, *J. Neurosci.Res.* **15**, 663-671.

Hughes, F.M. and Cidlowski, J.A. (1994) Apoptotic DNA degradation: evidence for novel enzymes, *Cell Death Differ.* **1**, 11-17.

Humke, E.W., Ni, J. and Dixit, V.M. (1998) ERICE, a novel FLICE-activatable caspase, *J. Biol. Chem.* **273**, 15702-15707.

Ichas, F. and Mazat, J.P. (1998) From calcium signaling to cell death: two conformations for the mitochondrial permeability transition pore. Switching from low- to high-conductance state, *Biochim. Biophys. Acta.* **1366**, 33-50.

Inohara, N., Ding, L., Chen, S. and Nunez, G. (1997) harakiri, a novel regulator of cell death, encodes a protein that activates apoptosis and interacts selectively with survival-promoting proteins Bcl-2 and Bcl-X(L), EMBO J. **16**, 1686-1694.

Inohara, N., Ekhterae, D., Garcia, I., Carrio, R., Merino, J., Merry, A., Chen, S., Nunez, G. (1998) Mtd, a novel Bcl-2 family member activates apoptosis in the absence of heterodimerization with Bcl-2 andBcl-XL, *J. Biol. Chem.* **273**, 8705-8710.

Ito, T., Deng, X., Carr B. and May, W.S. (1997) Bcl-2 phosphorylation required for anti-apoptosis function, *J. Biol. Chem.* **272**, 11671-11673.

Jacobson, M.D., Weil, M. and Raff, M.C. (1997) Programmed cell death in animal development, *Cell* **88**, 347-354.

Jurgensmeier, J.M., Xie, Z., Deveraux, Q,., Ellerby, L., Bredesen, D. and Reed, J.C. (1998) Bax directly induces release of cytochrome c from isolated mitochondria, *Proc. Natl. Acad. Sci. USA*. **95**, 4997-5002.

Kauffmann-Zeh, A., Rodriguez-Viciana, P., Ulrich, E., Gilbert, C., Coffer, P., Downward, J. and Evan, G. (1997) Suppression of c-Myc induced apoptosis by Ras signalling through PI(3)K and PKB, *Nature* **385**, 544-548.

Kelekar, A. and Thompson, C.B. (1998) Bcl-2-family proteins: the role of the BH3 domain in apoptosis, *Trends Cell Biol.* **8**, 324-330.

Kelekar, A., Chang, B.S., Harlan, J.E., Fesik, S.W. and Thompson, C.B. (1997) Bad is a BH3 domain-containing protein that forms an inactivating dimer with Bcl-XL, *Mol. Cell. Biol.* **17**, 7040-7046.

Kennedy, S.G., Wagner, A.J., Conzen, S.D., Jordan, J., Bellacosa, A., Tsichlis, P.N. and Hay, N. (1997) The PI3-kinase / Akt signaling pathway delivers an anti-apoptotic signal, *Genes Dev.* **11**, 701-713.

Kerr, J.F.R., Searle, J., Harmon, B.V. and Bishop, C.J. (1987) Apoptosis. In: Perspectives on mammalian cell death (ed. C.S. Potten), pp 93-119 Oxford University Press, Oxford, New York, Tokyo.

Kerr, J.F.R., Wyllie, A.H. and Currie, A.H. (1972) Apoptosis, a basic biological phenomenon with wider implications in tissue kinetics, *Br. J. Cancer* **26,** 239-245.

Kharbanda, S., Pandey, P., Ren, R., Mayer, B., Zon, L. and Kufe, D. (1995a) c-Abl activation regulates induction of the SEK1/stress-activated protein kinase pathway in the cellular response to 1-beta-D-arabinofuranosylcytosine, *J. Biol. Chem.* **270**, 30278-30281.

Kharbanda, S., Ren, R., Pandey, P., Shafman, T.D., Feller, S.M., Weichselbaum, R.R. and Kufe, D.W. (1995b) Activation of the c-Abl tryosine kinase in the stress response to DNA damaging agents, *Nature* **376**, 785-788.

Khwaja, A., Rodriguez-Viciana, P., Wennstrom, S., Warne, P. and Downward, J. (1997) Matrix adhesion and Ras transformation both activate a phosphoinositide 3-OH kinase and protein kinase B/Akt cellular survival pathway, *EMBO J.* **16**, 2783-2793.

Kinoshita, T., Shirouzu, M., Kamiya, A., Hashimoto, K., Yokoyama, S. and Miyajima, A. (1997) Raf/MAPK and rapamycin sensitive pathways mediate the anti-apoptotic function of p21Ras in IL-3 dependent hematopoietic cells, *Oncogene* **15**, 619-627.

Kinoshita, T., Yokota, T., Arai, K. and Miyajima, A. (1995) Suppression of apoptotic death in hematopoietic cells by signalling through the IL-3/GM-CSF receptors, *EMBO J.* **14**, 266-275.

Kluck, R.M., Bossy-Wetzel, E., Green. D.R. and Newmeyer, D.D. (1997) The release of cytochrome c from mitochondria: a primary site for Bcl-2 regulation of apoptosis, *Science*, **275**, 1132-1136.

Koliatsos, V.E., Cayouette, M.H., Berkemeier, L.R., Clatterbuck, R.E., Price, D.L. and Rosenthal, A. (1994) Neurotrophin4/5 is a trophic factor for mammalian facial motor neurons, *Proc. Natl. Acad. Sci. USA* **91**, 3304-3308.

Koseki, T., Inohara, N., Chen, S. and Nunez, G. (1998) ARC, an inhibitor of apoptosis expressed in skeletal muscle and heart that interacts selectively with caspases, *Proc. Natl. Acad. Sci. USA* **95**, 5156-60.

Koury, M.J. and Bondurant, M.C. (1990) Erythropoietin retards DNA breakdown and prevents programmed death in erythroid progenitor cells, *Science* **248**, 378-381.

Kozopas, K.M., Yang, T., Buchan, H.L., Zhou, P. and Craig, R.W. (1993) MCL1, a gene expressed in programmed myeloid cell differentiation, has sequence similarity to BCL2, *Proc. Natl. Acad. Sci. USA.* **90**, 3516-3520.

Kroemer, G., Dallaporta, B. and Resche-Rigon, M. (1998) The mitochondrial death/life regulator in apoptosis and necrosis, *Annu. Rev. Physiol.* **60**, 619-642.

Kuida, K., Haydar, T.F., Kuan, C.Y., Gu, Y.o.n.g., Taya, C., Karasuyama, H., Su, M.S.-.S., Rakic, P. and Flavell, R.A. (1998) Reduced apoptosis and cytochrome c mediated caspase activation in mice lacking caspase 9, *Cell* **94**, 325-337.

Kuida, K., Lippke, J.A., Ku, G., Harding, M.W., Livingston, D.J., Su, M.S. and Flavell, R.A. (1995) Altered cytokine export and apoptosis in mice deficient in interleukin-1 beta converting enzyme, *Science* **267**, 2000-2003.

Kuida, K., Zheng, T.S., Na, S., Kuan, C.-.Y., Yang, D., Karasuyama, H., Rakic, P. and Flavell, R.A. (1996) Decreased apoptosis in the brain and lethality in CPP32-deficient mice, *Nature* **384**, 368-372.

Kulik, G., Klippel, A. and Weber, M.J. (1997) Antiapoptotic signalling by the insulin like growth factor I receptor, phosphatidylinositol 3-kinase and Akt, *Mol Cell Biol* **17**, 1595-1606.

Kunstle, G., Leist, M., Uhlig, S., Revesz, L., Feifel, R., MacKenzie, A. and Wendel, A. (1997) ICE-protease inhibitors block murine liver injury and apoptosis caused by CD95 or by TNF alpha, *Immunol. Lett.* **55**, 5-10.

Lam, M., Dubyak, G., Chen, L., Nunez, G., Miesfeld, R.L. and Distelhorst, C.W. (1994) Evidence that bcl-2 represses apoptosis by regulating endoplasmic recticulum associated calcium fluxes, *Proc. Natl. Acad. Sci .USA.* **91**, 6569-6573.

Lazebnik, Y.A., Kaufmann, S.H., Desnoyers, S., Poirier, G.G. and Earnshaw, W.C. (1994) Cleavage of poly(ADP-ribose) polymerase by a proteinase with properties like ICE, *Nature* **371**, 346-347.

Leist, M., Single, B., Castoldi, A.F., Kuhnle, S. and Nicotera, P. (1997) Intracellular adenosine triphosphate (ATP) concentration: a switch in the decision between apoptosis and necrosis, *J. Exp. Med.* **185**, 1481-1486.

Lewin, G.R. (1996) Neurotrophins and the specification of neuronal phenotype, *Phiolos. Trans. R. Soc. Lond. B. Biol. Sci.* **351**, 405-411.

Lewinson, D. and Silbermann, M. (1992) Chondroclasts and endothelial cells collaborate in the process of cartilage resorption, *Anat. Rec.* **233**, 504-514.

Li, P., Allen, H., Banerjee, S., Franklin, S., Herzog, L., Johnston, C., McDowell, J., Paskind, M., Rodman, L., Salfield, J. and etal (1995) Mice deficient in IL-1B converting enzyme are defective in production of mature IL-1B and resistant to endotoxic shock, *Cell* **80**, 401-411.

Li, P., Nijhawan, D., Budihardjo, I., Srinivasula, S.M., Ahmad, M., Alnemri, E.S. and Wang, X. (1997) Cytochrome c and dATP-dependent formation of Apaf-1/caspase-9 complex initiates an apoptotic protease cascade, *Cell* **91**, 479-489.

Lin, E.Y., Orlofsky, A., Berger, M.S. and Prystowsky, M.B. (1993) Characterization of A1, a novel hemopoietic-specific early-response gene with sequence similarity to bcl-2, *J. Immunol.* **151**, 1979-1988.

Lin, Y. and Benchimol, S. (1995) Cytokines inhibit p53-mediated apoptosis but not p53-mediated G1 arrest, *Mol. Cell. Biol.* **15**, 6045-6054.

Lindsay, R.M. (1996) Role of neurotrophins and trk receptors in the development and maintenance of sensory neurons: an overview, *Phiolos. Trans. R. Soc. Lond. B. Biol. Sci.* **351**, 365-373.

Ling, Y.H., Tornos, C. and Perez-Soler, R. (1998) Phosphorylation of Bcl-2 is a marker of M phase events and not a determinant of apoptosis, *J. Biol. Chem.* **273**, 18984-18991.

Liston, P., Roy, N., Tamai, K., Lefebvre, C., Baird, S., Chertonhorvat, G., Farahani, R., Mclean, M., Ikeda, J.E., Mackenzie, A. and Korneluk, R.G. (1996) Supression of apoptosis in mammalian cells by NAIP and a related family of IAP genes, *Nature* **379**, 349-353.

Liu, J.P., Baker, J., Perkins, A.S., Robertson, E.J. and Efstratiadis, A. (1993) Mice carrying null mutations of the gènes encoding insulin like growth factor I (IGF-I) and type I IGF receptor (IGF-IR), *Cell* **75**, 59-72.

Lotem, J. and Sachs, L. (1995) A mutant p53 antagonizes the deregulated c-*myc*-mediated enhancement of apoptosis and decrease in leukemogenicity, *Proc. Natl. Acad. Sci. USA* **92**, 9672-9676.

Lowery, P.A. (1995) Hematopoietic stem cell cytokine response, *J. Cell Biochem.* **58**, 410-415.

Luciani, M.F. and Chimini, G. (1996) The ATP binding cassette transporter ABC1, is required for the engulfment of corpses generated by apoptotic cell death, *EMBO J.* **15**, 226-235.

Mancini, M., Anderson, B.O., Caldwell, E., Sedghinasab, M., Paty, P.B. and Hockenbery, D.M. (1997) Mitochondrial proliferation and paradoxical membrane depolarization during terminal differentiation and apoptosis in a human colon carcinoma cell line, *J. Cell Biol.* **138**, 449-469.

Mancini, M., Nicholson, D.W., Roy, S., Thornberry, N.A., Peterson, E.P., Casciola-Rosen, L.A. and Rosen, A. (1998) The caspase-3 precursor has a cytosolic and mitochondrial distribution: implications for apoptotic signaling, *J. Cell Biol.* **140**, 1485-1495.

Mandrell, K., Antonsson, B., Magnenat, E., Camps, M., Muda, M., Chabert, C. and et al (1997) Bcl-2 undergoes phosphorylation by c-Jun N-terminal kinase/stress activated protein kinase in the presence of the constitutively active GTP binding protein Rac1, *J. Biol. Chem.* **272**, 25238-25242.

Maraskovsky, E., O'Reilly, L.A., Teepe, M., Corcoran, L.M., Peschon, J.J. and Strasser, A. (1997) Bcl-2 can rescue T lymphocyte development in interleukin-7 receptor-deficient mice but not in mutant rag-1-/- mice, *Cell* **89**, 1011-1019.

Mariani, S.M., Matiba, B., Armandola, E.A. and Krammer, P.H. (1997) Interleukin 1 beta-converting enzyme related proteases/caspases are involved in TRAIL-induced apoptosis of myeloma and leukemia cells, *J. Cell Biol.* **137**, 221-229.

Marsters, S.A., Sheridan, J.P., Pitti, R.M., Brush, J., Goddard, A. and Ashkenazi, A. (1998) Identification of a ligand for the death-domain-containing receptor Apo3, *Curr. Biol.* **8**, 525-528.

Marsters, S.A., Sheridan, J.P., Pitti, R.M., Huang, A., Skubatch, M., Baldwin, D., Yuan, J., Gurney, A., Goddard, A.D., Godowski, P. and Ashkenazi, A. (1997) A novel receptor for Apo2L/TRAIL contains a truncated death domain, *Curr. Biol.* **7**, 1003-1006.

Martin, S.J., Reutelingsperger, C.P.M., McGahon, A.J., Radar, J.A., van Schie, R.C., LaFace, D.M. and Green D.R. (1995) Early redistribution of plasma membrane phosphatidylserine is a general feature of apoptosis regardless of the initiating stimulus: inhibition by overexpression of Bcl-2 and Abl, *J. Exp. Med.* **182**, 1545-1556.

Mathias, S., Dressier, K.A. and Kolesnick, R.N. (1991) Characterization of a ceramide activated protein kinase: stimulation by tumor necrosis factor alpha, *Proc. Natl. Acad. Sci. USA* **88**, 10009-10013.

Maundrell, K., Antonsson, B., Magnenat, E., Camps, M., Muda, M., Chabert, C., Gillieron, C., Boschert, U., Vial-Knecht, E., Martinou, J.C. and Arkinstall, S. (1997) Bcl-2 undergoes phosphorylation by c-Jun N-terminal kinase/stress-activated protein kinases in the presence of the constitutively active GTP-binding protein Rac1, *J. Biol. Chem.* **272**, 25238-25242.

McCarthy, J.V. and Cotter, T.G. (1997) Cell shrinkage: a role for potassium and sodium ion flux, *Cell. Death. Differ.* **4**, 756-770.

McKenna, S.L. and Cotter, T.G. (1997) Functional aspects of apoptosis in haematopoiesis and consequences of failure, *Adv. Cancer Res.* **71**, 121-164.

Minn, A.J., Velez, P., Schendel, S.L., Liang, H., Muchmore, S.W., Fesik, S.W., Fill, M. and Thompson, C.B. (1997) Bcl-XL forms an ion channel in sythetic lipid membranes, *Nature*, **385**, 353-357.

Minshall, C., Arkins, S., Freund, G.G. and Kelley, K.W. (1996) Requirement for phosphatidylinositol 3'kinase to protect hemopoietic progenitors against apoptosis depends upon the extracellular survival factor, *J. Immunol.* **156**, 939-947.

Miura, M., Zhu, H., Rotello, R., Hartwieg, E.A. and Yuan, J. (1993) Induction of apoptosis in fibroblasts by IL-1beta converting enzyme, a mammalian homologue of the C. elegans cell death gene ced-3, *Cell* **75**, 653-660.

Miyashita, T. and Reed, J.C. (1995) Tumor suppressor p53 is a direct transcriptional activator of the human bax gene, *Cell* **80**, 293-299.

Miyashita, T., Harigai, M., Hanada, M. and Reed, J.C. (1994) Identification of a p53-dependant negative response element in the *bcl-2* gene, *Cancer Res.* **54**, 3131-3135.

Mongkolsapaya, J., Cowper, A.E., Xu, X.N., Morris, G., McMichael, A.J., Bell, J.I. and Screaton, G.R. (1998) Lymphocyte inhibitor of TRAIL (TNF-related apoptosis-inducing ligand): a new receptor protecting lymphocytes from the death ligand TRAIL, *J. Immunol.* **160**, 3-6.

Morgan, S.E. and Kastan, M.B. (1997) p53 and ATM: Cell cycle, cell death, and cancer, *Adv. Cancer Res.* **71**, 1-25.

Morris, R.G., Duvall, E., Hargreaves, A.D. and Wyllie, A.H. (1984) Hoermoneinduced cell death. 2. Surface changes in thymocytes undergoing apoptosis, *Am. J. Pathol.* **115**, 426-436.

Muchmore, S.W., Sattler, M., Liang, H., Meadows, R.P., Harlan, J.E., Yoon, H.S., Nettesheim, D., Chang, B.S., Thompson, C.B., Wong, S.L., Ng, S-C, and Fesik, S.W. (1996) X-ray and NMR structure of human Bcl-xL, an inhibitor of programmed cell death, *Nature* **381,** 335-341.

Muzio, M., Stockwell, B.R., Stennicke, H.R., Salvesen, G.S. and Dixit V.M. (1998) An induced proximity model for caspase-8 activation, *J. Biol. Chem.* **273**, 2926-2930.

Nagata, S. (1997) Apoptosis by death factor, *Cell* **88**, 3553-65.

Nicholson, D.W. and Thornberry, N.A. (1997) Caspases: killer proteases, *Trends Biochem. Sci.* **22**, 299-306.

O'Connor, L., Strasser, A., O'Reilly, L.A., Hausmann, G., Adams, J.M., Cory, S. and Huang, D.C. (1998) Bim: a novel member of the Bcl-2 family that promotes apoptosis, *EMBO J.* **17**, 384-395.

O'Connor, R. (1998) Survival factors and apoptosis, *Adv.Biochem.Eng.Biotech.* **62**, 137-166.

Oberhammer, F., Wilson, J.W., Dive, C., Morris, I.D., Hickman, J.A., Wakeling, A.E., Walker, P.R. and Sikorska, M. (1993) Apoptotic death in epithelial cells: cleavage of DNA to 300 and/or 50kb fragments prior to or in the absence of internucleosomal fragmentation, *EMBO J.* **12**, 3670-3684.

Ockel, M., von-Schack, D., Schropel, A., Dechant, G., Lewin, G.R. and Barde, Y.A. (1996) Roles of neurotrophin-3 during early development of the peripheral nervous system, *Phiolos. Trans. R. Soc. Lond. B. Biol. Sci.* **351**, 383-387.

Oltvai, Z.N. and Korsmeyer, S.J. (1994) Checkpoints of dueling dimers foil death wishes, *Cell* **79**, 189-192.

Oltvai, Z.N., Milliman, C.L., Korsmeyer, S.J. (1993) Bcl-2 heterodimerizes in vivo with a conserved homolog, Bax, that accelerates programmed cell death, *Cell* 74, 609-619.

Oppenheim, R.W. (1991) Cell death during development of the nervous system, *Annu. Rev. Neurosci.* **14**, 453-501.

Orth, K., O'Rourke, K., Salvesen, G.S. and Dixit, V.M. (1996) Molecular ordering of mammalian CED-3/ICE-like proteases, *J. Biol. Chem.* **271**, 20977-20980.

Owen-Schaub, L.B., Zhang, W., Cusack, J.C., Angelo, L.S., Santee, S.M., Fujiwara, T., Roth, J.A., Deisseroth, A.B., Zhang, W.-.W., Kruzel, E. and etal (1995) Wild-type human p53 and a temperature-sensitive mutant induce Fas/APO-1 expression, *Mol Cell Biol* **15**, 3032-3040.

Pan, G., Ni, J., Yu, G., Wei, Y.F. and Dixit, V.M. (1998) TRUNDD, a new member of the TRAIL receptor family that antagonizes TRAIL signalling, *FEBS Lett.* **424**, 41-45.

Pan, G., O'Rourke, K. and Dixit, V.M. (1998) Caspase-9, Bcl-XL, and Apaf-1 form a ternary complex, *J. Biol. Chem.* **273**,5841-5845.

Pan, G., O'Rourke, K., Chinnaiyan, A.M., Gentz, R., Ebner. R., Ni, J. and Dixit, V.M. (1997) The receptor for the cytotoxic ligand TRAIL, *Science* **276**, 111-113.

Penfold, P.L. and Provis, J.M. (1986) Cell death in the development of the human retina: phagocytosis of pyknotic and apoptotic bodies by retinal cells, *Graefes. Arch. Clin. Exp. Ophthalmol.* **224**, 549-553.

Porter, A.G., Ng, P. and Janicke, R.U. (1997) Death substrates come alive, *BioEssays* **19**, 501-507.

Raff, M.C. (1992) Social controls on cell survival and cell death, *Nature* **356**, 397-400.

Rao, L., Debbas, M., Sabbatini, P., Hockenbery, D., Korsmeyer, S. and White, E. (1992) The adenovirus E1A proteins induce apoptosis which is inhibited by the E1B 19-kDa and Bcl-2 proteins, *Proc. Natl. Acad. Sci. USA* **89**, 7742-7746.

Reed, J.C. (1997) Double identity for proteins of the bcl-2 family, *Nature,* **387**, 773-776.

Reichel, M.B., Ali, R.R., D'Esposito, F., Clark, A.R., Luthert, P.J., Battacharya, S.S. and Hunt, D.M. (1998) High frequency of persistant hyperplastic primary vitreous and cataracts in p53 deficient mice, *Cell Death Differ.* **5**, 156-162.

Rodriguez-Tarduchy-G, Collins-MK, Garcia-I and Lopez-Rivas, A. (1992) Insulin like growth factor-I inhibits apoptosis in IL-3 dependent hemopoietic cells, *J. Immunol.* **149**, 535-540.

Rosse, T., Olivier, R., Monney, l., Rager, M., Conus, S., Fellay, I., Jansen, B. and Borner, C. (1998) Bcl-2 prolongs cell survival after Bax-induced release of cytochrome c, *Nature* **391**, 496-499.

Rothe, M., Pan, M.G., Henzel, W.J., Ayres, T.M. and Goeddel, D.V. (1995) The TNFR2-TRAF signaling complex contains two novel proteins related to baculoviral-

inhibitor of apoptosis proteins, *Cell* **83**, 1243-1252.
Rotonda, J., Nicholson, D.W., Fazil, K.M., Gallant, M., Gareau, Y., Labelle, M., Peterson, E.P., Rasper, D.M., Ruel, R., Vaillancourt, J.P., Thornberry, N.A. and BeckerJW (1996) The three dimensional structure of apopain/CPP32, a key mediator of apoptosis, *Nat. Struct. Biol.* **3**, 619-625.
Roy, N., Mahadevan, M.S., McLean, M., Shutler, G., Yaraghi, Z., Farahani, R., Baird, S., Besner-Johnston, A., Lefebvre, C., Kang, X., Salih, M., Aubry, A., Tamai, K., Guan, X., Ioannou, P., Crawford, T.O., de Jong, P.J., Surh, L., Ikeda, J.E., Korneluk, R.G. and MacKenzie, A. (1995) The gene for neuronal apoptosis inhibitory protein is partially deleted in individuals with spinal muscular atrophy, *Cell* **80**, 167-178.
Ruoslahti, E. and Reed, J.C. (1994) Anchorage dependance, Integrins, and apoptosis, *Cell* **77**, 477-478.
Ryan, J.J., Prochownik, E., Gottlieb, C.A., Apel, I.J., Merino, R., Nunez, G. and Clarke, M.F. (1994) c-myc and bcl-2 modulate p53 function by altering p53 subcellular trafficking during the cell cycle, *Proc. Natl. Acad. Sci. USA.* **91**, 5878-5882.
Ryeom, S.W., Sparrow, J.R. and Silverstein, R.L. (1996) CD36 participates in the phagocytosis of rod outer segments by retinal pigment epithelium, *J. Cell. Sci.* **109**, 387-395.
Sabbatini, P., Lin, J., Levine, A.J. and White, E. (1995) Essential role for p53-mediated treanscription in E1A-induced apoptosis, *Genes Dev.* **9**, 2184-2192.
Sah, V.P., Attardi, L.D., Mulligan, G.J., Williams, B.O., Bronson, R.T. and Jacks, T. (1995) A subset of p53-deficient embryos exhibit exencephaly, *Nature Genet.* **10**, 175-180.
Sakahira, H., Enari, M. and Nagata, S. (1998) Cleavage of CAD inhibitor in CAD activation and DNA degradation during apoptosis, *Nature* **391**, 96-99.
Savill, J., Dransfield, I., Hogg, N. and Haslett, C. (1990) Vitronectin receptor-mediated phagocytosis of cells undergoing apoptosis, *Nature* **343**, 170-173.
Savill, J., Hogg, N., Ren, Y. and Haslett, C. (1992) Thrombospondin cooperates with CD36 and the vitronectin receptor in macrophage recognition of neutrophils undergoing apoptosis, *J. Clin. Invest.* **90**, 1513-1522.
Schendel, S.L., Xie, Z., Montal, M.O., Matsuyama, S., Montal, M., Reed, J.C. (1997) Channel formation by antiapoptotic protein Bcl-2, *Proc. Natl. Acad. Sci. USA.* **94**, 5113-5118.
Sell, C., Baserga, R. and Rubin, R. (1995) Insulin like growth factor I (IGF-1) and the IGF-I receptor prevent etoposide induced apoptosis, *Cancer Res.* **55**, 303-306.
Sheridan, J.P., Marsters, S.A., Pitti, R.M., Gurney, A., Skubatch, M., Baldwin, D., Ramakrishnan, L., Gray, C.L., Baker, K., Wood, W.I., Goddard, A.D., Godowski, P. and Ashkenazi, A. (1997) Control of TRAIL-induced apoptosis by a family of signaling and decoy receptors, *Science* **277**, 818-821.
Smithgall, T.E. (1998) Signal transduction pathways regulating hematopoietic differentiation, *Pharmacological Reviews* **50**, 1-19.
Song, Q., Kees-Miller, S.P., Kumar, S., Zhang, N., Chan, D.W., Smith, G.C.M., Jackson, S.P., Alnemri, E.S., Litwack, G., Khanna, K.K. and Lavin, M.F. (1996) DNA dependent protein kinase catalytic subunit: a target for an ICE like protease in apoptosis, *EMBO J.* **15**, 3238-3246.
Srinivasula, S.M., Ahmad, M., Ottilie, S., Bullrich, F., Banks, S., Wang, Y., Fernandes-Alnemri, T., Croce, C.M., Litwack, G., Tomaselli, K.J., Armstrong,

R.C. and Alnemri, E.S. (1997) FLAME-1, a novel FADD-like anti-apoptotic molecule that regulates Fas/TNFR1-induced apoptosis, *J. Biol. Chem.* **272**, 18542-18545.

Sun, X.J., Wang, L.M., Zhang, Y., Yenush, L., Myers, M.J.J.r., Glasheen, E., Lane, W.S., Pierce, J.H. and White, M.F. (1995) Role of IRS-2 in insulin and cytokine signalling, *Nature* **377**, 173-177.

Susin, S.A., Zamzami, N., Castedo, M., Daugas, E., Wang, H.G., Geley, S., Fassy, F., Reed, J.C. and Kroemer, G. (1997) The central executioner of apoptosis: multiple connections between protease activation and mitochondria in Fas/APO-1/CD95- and ceramide-induced apoptosis, *J. Exp. Med.* **186**, 25-37.

Susin, S.A., Zamzami, N., Castedo, M., Hirsch, T., Marchetti, P., Macho, A., Daugas, E., Geuskens, M. and Kroemer, G. (1996) Bcl-2 inhibits the mitochondrial release of an apoptogenic protease, *J. Exp. Med.* **184**, 1331-1341.

Tartaglia, L.A. and Goeddel, D.V. (1992) Two TNF receptors, *Immunol. Today* **13**, 151-153.

Thome, M., Schneider, P., Hofmann, K., Fickenscher, H., Meinl, E., Neipel, F., Mattmann, C., Burns, K., Bodmer, J.L., Schroter, M., Scaffidi, C., Krammer, P.H., Peter, M.E. and Tschopp, J. (1997) Viral FLICE-inhibitory proteins (FLIPs) prevent apoptosis induced by death receptors, *Nature* **386**, 517-521.

Thornberry, N.A. and Lazebnik, Y. (1998) Caspases: Enemies within, *Science* **281**, 1312-1316.

Ting, A.T., Pimentel-Muinos, F.X. and Seed, B. (1996) RIP mediates tumor necrosis factor receptor 1 activation of NF-kappaB but not Fas/APO-1-initiated apoptosis, *EMBO J.* **15**, 6189-6196.

Uren, A.G., Pakusch, M., Hawkins, C.J., Puls, K.L. and Vaux, D.L. (1996) Cloning and expression of apoptosis inhibitory protein homologs that function to inhibit apoptosis and/or bind tumor necrosis factor receptor associated factors, *Proc. Natl. Acad. Sci. USA* **93**, 4974-4978.

Van de Craen, M., Van Loo, G., Pype, S., Van Criekinge, W., Van den brande, I., Molemans, F., Fiers, W., Declercq, W. and Vandenabeele, P. (1998) Identification of a new caspase homologue: caspase 14, *Cell Death Differ.* **5**, 838-846.

Vander Heiden, M.G, Chandel, N.S., Williamson, E.K., Schumacker, P.T. and Thompson, C.B. (1997) Bcl-xL regulates the membrane potential and volume homeostasis of mitochondria, *Cell* **91**, 627-637.

Verheij, M., Bose, R., Lin, X.H., Yao, B., Jarvis, W.D., Grant, S. and al, e.t. (1996) Requirement for ceramide initiated SAPK/JNK signalling in stress induced apoptosis, *Nature* **380**, 75-79.

Verhoven, B., Schengel, R.A. and Williamson, P. (1995) Mechanisms of phosphatidylserine exposure, a phagocyte recognition signal, on apoptotic T lymphocytes, *J. Exp. Med.* **182**, 1597-1601.

Walczak, H., Degli-Esposti, M.A., Johnson, R.S., Smolak, P.J., Waugh, J.Y., Boiani, N., Timour, M.S., Gerhart, M.J., Schooley, K.A., Smith, C.A., Goodwin, R.G. and Rauch, C.T. (1997) TRAIL-R2: a novel apoptosis-mediating receptor for TRAIL, *EMBO J.* **16**, 5386-5397.

Walker, N.P.C., Talanian, R.V., Brady, K.D., Dang, L.C., Bump, N.J., Ferenz, C.R., Franklin, S., Ghayur, T., Hackett, M.C., Hammil, L.D. and et al (1994) Crystal structure of the cysteine protease interleukin-1B converting enzyme: a (p20/p10)2 homodimer, *Cell* **78**, 343-352.

Walker, P.R., Pandey, S. and Sikorska, M. (1995) Degradation of chromatin in apoptotic cells, *Cell Death. Differ.* **2,** 97-104.

Wang, K., Yin, X.M., Chao, D.T., Milliman, C.L. and Korsmeyer, S.J. (1996) BID: a novel BH3 domain-only death agonist, *Genes Dev.* **10**, 2859-2869.

Wang, S., Miura, M., Jung, Y.-.k., Zhu, H., Li, E. and Yuan, J. (1998) Murine Caspase-11, an ICE-interacting protease is essential for the activation of ICE, *Cell* **92**, 501-509.

Wen, L.-.P., Madani, K., Martin, G.A. and Rosen, G.D. (1998) Proteolytic cleavage of Ras GTPase activating protein during apoptosis, *Cell Death Differ.* **5**, 729-734.

Widmann, C., Gibson, S. and Johnson, G.L. (1998) Caspase dependant cleavage of signaling proteins during apoptosis, *J. Biol. Chem.* **273**, 7141-7147.

Wiley, S.R., Schooley, K., Smolak, P.J., Din, W.S., Huang, C.P., Nicholl, J.K., Sutherland, G.R., Smith, T.D., Rauch, C., Smith, C.A., *et al* (1995) Identification and characterization of a new member of the TNF family that induces apoptosis, *Immunity* **3**, 673-682.

Wilson, K.P., Black, A.F., Thomson, J.A., Kim, E.E., Griffith, J.P., Navia, M.A., Murcko, M.A., Chambers, S.P., Aldape, R.A., Raybuck, S.A. and Livingston, D.L. (1994) Structure and mechanism of interleukin-1B converting enzyme, *Nature* **370**, 270-275.

Woo, M., Hakem, R., Soengas, M.S., Duncan, G.S., Shahinian, A., Kagi, D., Hakem, A., McCurrach, M., Khoo, W., Kaufman, S.A., Senaldi, G., Howard, T., Lowe, S.W. and Mak, T.W. (1998) Essential contribution of caspase 3/CPP32 to apoptosis and its associated nuclear changes, *Genes Dev.* **12**, 806-819.

Wu, Y.C. and Horvitz, H.R. (1998a) The C. elegans cell corpse engulfment gene ced-7 encodes a protein similar to ABC transporters, *Cell* **93**, 951-960.

Wu, Y.C. and Horvitz, H.R. (1998b) C. elegans phagocytosis and cell-migration protein CED-5 is similar to human DOCK180, *Nature* **392**, 501-504.

Xia, Z., Dickens, M., Raingeaud, J., Davis, R.J. and Greenberg, M.E. (1995) Opposing effects of ERK and JNK -p38 MAP kinases on apoptosis, *Science* **270**, 1326-1331.

Xue, D. and Horvitz, H.R. (1995) Inhibition of the Caenorhabditis elegans cell-death protease CED-3 by a CED-3 cleavage site in baculovirus p35 protein, *Nature* **377**, 248-251.

Yang, X., Khosravi-Far, R., Chang, H.Y. and Baltimore, D. (1997) Daxx, a novel Fas-binding protein that activates JNK and apoptosis, *Cell* **89**, 1067-1076.

Yeh, W.C., Pompa, J.L., McCurrach, M.E., Shu, H.B., Elia, A.J.,, Shahinian, A., Ng, M., Wakeham, A., Khoo, W., Mitchell, K., El-Deiry, W.S., Lowe, S.W., Goeddel, D.V. and Mak, T.W. (1998) FADD: essential for embryo development and signaling from some, but not all, inducers of apoptosis, *Science* **279**, 1954-1958.

Yonish-Rouach, E., Deguin, V., Zaitchouk, T., Breugnot, C., Mishal, Z., Jenkins, J.R. and May, E. (1995) Transcriptional activation plays a role in the induction of apoptosis by transiently transfected wild-type p53, *Oncogene* **11**, 2197-2205.

Yoshida, H., Kong, Y-Y., Yoshida, R., Elia, A.J., Haken, A., Hakem, R., Penninger, J.M. and Mak, T.W. (1998) Apaf-1 is required for mitochondrial pathways of apoptosis and brain development, *Cell*, **94**, 739-750.

Yu, B., Hailman, E. and Wright, S.D (1997) Lipopolysaccharide binding protein and soluble CD14 catalyze exchange of phospholipids, *J. Clin. Invest.* **99**, 315-324.

Yu, Y.T. and Snyder, L. (1994) Translation elongation factor Tu cleaved by a phage-exclusion system, *Proc. Natl. Acad. Sci. USA* **91**, 802-806.

Yuan, J., Shaham, S., Ledoux, S., Ellis, H.M. and Horvitz, H.R. (1993) The C.elegans cell death gene *ced-3* encodes a protein similar to mammalian interleukin-1 beta converting enzyme, *Cell* **75**, 641-652.

Zamzami, N., Susin, S.A., Marchetti, P., Hirsch, T., Gomez-Monterrey, I., Castedo, M. and Kroemer, G. (1996) Mitochondrial control of nuclear apoptosis, *J. Exp. Med.* **183**, 1533-1544.

Zha, J., Harada, H., Yang, E., Jockel, J. and Korsmeyer, S.J. (1996) Serine phosphorylation of death agonist BAD in response to survival factor results in binding to 14-3-3 not BCL-X, *Cell* **87**, 619-628.

Zhang, J., Cado, D., Chen, A., Kabra, N.H. and Winoto, A. (1998) Fas-mediated apoptosis and activation-induced T-cell proliferation are defective in mice lacking FADD/Mort1, *Nature* **392**, 296-300.

Zhivotosky, B., Orrenius, S., Brutugun, O.T. and Doskeland, S.O. (1998) Injected cytochrome-c induces apoptosis, *Nature* **391**, 449-450.

Zou, H., Henzel, W.J., Liu, X., Lutschg, A. and Wang, X. (1997) Apaf-1, a human protein homologous to C. elegans CED-4, participates in cytochrome-c-dependent activation of caspase-3, *Cell* **90**, 405-413.

INHIBITION OF APOPTOSIS IN MAMMALIAN CELL CULTURE
The Biotechnological Relevance of Limiting Cell Death

A. J. MASTRANGELO
The Johns Hopkins University
Department of Chemical Engineering
3400 North Charles Street
Baltimore, Maryland 21218 U.S.A.
E-mail: ajm@jhu.edu

1. Introduction

When J.F. Enders first used primate cells to produce poliomyelitis virus in 1949, a multi-billion dollar industry was born. The realization that viruses could be attenuated *in vitro* for use as vaccines initiated attempts to develop large scale cultures of mammalian cells. In the 1970's, two discoveries led to the expanded use of such cultures: (1) the advent of recombinant DNA technology meant that cell lines could be engineered to overexpress heterologous genes, and (2) the development of hybridoma cell lines provided a system capable of continuously secreting antibodies (Kohler and Milstein, 1975). Today, dozens of products including virus vaccines, antibodies, interferons, immunoregulators, hormones, and growth factors are manufactured by mammalian cells in culture; this number promises only to increase as our understanding of cellular processes grows.

The need for animal cell cultures arose due to inherent limitations in the well-established bacterial and yeast expression vehicles. Specifically, many therapeutics require post-translational modifications in order to possess biological activity, a requirement that the bacterial and yeast systems often could not fulfill. However, the culture of mammalian cells poses a significant technical challenge, and product yields in original processes were quite low. Early attempts to improve culture productivities focused on increasing cell densities by optimizing media formulations (Thomas, 1990) and altering bioreactor configurations (Bliem *et al.*, 1991). These manipulations led to vast improvements in attainable densities. By comparison, very little work was done to limit cell death in culture systems since it was assumed that the process was an accidental response to the harsh environment within the bioreactors. However, recent studies indicate that the primary mechanism of cell death in typical animal cell cultures is not a passive one. Instead, a large fraction of mammalian cells in culture actively participate in their own demise (Al-Rubeai *et al.*, 1990; Franěk and Dolníková, 1991a; Mercille and Massie, 1994a; Singh *et al.*, 1994), a process known as programmed cell death (PCD), or apoptosis.

Apoptosis was first described in a landmark paper by Kerr *et al.* (1972) almost 30 years ago. This mechanism of cell death is characterized by a series of well-known and easily recognizable morphological changes that include cell shrinkage, membrane blebbing, nuclear fragmentation, and finally disintegration of the cell into a series of membrane-enclosed apoptotic bodies (reviewed in Cummings *et al.*, 1997). The importance of PCD in biological systems was initially underestimated because the rapid phagocytosis of apoptotic bodies *in vivo* made the process difficult to identify (Gerschenson and Rotello, 1992). However, it is now believed that all mammalian cells possess the capacity to initiate their own death. This fact is particularly relevant in biotechnology, where a cell's immediate environment may consist of high shear stresses, low levels of nutrients and growth factors, and elevated concentrations of toxins, each of which has been shown to induce PCD. Given the active nature of death in culture processes, it is now recognized that ongoing attempts to increase cell culture productivities should include methods to inhibit apoptosis. In order to accomplish this, it is crucial that the biochemical mechanisms involved in apoptotic death be fully understood.

Apoptosis occurs as a series of events that can be subdivided into four distinct phases: initiation, signaling, effector, and degradation (Mastrangelo and Betenbaugh, 1998) (Figure 1). During the initial phase, a cell is exposed to a certain stimulus that launches the death program. In culture, this can be a buildup of toxins, the depletion of nutrients and/or growth factors, oxygen limitations, hydrodynamic stresses, or virus infections. Such an insult is processed by the cell through a number of individual paths that propel the cell toward death during the signaling phase of apoptosis. The next step of the apoptotic cascade begins when each of these pathways converge upon a single (or possibly a small number of) execution pathway(s). Included in this phase are the action of Bcl-2 family members, the mitochondrial permeability transition (PT), release of cytochrome c (cyt c) from the mitochondria, and activation of the caspase family of proteases. Finally, in the last phase of apoptosis, the activated caspases cleave a number of substrates that lead to the death and subsequent disintegration of the cell. *In vivo*, the resultant apoptotic bodies are phagocytosed by neighboring cells as well as professional phagocytes; *in vitro*, however, these cellular remnants lyse in a process known as secondary necrosis.

Techniques employed to block or delay PCD in cell culture typically focus on one of two methods: interfering with individual signaling cascades, or modulating the action of key molecules in the execution pathway. In the former case, apoptosis has been delayed in cell cultures by changing the amino acid composition of the culture medium (Franěk and Dolníková, 1991b; DiStefano *et al.*, 1996; Duval *et al.*, 1991) as well as by exposing cells to different antioxidants (Lin *et al.*, 1995; Malorni *et al.*, 1993; McGowan *et al.*, 1996). A more common approach, however, is to change a cell's ability to execute the final common pathway of apoptosis. Strategies employed to accomplish this include engineering cells to express anti-apoptosis genes (Itoh *et al.*, 1995; Mastrangelo *et al.*, 1996; Mercille *et al.*, 1999; Singh *et al.*, 1996; Simpson *et al.*, 1997; Suzuki *et al.*, 1997; Terada *et al.*, 1997), exposing cells to caspase inhibitors (Chiou and White, 1998; Duckett *et al.*, 1996), and blocking the mitochondrial membrane PT (Pastorino *et al.*, 1996; Zamzami *et al.*, 1996a). This chapter will attempt to review the work currently underway in these areas.

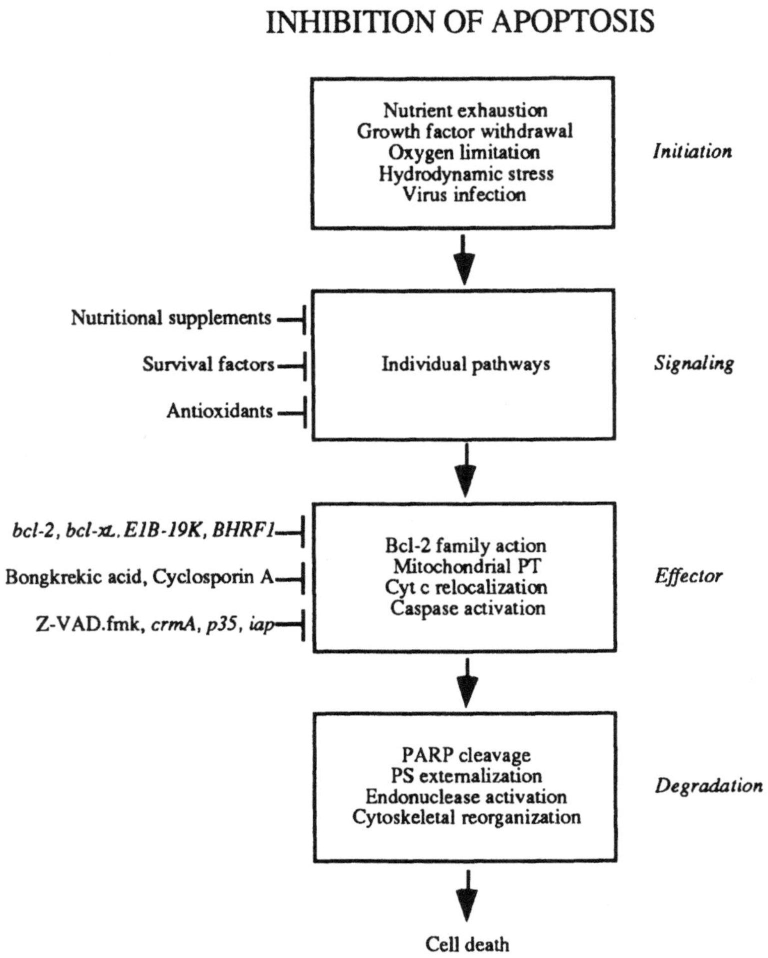

Figure 1. The apoptotic cascade. During the initiation phase, the cell is exposed to some stimulus which launches PCD. This insult, which in culture can be any of the stimuli noted, is processed by the cell through a number of individual pathways during the signaling phase. Attempts to stop the progression of cell death in culture during this phase typically utilize nutritional supplements, survival factors, or antioxidants. If apoptosis is not halted, these pathways converge into one common pathway during the effector phase. Included in this stage of apoptosis are the action of the Bcl-2 family members, the mitchondrial PT, relocalization of cyt c from the mitchondria, and caspase activation. These events can be blocked by a number of genetic and chemical methods, as described. If apoptosis is not suppressed, the effector phase leads into the degradation phase where a number of biochemical changes take place, such as PARP cleavage, exposure of phosphatidyl serine (PS) on the outer leaflet of the plasma membrane, endonulease activation and activity, and cytoskeletal reorganization and collapse, before the disintegration of the cell body.

2. Limiting Apoptosis During the Signaling Phase

In any cell culture application, the composition of the medium can have a significant impact on culture productivity. This was recognized almost 50 years ago when Eagle defined the growth requirements of HeLa and mouse L cells (Eagle, 1955). Upon further study, it became clear that different cell types have distinct nutritional requirements; in particular, amino acid utilization can vary dramatically from one cell line to the next (Eagle, 1959). This discovery led to the development of a number of other media formulations, each hoping to maximize cellular proliferation. More recently, it has become clear that not only can the medium composition influence cell growth, it can have a significant impact on the viability of a culture. The disappearance of a signaling molecule, the exhaustion of a nutrient, the accumulation of a toxin, or the absence of key survival and growth factors can all induce the PCD in animal cells, most likely through distinct signaling pathways. Apoptosis in cell culture systems could be limited, then, by modifying the cells' external environment, therefore blocking cell death along these different paths. Extending the life of a culture, even if only by a day or two, could lead to significant gains in product yield, thereby increasing the cost-effectiveness of a particular production scheme.

2.1 COMPLEMENTATION OF CULTURE MEDIUM WITH AMINO ACIDS

Many of the components added to media to stimulate cell growth also inhibit death indirectly. For instance, the simple addition of glucose to media formulations is actually a way of limiting PCD, since glucose deprivation leads to apoptosis (Mercille and Massie, 1994a). Similarly, a number of cell lines require growth and survival factors that can be easily provided by serum. Again, maintaining these cells on serum can be viewed as an attempt to mitigate apoptotic death (Cotter and Al-Rubeai, 1995). However, due to the cost and undefined nature of serum, there exist many ongoing studies aimed at identifying essential factors that would provide high cell viabilities while allowing for sufficient protein production in the absence of serum. Chung *et al.* (1997) demonstrated that hybridoma cells experiencing decreased serum levels can be saved from apoptosis by the addition of interleukin-6 (IL-6) to the medium. In fact, viabilities and viable cell densities of Sp2/0 cells grown in 1% serum supplemented with IL-6 were similar to the control cultures containing 10% serum.

Another method of limiting PCD in cell culture systems is the addition of amino acids to the medium, since amino acid deprivation can lead to high levels of apoptosis, particularly in hybridoma and myeloma cells (Mercille and Massie, 1994a; Singh *et al.*, 1994; Vomastek and Franěk, 1993). Several approaches have demonstrated that higher viabilities can be maintained simply by altering the amino acid composition of the medium (DiStefano *et al.*, 1996; Duval *et al.*, 1991; Franěk and Dolníková, 1991b). One such method involves analyzing the amino acid consumption rates of individual cell lines and subsequently altering the medium formulation to ensure that limitations in the amino acid supply are not experienced during the culture period (DiStefano *et al.*, 1996; Duval *et al.*, 1991). Complementation of hybridoma cell lines by such carefully determined amino acid mixtures has been shown to provide higher cell densities, leading to increases in monoclonal antibody production (Duval *et al.*, 1991). In a similar study, regular nutrient additions consisting of glucose, proteins, and amino acids to fed-batch cultures of myeloma cells delayed apoptosis by 24 to 48 hours and resulted in five- to ten-fold increases in final product concentrations (DiStefano *et al.*, 1996). The drawback to this approach, however, is the requirement for time-consuming, detailed studies on the metabolism of each cell line of interest.

A similar, but less exact, strategy is the general addition of amino acids to the culture medium. In a number of studies, F. Franěk's laboratory has demonstrated that the addition of amino acid mixtures to hybridoma cultures can delay apoptosis while increasing Ab yields. Early investigations indicated that the fortification of cell culture medium with conventional amino acid mixtures could increase the cell viability in hybridoma cultures, while elevating the total antibody production by a factor of almost three (Franěk and Dolníková, 1991b). Subsequent work focused on identifying the individual amino acids responsible for the protective effect and their role in apoptosis inhibition.

In a set of novel experiments, cells were maintained on "starvation medium", growth medium which was diluted by up to 90% with saline, that had been complemented with different amino acids. This allowed the identification of eight amino acids (histidine, glutamine, asparagine, proline, threonine, serine, alanine, and glycine) capable of delaying apoptosis (Franěk and Chládkova-Srámková, 1995; Franěk and Srámková, 1996a; Franěk and Srámková, 1996b). Because each of these amino acids is able to act as an apoptotic suppressor individually, it appears that they do not function solely as nutrients or protein precursors but also as signal molecules. Further evidence for this role is provided by the fact that not only are five of the eight "protective" amino acids considered non-essential in terms of mammalian cell nutrition (Eagle, 1959), but other non-coded amino acids and a non-metabolizable analog were also able to inhibit apoptosis and allow for cell growth in starving cultures (Franek and Chládkova-Srámková, 1995). In other words, fortification of the medium with these substrates appears to block apoptosis along some signaling pathway (Franěk and Srámková, 1996a; Franěk and Srámková, 1996b).

It is important to note, however, that the potential capacity of amino acids to act as survival signals is not a unique function in their ability to regulate apoptosis; attention must also be paid to their role as protein precursors. The deprivation of any single amino acid from hybridoma cultures has been shown to lead to apoptotic death (Simpson *et al.*, 1998). Cell death in this case is most likely due to the inability of the cells to manufacture certain proteins in the absence of key amino acids, since inhibition of protein synthesis can induce apoptosis in both hybridoma and myeloma cell lines (Perreault and Lemieux, 1993; Mercille and Massie, 1994a). Complementation of culture medium with amino acids may therefore be useful in delaying PCD, particularly in hybridoma and myeloma cultures. However, the exhaustion of an essential amino acid will not always lead to the rapid induction of apoptosis (Perreault and Lemieux, 1993). Indeed, a number of commercially relevant mammalian cell lines will respond to this type of deprivation by entering a period of growth arrest in which the cell can still be saved from death, though apoptosis will most likely ensue if the limitation is sustained (Petronini *et al.*, 1996). It is clear that the medium formulation will have an enormous impact on both the levels of cell death and, consequently, the productivity of any culture.

2.2 UTILIZATION OF ANTIOXIDANTS TO LIMIT APOPTOSIS

Reactive oxygen species (ROS) are another important class of molecules involved in the early stages of PCD. Though their role in the apoptotic cascade was initially unclear, evidence now suggests that the participation of ROS in cell death is limited to certain signal transduction pathways either directly, as signaling molecules, or indirectly, as modulators of the cellular redox potential (Jacobson, 1996; Mignotte and Vayssiere, 1998). ROS can also be generated in the late stages of apoptosis (Degli Esposti and McLennan, 1998; Kroemer *et al.*, 1997), although it is their capacity to act along certain signaling pathways that makes them attractive targets for apoptosis inhibition.

Since antioxidants possess the ability to neutralize ROS, they have been employed to effectively limit apoptosis in a number of studies (Greenlund *et al.*, 1995; Wolfe *et al.*, 1994; Wong *et al.*, 1989). In particular, N-acetylcysteine (NAC) and pyrrolidine dithiocarbamate (PDTC), two thiol-containing antioxidant reagents, have been utilized to protect a variety of cell lines from numerous insults. For example, treatment with NAC extended the survival of human neurons (Talley *et al.*, 1995), U937 cells (Cossarizza *et al.*, 1995), and murine L929 cells (Mayer and Noble, 1994) upon exposure to tumor necrosis factor alpha (TNF-α). Additionally, PDTC has been shown to effectively delay death in HL-60 cells in response to such apoptosis-inducing stimuli as etoposide, ultraviolet irradiation, camptothecin, and actinomycin D (Bessho *et al.*, 1994; McGowan *et al.*, 1996).

These reagents can also be used to limit cell death in response to virus infections. NAC can inhibit HIV-induced apoptosis in U937 cells (Malorni *et al.*, 1993), while both NAC and PDTC have been shown to suppress cell death in AT3 and N18 cells in response to a Sindbis virus (SV) infection (Lin *et al.*, 1995). Furthermore, when used to suppress apoptosis during a recombinant viral infection, NAC was able to extend cellular lifespans while allowing for the increased and extended production of a model protein (Mastrangelo *et al.*, unpublished results).

The capacity to block PCD in virally infected cells is important in culture processes for several reasons. First, animal cells are currently utilized in the generation of a number of attenuated virus vaccines and gene therapy vectors. Providing cells with a defense against the apoptotic death associated with these production schemes could allow for more viral replication cycles and therefore higher product titers. Secondly, the use of viral expression systems is a common method of quickly producing high levels of a product of interest. Delaying cell death could increase productivities while perhaps simplifying purification trains, since lower levels of apoptosis could lead to lower levels of cell debris.

However, since antioxidants function in the signaling pathways of apoptosis, they are not always able to inhibit cell death. For example, PDTC offered rat thymocytes no protection from etoposide-induced apoptosis (Slater *et al.*, 1995). Moreover, both NAC and PDTC have been shown to induce PCD in at least one instance (Tsai *et al.*, 1996). The ability of such reagents to protect cells, therefore, must be tested for each system of interest.

3. Modulation of the Effector Phase of Apoptosis

Though inhibiting apoptosis during the signaling phase can be quite effective in certain instances, many of the strategies employed to do so are cell type- and stimulus-specific.

A more practical method of blocking PCD may be to change the action of the final execution pathway. Such an approach could have wide-ranging applications in the cell culture industry where cells are exposed to a variety of insults, each utilizing a unique signal transduction pathway. Interfering with an event common to most cases of PCD might therefore provide a stronger defense against apoptotic death, allowing for increased product yields.

Despite the fact that much of the apoptotic cascade remains unclear, intense investigation has provided a preliminary description of the effector phase. Though still ambiguous, certain key events are believed to occur at this stage of apoptosis (Figure 2):

(i) The mitochondrial PT and release of cyt c from the mitochondria: It is well documented that the mitochondria play a central role in the execution of apoptosis. Two major changes associated with this organelle are the formation of a PT pore, typically inducing the collapse of the membrane potential ($\Delta\Psi_m$) (Zamzami *et al.*, 1996a), and the relocalization of cyt c from the intramembrane space of the mitochondria to the cytosol. Whether or not these two events act in tandem is currently a matter of debate. In some cases, the formation of the PT pore seems to be sufficient for apoptosis to occur (Marzo *et al.*, 1998), while in others, cyt c release is necessary for cell death. The two may be linked, for in at least one instance the PT was required for cyt c relocalization (Bradham *et al.*, 1998). However, this is not always the case as the release of cyt c from the mitochondria has been documented both before and in the absence of a drop in $\Delta\Psi_m$ (Bossy-Wetzel *et al.*, 1998; Vander Heiden *et al.*, 1997). It does remain possible, though, that the PT rapidly opens and closes allowing a repeated re-establishment of $\Delta\Psi_m$ (Ichas and Mazat, 1998) while permitting the relocalization of cyt c. Once released into the cytosol, cyt c is able to assist in the activation of certain caspases.

(ii) Caspase activation: Caspases, a family of cysteine proteases with aspartate specificity, are often called the "executioners" of apoptosis due to their role in the destruction of the cell. This family was first implicated in apoptosis when the sequence homology between caspase-1, also known as interleukin-1β-converting enzyme (ICE), and the *Caenorhabditis elegans* gene product CED-3 was noted (Yuan *et al.*, 1993). To date, 12 other members of this family of cysteine proteases have been discovered with varying, and sometimes redundant, roles in PCD. Caspases are synthesized in an inactive pro-form and can accumulate in the cell as dormant molecules. When cell death signals are received, "initiator" caspases such as caspases-8 and -9 are activated by several mechanisms: pro-caspase-8 is cleaved into caspase-8 during the signaling cascade associated with receptor-mediated apoptosis, while the release of cyt c from the mitochondria assists in the activation of caspase-9. Due to the autocatalytic nature of this family, the activation of these initiators leads to a cascade, or perhaps an amplification cycle, wherein the "effector" caspases, numbers 3, 6, and 7, are activated. These footsoldiers of apoptosis are responsible for cleaving a number of cellular substrates, including poly(ADP-ribose) polymerase (PARP) and nuclear lamins, that lead to the breakdown of the cell body (Thornberry and Lazebnik, 1998).

(iii) Action of the Bcl-2 family of proteins: With its ability to suppress apoptosis in a number of instances, *bcl-2* has been, perhaps, the most widely studied apoptosis regulator. Since the discovery of this proto-oncogene in follicular B-cell lymphoma (Tsujimoto *et al.*, 1984), over a dozen related genes have been found, some of which encode proteins that inhibit apoptosis and others whose products promote the process.

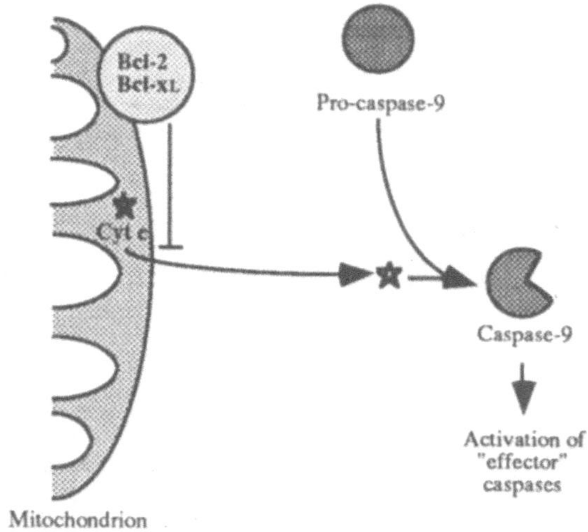

Figure 2. The effector phase of apoptosis. During the "final common pathway" of apoptosis, cyt c relocalizes from the mitochondria to the cytosol where it assists in activating caspase-9, an event which leads to the activation of a number of downstream caspases. Attempts to prevent this may utilize Bcl-2 or Bcl-x_L which somehow prevent the release of cyt c. See text for further detail.

Unfortunately, Bcl-2's specific mechanism of action remains elusive, though certain functions of this protein are well described. For instance, it is clear that Bcl-2 can heterodimerize with Bax, a related death promoter, in an effort to render Bax useless (Oltvai *et al.*, 1993). Another important function of this anti-apoptotic protein is its ability to prevent the release of cyt c from the mitochondria, thereby preventing caspase activation in certain instances (Kluck *et al.*, 1997; Yang *et al.*, 1997). However, this is apparently not its only *raison d'être* as it is also able to protect cells downstream of cyt c release (Rossé *et al.*, 1998; Zhivotovsky *et al.*, 1998). With the ability to halt the progression of apoptosis at more than one location in the effector pathway, Bcl-2 and its anti-death relatives are prime candidates for inhibiting PCD in large scale culture applications. Indeed, overexpression of *bcl-2* has been the most commonly utilized strategy of limiting apoptosis in biotechnological applications to date.

Interfering with any of these conserved steps in the effector phase of PCD could significantly decrease levels of cell death in a culture application, where cells may be exposed to a variety of apoptotic stimuli.

3.1 BLOCKING THE MITOCHONDRIAL PERMEABILITY TRANSITION

The overwhelming importance of the mitochondria in apoptosis regulation was not fully appreciated until recently; it is now clear, however, that these organelles play a central role in the progression of PCD. As previously discussed, two significant events associated with the mitochondria during apoptotic death are the formation of the PT pore and the release of cyt c into the cytosol. Since these two steps are crucial for many instances of apoptosis, inhibiting either event could have wide ranging applications in

TABLE 1. Agents capable of inhibiting the permeability transition of the mitochondria.

Inhibitors of Mitochondrial PT
Cyclosporin A
Bongkrekic acid
Chloromethyl-X-rosamine
Calpain inhibitors
Phosphotyrosine
Calcium chelators

the cell culture industry. Unfortunately, the exact mechanism of cyt c release is as of yet unknown, which has hindered the development of strategies to prevent this occurrence. However, research has indicated that overexpression of Bcl-2 and Bcl-x_L are able to block this process (Kluck *et al.*, 1997; Vander Heiden *et al.*, 1997; Yang *et al.*, 1997). As more insight is gained into the mode of action of cyt c relocalization, other methods to block this process will likely arise. Until then, though, it may be useful to limit PCD by inhibiting the mitochondrial PT, since a number of reagents have demonstrated this ability (reviewed in Zamzami *et al.*, 1997) (Table 1).

Of the many agents known to block the PT of the mitochondria, the two most commonly employed to delay apoptosis are cyclosporin A (CsA) and bongkrekic acid (BA). The former of these compounds has been shown to effectively inhibit apoptotic death in response to low levels of thapsigargin (Waring and Beaver, 1996), TNF-α (Bradham *et al.*, 1998; Pastorino *et al.*, 1996), and anoxia (Pastorino *et al.*, 1993). Furthermore, a CsA derivative, N-methyl-4-Val-cyclosporin A can limit PCD induced by nitric oxide (Hortelano *et al.*, 1997) as well as dexamethasone (Zamzami *et al.*, 1996a). It is important to note, however, that some studies indicate that the effects of CsA on PT are transitory, capable of reversing within an hour (Zamzami *et al.*, 1996b). Though significant, such modest extensions in cell lifetimes may not be practical in biotechnological applications where the cost of both the PT inhibitor and the purification efforts required to rid the final product of such an additive may no longer offset any profits gained from increased yields. Instead, it may be appropriate to utilize BA, which is a more potent inhibitor of the mitochondrial PT, in order to further extend the culture lifetimes.

The anti-apoptotic properties of BA, a toxic antibiotic produced by the microorganism *Pseudomonas cocovenenans*, were first noted in G. Kroemer's laboratory (Zamzami *et al.*, 1996a). In the time since, it has been used to limit PCD in thymocytes upon exposure to dexamethasone, irradiation, and etoposide (Marchetti *et al.*, 1996), as well as nitric oxide (Hortelano *et al.*, 1997). BA has also been shown to protect neuroblastoma cells from treatment with doxorubicin and betulinic acid (Fulda *et al.*, 1998) and L929 fibroblasts from TNF-α-induced apoptosis (Hehner *et al.*, 1998). Additionally, it is able to delay PCD in response to virus infections; a rat carcinomal cell line infected with an SV vector in the presence of BA maintained higher viabilities and survived 24 hours longer than untreated controls (Mastrangelo *et al.*, unpublished results).

As the significance of the mitochondria in apoptosis has only recently been realized, few strategies have been developed to inhibit its action in the death cascade. However, it is likely that many more methods of blocking the mitochondrial PT, as well as preventing the relocalization of cyt c, will be developed as our understanding of the mechanisms behind these events expands.

3.2 INHIBITION OF CASPASE ACTIVITY

Another effective method of blocking apoptosis is to inhibit the activity of the caspases. Evidence for the significance of these proteases in apoptosis came with the discovery that their overexpression can lead to cell death, whereas in many cases, their inhibition can prevent PCD (reviewed in Zhivotovsky *et al.*, 1996). A further indication of their importance in PCD is the fact that a number of viruses encode proteins which are able to inhibit caspase activity, thereby delaying cell death and allowing for the completion of the virus life cycle (Hardwick, 1998). Three different classes of such viral inhibitors have been discovered: CrmA, p35, and the IAP (inhibitors of apoptosis) family of proteins, all of which have been utilized to limit apoptosis in a number of cell lines. In addition, the use of synthetic peptide inhibitors of caspases is an effective alternative for blocking PCD. Each of these products holds great promise as apoptosis suppressors during the large scale cultivation of mammalian cell.

3.2.1 crmA

Though known to effectively inhibit caspase-1, other caspase targets of CrmA remain controversial. *In vitro* studies have suggested its ability to inhibit caspases-4 and -8, though not caspases-2 or -3 (Datta *et al.*, 1997; Dobbelstein and Shenk, 1996) . Affinity studies further indicated that CrmA most likely does not inhibit the effector caspases, caspases-3, -6, or -7, *in vivo* (Zhou *et al.*, 1997), though Tewari *et al.* (1995) demonstrated that it can indeed inhibit caspase-3, both *in vitro* and *in vivo*. Whatever its target caspases, it is clear that *crmA* can limit apoptosis in a number of experimental systems.

When microinjected into chicken dorsal root ganglion neurons, *crmA* is able to protect against apoptosis induced by nerve growth factor deprivation (Gagliardini *et al.*, 1994). Consistent with its ability to inhibit caspases-1 and -8, overexpression of *crmA* can block PCD in human breast cancer, myeloid leukemia, neuroblastoma, and epithelioid carcinoma cells in response to TNF-α treatment (Datta *et al.*, 1997; Enari *et al.*, 1995; Miura *et al.*, 1995; Talley *et al.*, 1995; Tewari and Dixit, 1995), an apoptotic stimulus that utilizes the receptor-mediated pathway involving these caspases. Additionally, expression of this gene is able to protect baby hamster kidney (BHK) and N18 cells resulting from virus-induced death (Nava *et al.*, 1998) and rat fibroblasts from apoptosis associated with serum deprivation (Wang *et al.*, 1994).

3.2.2 p35

A broad spectrum caspase inhibitor (Hardwick, 1998), the baculovirus p35 protein is able to limit apoptosis in a number of different systems, spanning both invertebrate and vertebrate species. In the nematode *C. elegans*, expression of *p35* prevents death in developing larvae and can rescue larval mutants lacking the anti-apoptosis *ced-9* gene (Sugimoto *et al.*, 1994). Illustrating its potential for use in biotechnological applications, p35 can protect Sf-21 insect cells from apoptosis induced by actinomycin

D (Crook *et al.*, 1993; Clem and Miller, 1994) as well as by infection with a mutant baculovirus lacking a *p35* gene (Cartier *et al.*, 1994). The overexpression of this gene has also limited apoptosis in a number of mammalian cell lines, including human breast cancer cells exposed to thapsigargin (Qi *et al.*, 1997) and TNF-α (Beidler *et al.*, 1995). Especially relevant to industrial applications is p35's ability to protect mammalian neurons from glucose and serum withdrawal (Rabizadeh *et al.*, 1993), two insults common in large scale animal cell culture. In addition, expression of *p35* from an SV vector is able to protect BHK cells from virus-induced apoptosis (Clem *et al.*, 1996).

3.2.3 IAP

A second class of caspase inhibitors is encoded by the baculoviruses, further illustrating the importance of these proteases in apoptosis. The IAP proteins are a large family and, unlike p35 and CrmA, have mammalian counterparts. Three homologues have been found in divergent baculoviruses, Op-IAP, Cp-IAP, and the non-functional Ac-IAP; additionally, the human NAIP (neuronal apoptosis inhibitor protein), xIAP (hILP, MIHA), IAP-1 (cIAP-1, MIHB, hIAP-2), and IAP-2 (cIAP-2, MIHC, hIAP-1), as well as the *Drosophila melanogaster* dIAP, all show some sequence similarity. Though their specific caspase target(s) remains unclear, IAPs can limit apoptosis induced by a wide variety of stimuli and consequently may be useful in large scale animal cell cultures.

Both Op-IAP and Cp-IAP are able to protect cultured insect cells from apoptosis caused by exposure to actinomycin D (Clem and Miller, 1994; McLachlin and Miller, 1997) as well as infection with mutant baculoviruses lacking *p35* and *iap* genes (Seshagiri and Miller, 1997). Additionally, Op-*iap* expressed from an SV vector can protect both BHK and N18 cells from virus-induced PCD (Duckett *et al.*, 1996). The human homologues also show great promise as caspase inhibitors. One or more of these genes has been shown to protect Chinese hamster ovary (CHO) and 293 cells, two industrially relevant cell lines, from apoptosis induced by stimuli as diverse as serum withdrawal (Duckett *et al.*, 1996; Liston *et al.*, 1996), staurosporine treatment (Miller, 1997), exposure to menadione (Liston *et al.*, 1996), and SV infection (reviewed in Miller, 1997). Overexpression of these genes could therefore have numerous biotechnological applications since growth factor deprivation, exposure to toxins, and virus infections are common insults in many culture processes.

3.2.4 Peptide Inhibitors

In addition to overexpressing viral caspase inhibitors, another method of interfering with members of this protease family is through the use of synthetic peptide-based compounds designed to inhibit specific caspases. A number of these reagents have been created, including N-benzyloxycarbonyl-Val-Ala-Asp-fluoromethylketone (Z-VAD.fmk), acetyl-Tyr-Val-Ala-Asp-chloromethylketone (Y-VAD.cmk), Boc-aspartyl-fluoromethylketone (BD.fmk), and acetyl-Asp-Glu-Val-Asp-aldehyde (Ac-DEVD-CHO) (Enari *et al.*, 1995; Pervaiz *et al.*, 1998; Sarin *et al.*, 1996; Schafer *et al.*, 1998).

Y-VAD.cmk, BD.fmk, and Ac-DEVD-CHO have been utilized to block apoptosis in response to insults such as serum withdrawal (Schafer *et al.*, 1998), treatment with anticancer agents (Pervaiz *et al.*, 1998), and anti-Fas antibody (Enari *et al.*, 1995), as well as IL-2 withdrawal, irradiation, and exposure to dexamethasone and etoposide (Sarin *et al.*, 1996). However, it is Z-VAD.fmk that has received the most attention in apoptosis suppression, having been utilized to limit cell death in a variety of applications. Both breast cancer cells and WEH17.2 cells are protected from thapsigargin-induced apoptosis by this irreversible tripeptide inhibitor (McColl *et al.*, 1998; Qi *et al.*, 1997), which is also able to limit PCD in thymocytes in response to dexamethasone treatment, irradiation, and exposure to etoposide (Beyette *et al.*,, 1998; Sarin *et al.*, 1996). Treatment with Z-VAD.fmk can rescue PC12 cells undergoing trophic factor deprivation (Gagliardini *et al.*, 1994) and T cell hybridomas exposed to either ceramide or the ROS donor ter-butylhydroperoxide (Susin *et al.*, 1996). It is the ability of this halomethyl ketone to limit virus-induced apoptosis, however, that makes it an attractive compound for biotechnological use.

Z-VAD.fmk can limit apoptosis in response to Semliki Forest virus (SFV) infection, though a greater level of protection was observed when this treatment was paired with Bcl-2 overexpression (Grandgirard *et al.*, 1998). It can also protect Vero cells from rubella virus-induced PCD (Pugachev and Frey, 1998) and BHK and N18 cells from death upon SV infection (Nava *et al.*, 1998). Interestingly, when used to suppress apoptosis in HeLa cells infected with a mutant adenovirus, Z-VAD.fmk not only inhibited cell death, but allowed for an increase in virus production (Chiou and White, 1998). This could have applications in the vaccine industry where mammalian cells are often used to produce attenuated virus strains. In a similar study, Z-VAD.fmk delayed apoptosis induced by an SV virus vector and allowed for a significant increase in yield of a model recombinant protein (Mastrangelo *et al.*, unpublished results).

The ability to inhibit a species as central to the apoptotic cascade as caspases is of great interest to biotechnologists, as this methodology may be exploited to extend cell lifetimes as a means of increasing culture productivities.

3.3 OVEREXPRESSION OF PROTECTIVE BCL-2 FAMILY MEMBERS

Perhaps the most widely used method of inhibiting apoptosis in cell lines of industrial interest to date is the overexpression of *bcl-2*. The ability of this proto-oncogene to limit PCD was first demonstrated when Vaux *et al.* (1988) showed that an IL-3 dependent myeloid cell line could survive in the absence of IL-3 upon *bcl-2* overexpression. Further studies revealed *bcl-2*'s ability to protect numerous cell lines from insults as diverse as *c-myc* overexpression (Bissonette *et al.*, 1992), UV irradiation (Martin *et al.*, 1995), exposure to TNF-α (Fernandez *et al.*, 1994), ROS (Hockenberry *et al.*, 1993), virus infections (Levine *et al.*, 1993), heat shock (Strasser and Anderson, 1995), calcium (Baffy *et al.*, 1993), and treatment with chemotherapeutic agents (Miyashita and Reed, 1992). Additionally, other Bcl-2 family members, including Bcl-x_L and the viral homologues E1B-19K, BHRF1, and KSbcl-2, have been shown to limit apoptosis in response to TNF-α treatment (Hashimoto *et al.*, 1991; White *et al.*, 1992), gamma irradiation (McCarthy *et al.*, 1996), growth factor deprivation (Boise *et al.*, 1993; Martinou *et al.*, 1995; Takayama *et al.*, 1994), serum depletion (Henderson *et al.*, 1993), exposure to cytotoxic drugs (McCarthy *et al.*, 1996), and virus infections (Cheng *et al.*, 1997a; Ink *et al.*, 1995; Tarodi *et al.*, 1994).

The importance of *bcl-2* to the biotechnology industries was not fully realized, however, until the early 1990's when Al-Rubeai *et al.* (1990) demonstrated the occurrence of apoptosis in hybridoma cultures. Subsequent research by this group and others indicated that apoptosis is often a result of insults commonly experienced during large scale cultivation of mammalian cells (Al-Rubeai *et al.*, 1995; Franěk and Dolníková, 1991a; Mercille and Massie, 1994a and b; Singh *et al.*, 1994). In recent years, the ability of *bcl-2* and its family members to protect cells from death-inducing stimuli during culture applications has been investigated, and is discussed below.

3.3.1 Nutrient Exhaustion

Cells in culture, especially those at the end of a batch process, often experience limitations in nutrient supplies. In a number of applications, this deprivation leads to apoptosis, especially upon the exhaustion of key nutrients such as glutamine and glucose (Mercille and Massie, 1994a: Singh *et al.*, 1994). However, the overexpression of anti-apoptosis genes can often extend the viable culture period, sometimes giving rise to increases in product yields.

Early work by Itoh *et al.* (1995) demonstrated the utility of *bcl-2* in hybridoma cultures. When stably expressed in 2E3 cells, this gene not only allowed for a 45% higher maximum cell density than that of control cells, it delayed the initiation of apoptosis by two days and prolonged the entire culture period by four days, all of which gave rise to a four-fold increase in antibody (Ab) yields. Additionally, coexpression of *bag-1* and *bcl-2* in this cell line further increased cell survival as well as Ab production (Suzuki *et al.*, 1997). In similar studies in TB/C3 hybridoma and Burkitt's lymphoma cells, the overexpression of *bcl-2* extended culture duration and, in the case of the hybridomas, allowed for a 40% increase in Ab titers (Simpson *et al.*, 1997; Singh *et al.*, 1996). It should be noted, however, that increases in cell survival do not always translate into gains in productivity. Mercille *et al.* (1999) observed that NS/0 myeloma cells expressing the *bcl-2* homologue *E1B-19K* did not generate any increase in Ab yield even though they were resistant to apoptosis induced by nutrient exhaustion during a batch culture. It is possible, though, that the failure to increase productivity in this particular system is related to a specific interaction of *E1B-19K* with this cell type as opposed to a more general phenomenon. Regardless, extension of cell viability may minimize the cell debris associated with a particular process, thereby simplifying product purification and leading to a more cost-effective production scheme.

3.3.2 Growth Factor Withdrawal

As mentioned previously, removal of serum will induce apoptosis in a number of cell lines, including hybridomas and plasmocytomas (Singh *et al.*, 1994). However, from a bioprocessing standpoint, the use of serum in large scale cultures has several drawbacks (Butler, 1987):

(i) It is expensive and can account for up to 80% of the total media cost.

(ii) The undefined nature of serum allows for variability between batches which can translate into product inconsistency between cultures.

(iii) The protein content in serum is quite high, especially when compared to product concentration; this can lead to complex and expensive purification schemes.

(iv) There are regulatory issues involved when using serum since it may contain adventitious agents such as viruses, mycoplasma, and prion-like particles, which necessitates extensive and costly prescreening operations.

Recently, there has been considerable interest in developing methods to adapt cells to serum-free conditions while maintaining adequate viability levels. Several researchers have demonstrated the utility of Bcl-2 in such systems.

In an early study, the Bcl-2 homologue BHRF1 was shown to protect human B cells from apoptosis induced by serum deprivation (Henderson *et al.*, 1993). Additionally, more recent work has indicated that COS-1 cells stably transfected with *bcl-2* show enhanced survival as compared to mock transfectants when cultured in low serum conditions. In fact, the viable cell number of the *bcl-2* transfectants was nine fold higher than that of the controls at the end of a nine day culture period (Suzuki *et al.*, 1997) . Other cell lines have also benefited from *bcl-2* overexpression in a serum-starved environment. TB/C3 hybridomas engineered to stably express *bcl-2* adapted to serum-free medium more rapidly than control cells (three weeks versus two months), and were able to survive in a serum-free batch culture for extended periods (Fassnacht *et al.*, 1998). Furthermore, the coexpression of *bag-1* and *bcl-2* in 2E3 hybridomas prolonged cultures by two days while increasing the rate of Ab production by three-fold (Terada *et al.*, 1997).

3.3.3 Oxygen Limitations

The delivery of sufficient oxygen levels to large scale cultures can be difficult to achieve since oxygen is sparingly soluble in aqueous media and is rapidly consumed by proliferating cells. Methods capable of providing adequate amounts of oxygen may, unfortunately, cause significant physical damage to the cells. Sparging cultures, for example, can lead to high levels of cell death during bubble disengagement at the gas-liquid interface (Jobses *et al.*, 1991). However, the development of less intensive approaches capable of efficiently supplying oxygen to large scale cultures has been difficult. An indirect route to mitigate cell damage associated with procedures such as sparging may be to provide cells with less oxygen, although many cell types, including hybridomas, respond to limitations in oxygen by activating the apoptotic cascade (Mercille and Massie, 1994b). Modifying cells to better withstand oxygen deprivation could therefore increase product yields by limiting the problems associated with culture oxygenation. Indeed, overexpression of *bcl-2* and *bcl-x_L* has demonstrated the ability to suppress apoptosis induced by hypoxia (Jacobson and Raff, 1995; Shimizu *et al.*, 1995). Protection from PCD in such environments could have industrial implications, since commercially relevant cell lines can also be saved from anaerobic conditions using Bcl-2 (Simpson *et al.*, 1997; Singh *et al.*, 1997).

3.3.4 Hydrodynamic Stresses

Another common problem typically associated with the large scale culture of animal cells is the occurrence of apoptosis resulting from the stressful hydrodynamic conditions often found within bioreactors (Al-Rubeai *et al.*, 1995). Since agitation of cultures is necessary to ensure homogeneity, methods to limit cell death under conditions of high shear stresses could be worthwhile, and may lead to gains in culture productivities. The ability of Bcl-2 to protect cells from moderate to high levels of shear has been demonstrated in several instances. In one example, a lymphoblastoid cell line stably expressing *bcl-2* was capable of growing in suspension without prior adaptation, possibly indicating the capacity of this gene to protect from PCD due to hydrodynamic forces (Singh *et al.*, 1996). In another study, *bcl-2* extended the culture period of a murine hybridoma cell line by 24 hours following exposure to high shear stresses, while allowing for a maximum viable cell number almost four times higher than that of the controls. Even under mild shear conditions, when the maximum viable densities of the two cultures were similar, the overexpression of *bcl-2* extended the culture period by 100 hours (Perani *et al.*, 1998). Shear stresses are not, however, always associated with increases in apoptosis levels. Indeed, low levels of shear have actually been shown to limit PCD induced by both growth factor withdrawal and incubation with TNF-α in an endothelial cell culture (Dimmeler *et al.*, 1996). This suggests the possible existence of an optimal level of shear stress that could permit sufficient culture mixing while suppressing apoptosis.

3.3.5 Virus Infections

It is well established that virus infections lead to apoptosis in a number of cell lines (reviewed in Hardwick, 1997). Viral vector-induced apoptosis, however, is an important limitation in the biotechnology industries where the viral expression systems are used for the rapid production of high levels of recombinant proteins. Encouragingly, it has been demonstrated in a number of instances that Bcl-2 family members are able to limit the cell death response to viruses.

When expressed from SV vectors, *bcl-2*, *bcl-*x_L, *BHRF1*, and *KSbcl-2* were each able to limit apoptosis in BHK cells resulting from the virus infections (Cheng *et al.*, 1997a). Furthermore, stably transfected cell lines expressing *bcl-2* are able to suppress PCD in response to Japanese encephalitis virus (Liao *et al.*, 1997), SV (Levine *et al.*, 1993), and SFV (Scallan *et al.*, 1997). In one instance, the overexpression of *bcl-2* in a rat prostate carcinomal cell line (AT3) resulted in the establishment of a persistent SV infection (Levine *et al.*, 1993).

The development of cell lines stably expressing Bcl-2 family members has also allowed for extended survival in cells infected with recombinant viruses. For example, RIN cells engineered to overexpress *bcl-2* were protected from apoptosis induced by an SFV vector containing the *lacZ* gene, exhibiting viabilities 40% higher than those of control cells after 72 hours (Lundstrom *et al.*, 1997). In a similar study, AT3-*bcl2* cells were completely resistant to apoptosis induced by an SV vector, maintaining viabilities greater than 80% over a ten day culture period, while AT3-*neo* control cells completely lost viability by three days post-infection. Though the control cell line did initially produce more of the model protein chloramphenicol acetyltransferase (CAT), the repeated infection of AT3-*bcl2* cells allowed for extended and amplified CAT production, giving rise to CAT levels far exceeding those obtained from the AT3-*neo* culture (Mastrangelo *et al.*, 1996).

Insect cells can also benefit from the anti-apoptotic properties of *bcl-2*. Overexpression of this gene in Sf-9 cells can result in an extended lifespan following baculovirus infection (Alnemri *et al.*, 1992). However, when coinfected with a baculovirus expressing Bcl-2 and another expressing either β-galactosidase or tissue plasminogen activator (tPA), these cells showed no increases in survival or recombinant protein production as compared to cells infected with only the latter virus. In contrast, High Five cells coinfected with baculoviruses carrying the genes for Bcl-2 and tPA demonstrated significant enhancements in survival as well as in tPA production (Mitchell-Logean and Murhammer, 1997).

3.3.6 Other Culture Modes

Most of the culture systems employing Bcl-2 family members discussed up to this point have been batch processes where cells follow a well-established growth pattern comprised of four distinct phases. An initial lag period is followed by the log phase in which cells grow exponentially. Once the maximum cell density has been reached, a value that depends on not only the cell line, but also the bioreactor configuration, culture medium, and other culture conditions, cells will enter a stationary phase where the density remains fairly static. The length of this phase is largely dependent on cell type. In the case of hybridoma cells, for instance, no stationary phase is observed and cells transition from exponential growth to the decline phase of the culture. This is particularly unfortunate from a biotechnological perspective, since a negative correlation between growth rate and specific productivity exists in these cells. Without a stationary phase, potential gains in product yields are lost since a non-proliferating state could lead to significantly higher product titers. However, a number of methods now exist for arresting cell growth, allowing for a pseudo-stationary phase where culture productivities may increase. Common strategies include the use of osmolytes, as well as treating cultures with thymidine and other cytostatic agents.

Altering culture osmolarity. The osmolarity of a culture can have a significant impact on product yield; in hybridoma cultures, for instance, increased osmolarity can lead to enhanced productivity (Oh *et al.*, 1993; Ozturk and Palsson, 1991; Reddy and Miller, 1994). Unfortunately, hypertonic conditions promote PCD in a number of cell lines (Bortner and Cidlowski, 1996; Matthews and Feldman, 1996; Qin *et al.*, 1997), making the overexpression of anti-apoptosis genes an attractive strategy for limiting levels of cell death while improving product yields.

In one experimental system, an increase in culture osmolarity lowered the maximum cell densities of TB/C3 hybridoma cells. However, stable expression of *bcl-2* in these cells resulted in higher cell numbers as compared to the controls: maximums cell densities of this cell line were 33% higher at 350 mOsm and 100% higher at 400 mOsm. Additionally, unlike the mock transfectants, the *bcl-2* cell line was able to adjust to conditions of high osmolarity (400 mOsm). Though no comparison in Ab yields could be made at this osmolarity between the TB/C3-*bcl2* line and the controls (since the latter did not survive this transition), the *bcl-2* transfectants were shown to be about twice as productive at this higher osmolarity than at normal conditions (Perani 1998).

Treatment of cultures with cytostatic agents. Utilization of excess thymidine and other cytostatic agents is another method of inducing growth arrest in cells. This strategy, while providing higher yields for a short time, typically leads to high levels of apoptosis (Singh *et al.*, 1994). The overexpression of *bcl-2* in murine hybridomas treated with thymidine has been shown to extend culture duration while increasing specific Ab productivities (Simpson *et al.*, 1997). Furthermore, when *bcl-2* and *bag-1* were used in combination, 2E3 hybridoma cultures survived three days longer than controls and one day longer than cells expressing only *bcl-2*, giving rise to more than twice as much Ab overall than mock transfectants (Terada *et al.*, 1997).

In another study, Mercille and Massie (1998) treated NS/0 myelomas with a variety of inhibitors of cell proliferation including NaCl, thymidine, and OptiMab™. Cells engineered to overexpress the adenovirus *E1B-19K* gene consistently maintained higher viabilities as well as increased specific productivities than control cells under these conditions. Furthermore, the use of OptiMab™ provided a four-fold increase in volumetric Ab productivity, indicating the value of these methods for enhancing the production of monoclonal antibodies.

Metabolic engineering methods. Rather than treating cultures with agents to inhibit proliferation, Fussenegger *et al.* (1998) have demonstrated the utility of multigenic transfections to arrest cell growth while limiting apoptotic death. In this work, CHO cells were transfected with one of two tricistronic vectors, the second of which contained the genes for the cyclin-dependent kinase inhibitor p27, the anti-apoptotic Bcl-x_L, and a model heterologous protein, secreted alkaline phosphatase (SEAP). In order to manipulate expression levels, all three genes were placed under the control of a tetracycline-responsive promoter. When gene expression was induced, p27 blocked the progression of the cell cycle while Bcl-x_L function in an unclear manner to allow for increased SEAP expression levels. The coordination of all three events provided for very high product yields. In fact, CHO cells stably transfected with this construct showed a 30-fold increase in SEAP expression relative to the controls.

Perfusion systems. Another strategy of increasing culture productivity without actually arresting cell growth is the use of perfusion cultures. In continuous systems such as these, the specific cellular growth rate can be altered by changing the culture dilution rate. Perfusion systems have several advantages over other culture modes including higher cell densities, increased volumetric productivity, and rapid removal of waste products from the culture environment (Prior *et al.*, 1989). However, high levels of apoptosis are often associated with such processes (Mercille *et al.*, 1994), leading to decreases in attainable product yields. Engineering cells to make them resistant to apoptosis could therefore have a significant impact on culture productivities. Indeed, NS/0 cells overexpressing *E1B-19K* that were maintained at a slow growth rate during perfusion culture exhibited a three-fold decrease in death rate and subsequent increase in Ab production as compared to the control cells. Moreover, the increase in volumetric productivities when comparing the perfusion systems with batch processes was more than seven-fold for the *E1B-19K* cells, compared to only about three-fold in the NS/0 controls (Mercille and Massie, 1999). Similar studies utilizing TB/C3 hybridoma cells demonstrated that overexpression of *bcl-2* could increase the maximum cell density by roughly two-fold in two different perfusion systems, though no increase in culture productivity was seen in either case (Bierau *et al.*, 1998).

3.3.7 When Bcl-2 Fails

Though *bcl-2* is able to protect a variety of cell lines from numerous insults, there are examples in which *bcl-2* fails to suppress PCD (reviewed in Reed, 1994). In one instance of biotechnological relevance, the overexpression of *bcl-2* in NS/0 myeloma cells was unable to provide any protection from apoptosis induced upon entry into the decline phase of a batch culture. Additionally, these *bcl-2* transfectants were no more resistant to PCD induced by serum withdrawal than control cells (Murray *et al.*, 1996). The inability of *bcl-2* to protect these cells may have been due to the fact that Bcl-x_L, a close relative of Bcl-2, is endogenously expressed in this cell line; its presence, therefore, may have made Bcl-2 essentially redundant and, consequently, ineffective in apoptosis resistance.

Another possible mechanism of Bcl-2 failure is linked to the activity of caspase-3. Recent studies have demonstrated the ability of this caspase to cleave both Bcl-2 and Bcl-x_L, not only rendering them unable to protect against PCD, but converting them into pro-death proteins capable of potently inducing apoptosis (Cheng *et al.*, 1997b; Clem *et al.*, 1998; Grandgirard *et al.*, 1998). The cleavage sites for both proteins are found in a nonconserved loop domain unnecessary for anti-apoptotic activity. Therefore, the proteolytic inactivation of Bcl-2 and Bcl-x_L can be prevented, and their protective effects maintained, by genetic mutations that result in the deletion of one of the cleavage sites or removal of the loop domain from the protein. Indeed, the overexpression of loop deletion mutations can protect cells as well as or better than the wild type proteins in response to apoptotic stiumuli such as IL-3 deprivation, SV infection and serum withdrawal (Chang *et al.*, 1997; B. Figueroa, personal communication).

4. Concluding Remarks

This chapter has presented a number of methods by which apoptosis can be inhibited *in vitro*, during the cultivation of mammalian cell lines. Further development of several of these strategies, especially in regard to caspase inhibition, use of Bcl-2 deletion mutants, and multigenic transfections, will likely enhance our ability to suppress PCD in culture, thereby leading to more efficient bioprocesses.

It is important to remember, however, that other approaches can be utilized to limit cell death in culture, approaches that were once regarded as the only means of doing so. Effort must continually be made to improve bioreactor configurations, including culture oxygenation and agitation techniques, to optimize media formulations, especially in the absence of serum, and to characterize alternative culture modes such as fed-batch and perfusion operations. Insights gained in each of these areas will assuredly, when combined with those elaborated upon above, allow for the creation of a more stable environment for cells, one which will provide higher product yields, leading to more cost-effective production schemes.

5. Acknowledgments

The author would like to thank Bernard Massie and Mohammed Al-Rubeai for sharing unpublished data, as well as Michael J. Betenbaugh and George A. Oyler for critical review of the manuscript. Work on this topic by the author has been funded by the National Science Foundation (Grant number BES-9520248).

6. References

Al-Rubeai, M., Mills, D., and Emery, A.N. (1990). Electron microscopy of hybridoma cells with special regard to monoclonal antibody production. *Cytotechnology* **4**, 13-28.

Al-Rubeai, M., Singh, R.P. , Goldman, M.H., and Emery, A.N. (1995). Death mechanisms of animal cells in conditions of intensive cell-culture systems. *Biotechnol. Bioeng.* **45**, 463-472.

Alnemri, E. S., Robertson, N.M. , Fernandes, T.F., Croce, C.M., and Litwack, G. (1992). Overexpressed full-length human BCL-2 extends the survival of baculovirus-infected Sf9 insect cells. *Proc. Natl. Acad. Sci. USA* **89**, 7295-7299.

Baffy, G., Miyashita, T., Williamson, J.R., and Reed, J.C. (1993). Apoptosis induced by withdrawal of interleukin-3 (IL-3) from an IL-3-dependent hematopoietic cell line is associated with repartitioning of intracellular calcium and is blocked by enforced Bcl-2 oncoprotein production. *J. Biol. Chem.* **268**, 6511-6519.

Beidler, D. R., Tewari, M., Friesen, P.D., Poirier, G., and Dixit, V.M. (1995). The baculovirus p35 protein inhibits Fas- and tumor necrosis factor-induced apoptosis. *J. Biol. Chem.* **270**, 16526-16528.

Bessho, R., Matsubara, K., Kubota, M., Kuwakado, K., Hirota, H.,Wakazono, Y., Lin, Y.W., Okuda, A., Kawai, M., Nishikomori, R., and Heike, T. (1994). Pyrrolidine dithiocarbamate, a potent inhibitor of nuclear factor kappa B (NF-kappa B) activation, prevents apoptosis in human promelocytic leukemia HL-60 cells and thymocytes. *Biochem. Pharmacol.* **48**, 1883-9.

Beyette, J., Mason, G.G., Murray, R.A., Cohen, G.M., and Rivett, A.J. (1998). Proteasome activities decrease during dexamethasone-induced apoptosis of thymocytes. *Biochem. J.* **332**, 315-320.

Bierau, H., Perani, A., Al-Rubeai, M., and Emery, A.N. (1998). A comparison of intensive cell culture bioreactors operating with Hybridomas modified for inhibited apoptotic response. *J. Biotechnol.* **62**, 195-207.

Bissonnette, R. P., Echeverri, F., Mahboubi, A., and Green, D.R. (1992). Apoptotic cell death induced by *c-myc* is inhibited by *bcl-2*. *Nature* **359**, 552-556.

Bliem, R., Konoptsky, K., and Katinger, H. (1991). Industrial animal cell reactor systems: Aspects of selection and evaluation. In A. Fiechter (ed.), *Advances in Biochemical Engineering/Biotechnology* Vol. 44, Springer-Verlag, Berlin, pp.1-26.

Boise, L. H., Gonzalez-Garcia, M., Postema, C.E., Ding, L., Lindsten, T., Turka, L.A., Mao, X., Nunez, G., and Thompson, C.B. (1993). *bcl-x*, a *bcl-2* related gene that functions as a dominant regulator of apoptotic cell death. *Cell* **74**, 598-608.

Bortner, C. D., and Cidlowski, J.A. (1996). Absence of volume regulatory mechanisms contributes to the rapid activation of apoptosis in thymocytes. *Am. J. Physiol.* **271**, C950-961.

Bossy-Wetzel, E., Newmeyer, D.D., and Green, D.R. (1998). Mitochondrial cytochrome c release in apoptosis occurs upstream of DEVD-specific caspase activation and independently of mitochondrial transmembrane depolarization. *EMBO J.* **17**, 37-49.

Bradham, C. A., Zian, T., Streetz, K., Trautwein, C., Brenner, D.A., and Lemasters, J.J. (1998). The mitochondrial permeability transition is required for tumor necrosis factor alpha-mediated apoptosis and cytochrome c release. Mol. *Cell. Biol.* **18**, 6353-6364.

Butler, M. (1987) *Animal Cell Technology: Principles and Products*, Open University Press, London.

Cartier, J. L., Hershberger, P.A., and Friesen, P.D. (1994). Suppression of apoptosis in insect cells stably transfected with Baculovirus *p35*: dominant interference by N-terminal sequences $p35^{1-76}$. *J. Virol.* **68**, 7728-7737.

Chang, B.S., Minn, A.J., Muchmore, S.W., Fesik, S.W., and Thompson, C.B. (1997). Identification of a novel regulatory domain in Bcl-x_L and Bcl-2. *EMBO J.* **16**, 968-977.

Cheng, E. H.-Y., Nicholas, J., Bellows, D.S., Hayword, G.S., Guo, H-G., Reitz, M.S., and Hardwick, J.M. (1997a). A Bcl-2 homolog encoded by Kaposi sarcoma-associated virus, human herpesvirus 8, inhibits apoptosis but does not heterodimerize with Bax or Bak. *Proc. Natl. Acad. Sci. USA* **94**, 690-694.

Cheng, E. H.-Y., Kirsch, D.G., Clem, R.J., Ravi, R., Kastan, M.B., Bedi, A., Ueno, K., and Hardwick, J.M. (1997b). Conversion of Bcl-2 to a Bax-like death effector by caspases. *Science* **278**, 1966-1968.

Chiou, S. K., and White, E. (1998). Inhibition of ICE-like proteases inhibits apoptosis and increases virus production during adenovirus infection. *Virology* **244**, 108-118.

Chung, J. D., Zabel, C., Sinskey, A.J., and Stephanopoulos, G. (1997). Extension of Sp2/0 hybridoma cell viability through interleukin-6 supplementation. *Biotechnol. Bioeng.* **55**, 439-446.

Clem, R. J., and Miller, L.K. (1994). Control of programmed cell death by the baculovirus genes *p35* and *iap*. *Mol. Cell. Biol.* **14**, 5212-5222.

Clem, R. J., Hardwick, J.M., and Miller, L.K. (1996). Anti-apoptotic genes of baculoviruses. *Cell Death Diff.* **3**, 13-20.

Clem, R. J., Cheng, E. H.-Y., Karp, C.L., Kirsch, D.G., Ueno, K., Takahashi, A., Kastan, M.B., Griffin, D.E., Earnshaw, W.C., Veliuona, M.A., and Hardwick, J.M. (1998). Modulation of cell death by Bcl-xL through caspase interaction. *Proc. Natl. Acad. Sci. USA* **95**, 554-559.

Cossarizza, A., Franceschi, C., Monti, D., Salvioli, S., Bellesia, E., Rivabene, R., Biondo, L., Rainaldi, G., Tinari, A., and Malorni, W. (1995). Protective effect of N-acetylcysteine in tumor necrosis factor-alpha-induced apoptosis in U937 cells: the role of mitochondria. *Exp. Cell Res.* **220**, 232-240.

Cotter, T. G., and Al-Rubeai, M. (1995). Cell death (apoptosis) in cell culture systems. *Trends Biotechnol.* **13**, 150-154.

Crook, N. E., Clem, R.J., and Miller, L.K. (1993). An apoptosis-inhibiting baculovirus gene with a zinc finger-like motif. *J. Virol.* **67**, 2168-2174.

Cummings, M. C., Winterford, C.M. , and Walker, N.I. (1997). Apoptosis. *Am. J. Surg. Pathol.* **21**, 88-101.

Datta, R., Kojima, H., Banach, D., Bump, N.J., Talanian, R.V., Alnemri, E.S., Weichselbaum, R.R., Wong, W.W., and Kufe, D.W. (1997). Activation of a CrmA-insensitive, p35-sensitive pathway in ionizing radiation-induced apoptosis. *J. Biol. Chem.* **272**, 1965-1969.

Degli Esposti, M., and McLennan, H. (1998). Mitochondria and cells produce reactive oxygen species in virtual anaerobiosis: relevance to ceramide-induced apoptosis. *FEBS Lett.* **430**, 338-342.

Dimmeler, S., Haendeler, J., Rippmann, V., Nehls, M., and Zeiher, A.M. (1996). Shear stress inhibits apoptosis of human endothelial cells. *FEBS Lett.* **399**, 71-74.

DiStefano, D. J., Mark, G.E., and Robinson, D.K. (1996). Feeding of nutrients delays apoptotic death in fed-batch cultures of recombinant NS0 myeloma cells. *Biotechnol. Lett.* **18**, 1067-1072.

Dobbelstein, M., and Shenk, T. (1996). Protection against apoptosis by the vaccinia virus SPI-2 (B13R) gene product. *J. Virol.* **70**, 6479-6485.

Duckett, C. S., Nava, V.E. , Gedrich, R.W., Clem, R.J., VanDongen, J.L., Gilfillan, M.C., Shiels, H., Hardwick, J.M., and Thompson, C.B. (1996). A conserved family of cellular genes related to the baculovirus *iap* gene and encoding apoptosis inhibitors. *EMBO J.* **15**, 2685-2694.

Duckett, C. S., Li, F., Wang, Y., Tomaselli, K.J., Thompson, C.B., and Armstrong, R.C. (1998). Human IAP-like protein regulates programmed cell death downstream of Bcl-xL and cytochrome c. *Mol. Cell. Biol.* **18**, 608-615.

Duval, D., Demangel, C., Munier-Jolain, K., Miossec, S., and Geahel, I. (1991). Factors controlling cell proliferation and antibody production in mouse hybridoma cells: I. Influence of the amino acid supply. *Biotechnol. Bioeng.* **38**, 561-570.

Eagle, H. (1955). Nutrition needs of mammalian cells in tissue culture. *Science* **122**, 501-504.

Eagle, H. (1959). Amino acid metabolism in mammalian cell cultures. *Science* **130**, 432-437.

Enari, M., Hug, H., and Nagata, S. (1995). Involvement of an ICE-like protease in Fas-mediated apoptosis. *Nature* **375**, 78-81.

Fassnacht, D., Rössing, S., Franěk, F., Al-Rubeai, M., and Pörtner, R. (1998). Effect of *bcl-2* expression on hybridoma cell growth in serum-supplemented, protein-free and diluted media. *Cytotechnology* **26**, 219-225.

Fernandez, A., Marin, M.C., McDonnell, T.J., and Ananthaswamy, H.N. (1994). Differential sensitivity of normal and Ha-ras-transformed C3H mouse embryo fibroblasts to tumor necrosis factor: induction of *bcl-2*, *c-myc*, and manganese superoxide dismutase in resistant cells. *Oncogene* **9**, 2009-2017.

Franěk, F., and Dolníková, J. (1991a). Nucleosomes occuring in protein-free hybridoma cell cultures. Evidence for programmed cell death. *FEBS Lett.* **248**, 285-287.

Franěk, F., and Dolníková, J. (1991b). Hybridoma growth and monoclonal antibody production in iron-rich protein-free medium: effect of nutrient concentration. *Cytotechnology* **7**, 33-38.

Franěk, F., and Chládkova-Srámková, K. (1995). Apoptosis and nutrition: involvement of amino acid transport system in repression of hybridoma cell death. *Cytotechnology* **18**, 113-117.

Franěk, F., and Srámková, K. (1996a). Cell suicide in starving hybridoma culture: survival-signal effect of some amino acids. *Cytotechnology* **21**, 81-89.

Franěk, F., and Srámková, K. (1996b). Cell suicide in starving hybridoma culture: survival-signal effecto fo some amino acids. *Cytotechnology* **21**, 81-89.

Fulda, S., Susin, S.A., Kroemer, G., and Debatin, K.M. (1998). Molecular ordering of apoptosis induced by anticancer drugs in neuroblastoma cells. *Cancer Res.* **58**, 4453-4460.

Fussenegger, M., Schlatter, S., Dätwyler, S., Mazur, X., and Bailey, J.E. (1998). Controlled proliferation by multigene metabolic engineering enhances the productivity of Chinese hamster ovary cells. *Nat. Biotech.* **16**, 468-472.

Gagliardini, V., Fernandez, P.-A., Lee, R.K.K., Drexler, H.C.A., Rotello, R.J., Fishman, M.C., and Yuan, J. (1994). Prevention of vertebrate neuronal death by the *crmA* gene. *Science* **262**, 826-828.

Gerschenson, L. E., and Rotello, R. J. (1992). Apoptosis: a different type of cell death. *FASEB J.* **6**, 2450-2455.

Grandgirard, D., Studer, E., Monney, L., Belser, T., Fellay, I., Borner, C., and Michel, M.R. (1998). Alphaviruses induce apoptosis in Bcl-2-overexpressing cells: evidence for a caspase-mediated proteolytic inactivation of Bcl-2. *EMBO J.* **17**, 1268-1278.

Greenlund, L. J. S., Deckwerth, T.L., and Johnson, E.M. Jr. (1995). Superoxide dismutase delays neuronal apoptosis: a role for reactive oxygen species in programmed neuronal cell death. *Neuron* **14**, 303-315.

Hardwick, J. M. (1997). Virus-induced apoptosis. *Adv. Pharmacol.* **41**, 295-336.

Hardwick, J. M. (1998). Viral interference with apoptosis. *Semin. Cell Dev. Biol.* **9**, 339-349.

Hashimoto, S., Ishii, A., and Yonehara, S. (1991). The E1b oncogene of adenovirus confers cellular resistance to cytotoxicity of tumor necrosis factor and monoclonal anti-Fas antibody. *Int. Immunol.* **17**, 2835-2843.

Hehner, S. P., Hofmann, T.G., Ratter, F., Dumont, A., Droge, W., and Schmitz, M.L. (1998). Tumor necrosis factor-alpha-induced cell killing and activation of transcription factor NF-kappaB are uncoupled in L929 cells. *J. Biol. Chem.* **273**, 18117-18121.

Henderson, S., Huen, D., Rowe, M., Dawson, C., Johnson, G., and Rickinson, A. (1993). Epstein-Barr virus-coded BHRF1 protein, a viral homologue of Bcl-2, protects human B cells from programmed cell death. *Proc. Natl. Acad. Sci. USA* **90**, 8479-8483.

Hockenbery, D. M., Oltvai, Z.N., Yin, X.-M., Milliman, C.L., and Korsmeyer, S.J. (1993). Bcl-2 functions in an antioxidant pathway to prevent apoptosis. *Cell* **75**, 241-251.

Hortelano, S., Dallaporta, B., Zamzami, N., Hirsch, T., Susin, S.A., Marzo, I., Bosca, L., and Kroemer, G. (1997). Nitric oxide induces apoptosis via triggering mitochondrial permeability transition. *FEBS Lett.* **410**, 373-377.

Ichas, F., and Mazat, J.P. (1998). From calcium signaling to cell death: two conformations for the mitochondrial permeability transition pore. Switching from low- to high-conductance state. *Biochim. Biophys. Acta* **1366**, 33-50.

Ink, B. S., Gilbert, C.S., and Evan, G.I. (1995). Delay of vaccinia virus-induced apoptosis in nonpermissive Chinese hamster ovary cells by the cowpox virus *CHOhr* and adenovirus *E1B 19K* genes. *J. Virol.* **69**, 661-668.

Itoh, Y., Ueda, H., and Suzuki, E. (1995). Overexpression of *bcl-2*, apoptosis suppressing gene: prolonged viable culture period of hybridoma and enhanced antibody production. *Biotech. Bioeng.* **48**, 118- 122.

Jacobson, M. D., and Raff, M.C. (1995). Programmed cell death and Bcl-2 protection in very low oxygen. *Nature* **374**, 814-816.

Jacobson, M. D. (1996). Reactive oxygen species and programmed cell death. *Trends Biochem. Sci.* **21**, 83-86.

Jobses, I., Marten, D., and Tramper, J. (1991). Lethal events during gas sparging in animal cell culture. *Biotechnol. Bioeng.* **37**, 484-490.

Kerr, J. F. R., Wyllie, A.H., and Currie, A.R. (1972). Apoptosis: a basic biological phenomenon with wide-ranging implications in tissue kinetics. *Br. J. Cancer* **26**, 239-257.

Kluck, R. M., Bossy-Wetzel, E., Green, D.R., and Newmeyer, D.D. (1997). The release of cytochrome c from mitochondria: a primary site for Bcl-2 regulation of apoptosis. *Science* **275**, 1132-1136.

Kohler, G., and Milstein, C. (1975). Continuous cultures of fused cells secreting antibody of predefined specificity. *Nature* **256**, 495-497.

Kroemer, G., Zamzami, N., and Susin, S.A. (1997). Mitochondrial control of apoptosis. *Immunol. Today* **18**, 44-51.

Levine, B., Huang, Q., Isaacs, J.T., Reed, J.C., Griffin, D.E., and Hardwick, J.M. (1993). Conversion of lytic to persistent alphavirus infection by the *bcl-2* cellular oncogene. *Nature* **361**, 739-742.

Liao, C. L., Lin, Y.L., Wang, J.J., Huang, Y.L., Yeh, C.T., Ma, S.H., and Chen, L.K. (1997). Effect of enforced expression of human *bcl-2* on Japanese encephalitis virus-induced apoptosis in cultured cells. *J. Virol.* **71**, 5963-5971.

Lin, K.-I., Lee, S.-H., Narayanan, R., Baraban, J.M., Hardwick, J.M., and Ratan, R.R. (1995). Thiol agents and *bcl-2* identify an alphavirus-induced apoptotic pathway that requires activation of the transcription factor NF-kappa B. *J. Cell Biol.* **131**, 1149-1161.

Liston, P., Roy, N., Tamai, K., Lefebvre, C., Baird, S., Cherton-Horvat, G., Farahani, R., McLean, M., Ikeda, J, MacKenzie, A., and Korneluk, R.G. (1996). Suppression of apoptosis in mammalian cells by NAIP and a related family of IAP genes. *Nature* **379**, 349-353.

Lundstrom, K., Pralong, W., and Martinou, J.-C. (1997). Anti-apoptotic effect of Bcl-2 overexpression in RIN cells infected with Semliki Forest virus. *Apoptosis* **2**, 189-191.

Malorni, W., Rivabene, R., Santini, M.T., and Donelli, G. (1993). N-acetylcysteine inhibits apoptosis and decreases viral particles in HIV-chronically infected U937 cells. *FEBS Lett.* **327**, 75-8.

Marchetti, P., Castedo, M., Susin, S.A., Zamzami, N., Hirsch, T., Haeffner, A., Hirsch, F., Geuskens, M., and Kroemer, G. (1996). Mitochondrial permeability transition is a central coordinating event of apoptosis. *J. Exp. Med.* **184**, 1155-1160.

Martin, S. J., Newmeyer, D.D., Mathias, S., Farschon, D.W., Wang, H.G., Reed, J.C., Kolesnick, R.N., and Green, D.R. (1995). Cell-free reconstitution of Fas-, UV radiation- and ceramide-induced apoptosis. *EMBO J.* **14**, 5191-5200.

Martinou, I., Fernandez, P.-A., Missotten, M., White, E., Allet, B., Sadoul, R., and Martinou, J.-C. (1995). Viral proteins E1B19K and p35 protect sympathetic neurons from cell death induced by NGF deprivation. *J. Cell Biol.* **128**, 201-208.

Marzo, I., Brenner, C., and Kroemer, G. (1998). The central role of the mitochondrial megachannel in apoptosis: evidence obtained with intact cells, isolated mitochondria, and purified protein complexes. *Biomed. Pharmacother.* **52**, 248-251.

Mastrangelo, A. J., Hardwick, J.M., and Betenbaugh, M.J. (1996). Bcl-2 inhibits apoptosis and extends recombinant protein production in cells infected with Sindbis viral vectors. *Cytotechnology* **22**, 169-178.

Mastrangelo, A. J., and Betenbaugh, M.J. (1998). Overcoming apoptosis: new methods for improving protein expression systems. *Trends Biotechnol.* **16**, 88-95.

Matthews, C. C., and Feldman, E.L. (1996). Insulin-like growth factor I rescues SH-SY5Y human neuroblastoma cells from hypertonic induced programmed cell death. *J. Cell. Physiol.* **166**, 323-331.

Mayer, M., and Noble, M. (1994). N-Acetyl-L-cysteine is a pluripotent protector against cell death and enhancer of trophic factor-mediated cell survival in vitro. *Proc. Natl. Acad. Sci. USA* **91**, 7496-7500.

McCarthy, N. J., Hazlewood, S.A., Huen, D.S., Rickinson, A.B., and Williams, G.T. (1996). The Epstein-Barr virus gene BHRF1, a homologue of the cellular oncogene Bcl-2, inhibits apoptosis induced by gamma radiation and chemotherapeutic drugs, in S. Gupta and J.J. Cohen (eds.) *Mechamisms of Lymphocyte Activation and Immune Regulation VI: Cell Cycle and Programmed Cell Death in the Immune System* , Plenum Press, New York, pp. 83-97.

McColl, K. S., He, H., Zhong, H., Whitacre, C.M., Berger, N.A., and Distelhorst, C.W. (1998). Apoptosis induction by the glucocorticoid hormone dexamethasone and the calcium-ATPase inhibitor thapsigargin involves Bcl-2 regulated caspase activation. *Mol. Cell. Endocrinol.* **139**, 229-238.

McGowan, A. J., Fernandes, R.S., Samali, A., and Cotter, T.G. (1996). Anti-oxidants and apoptosis. *Biochem. Soc. Trans.* **24**, 229-33.

McLachlin, J. R., and Miller, L.K. (1997). Stable transformation of insect cells to coexpress a rapidly selectable marker gene and an inhibitor of apoptosis. *In Vitro Cell. Dev. Biol.* **33**, 575-579.

Mercille, S., Johnson, M., Lemieux, R., and Massie, B. (1994). Filtration-based perfusion of hybridoma cultures in protein-free medium: reduction of membrane fouling by medium supplementation with DNaseI. *Biotechnol. Bioeng.* **43**, 833-846.

Mercille, S., and Massie, B. (1994a). Induction of apoptosis in nutrient-deprived cultures of hybridoma and myeloma cells. *Biotechnol. Bioeng.* **44**, 1140-1154.

Mercille, S., and Massie, B. (1994b). Induction of apoptosis in oxygen-deprived cultures of hybridoma cells. *Cytotechnology* **15**, 117-128.

Mercille, S., and Massie, B. (1998). Apoptosis-resistant NS/0-E1B-19K myeloma cells exhibit increased viability and chB43 monoclonal antibody productivity using cell cycle modulators. *Cytotechnology* In Press.

Mercille, S., and Massie, B. (1999). Apoptosis-resistant E1B-19K-expressing NS/0 myeloma cells exhibit increased viability and chimeric antibody productivity under perfusion culture. *Biotechnol. Bioeng.* In Press.

Mercille, S., Jolicoeur, P., Gervais, C., Paquette, D., Mosser, D.D., and Massie, B. (1999). Dose-dependent reduction of apoptosis in nutrient-limited cultures of NS/0 myeloma cells transfected with the E1B-19K adenoviral gene. *Biotechnol. Bioeng.* In Press.

Mignotte, B., and Vayssiere, J.-L. (1998). Mitochondria and apoptosis. *Eur. J. Biochem.* **252**, 1-15.

Miller, L. K. (1997). Baculovirus interaction with host apoptotic pathways. *J. Cell. Physiol.* **173**, 178-182.

Mitchell-Logean, C., and Murhammer, D.W. (1997). bcl-2 expression in *Spodoptera frugiperda* Sf-9 and *Trichoplusia ni* BTI-Tn-5B1-4 insect cells: effect on recombinant protein expression and cell viability. *Biotechnol. Bioeng.* **56**, 380-388.

Miura, M., Friedlander, R.M., and Yuan, J. (1995). Tumor necrosis factor-induced apoptosis is mediated by a Crm-A-sensitive cell death pathway. *Proc. Natl. Acad. Sci. USA* **92**, 8318-8322.

Miyashita, T., and Reed, J.C. (1992). *bcl-2* gene transfer increases relative resistance of S49.1 and WEH17.2 lymphoid cells to cell death and DNA fragmentation induced by glucocorticoids and multiple chemotherapeutic drugs. *Cancer Res.* **52**, 5407-5411.

Murray, K., Ang, C.-E., Gull, K., Hickman, J.A., and Dickson, A.J. (1996). NSO myeloma cell death: influence of *bcl-2* overexpression. *Biotechnol. Bioeng.* **51**, 298-304.

Nava, V. E., Rosen, A., Veliuona, M.A., Clem, R.J., Levine, B., and Hardwick, J.M. (1998). Sindbis virus induces apoptosis through a caspase-dependent, CrmA-sensitive pathway. *J. Virol.* **72**, 452-459.

Oh, S. K. W., Vig, P., Chua, F.K.F, Teo, W.K., and Yap, M.S.G. (1993). Substantial overproduction of antibodies by applying osmotic pressure and sodium butyrate. *Biotechnol. Bioeng.* **42**, 601-610.

Oltvai, Z. N., Milliman, C.L., and Korsmeyer, S.J. (1993). Bcl-2 heterodimerizes in vivo with a conserved homolog, Bax, that accelerates programed cell death. *Cell* **74**, 609-619.

Ozturk, S. S., and Palsson, B.O. (1991). Effect of medium osmolarity on hybridoma growth, metabolism and antibody production. *Biotechnol. Bioeng.* **37**, 989-993.

Pastorino, J. G., Synder, J.W., Serroni, A., Hoek, J.B., and Farber, J.L. (1993). Cyclosporin and carnitine prevent the anoxic death of cultured hepatocytes by inhibiting the mitochondrial permeability transition. *J. Biol. Chem.* **268**, 13791-13798.

Pastorino, J. G., Simbula, G., Yamamoto, K., Glascott, P.A. Jr., Rothman, R.J., and Farber, J.L. (1996). The cytotoxicity of tumor necrosis factor depends on induction of the mitochondrial permeability transition. *J. Biol. Chem.* **271**, 29792-29798.

Perani, A., Singh, R.P., Chauhan, R., and Al-Rubeai, M. (1998). Variable functions of bcl-2 in mediating bioreactor stress-induced apoptosis in hybridoma cells. *Cytotechnology* In Press.

Perreault, J., and Lemieux, R. (1993). Essential role of optimal protein synthesis in preventing the apoptotic death of cultured B cell hybridomas. *Cytotechnology* **13**, 99-105.

Pervaiz, S., Hirpara, J.L., and Clement, M.V. (1998). Caspase proteases mediate apoptosis induced by anticancer agent preactivated MC540 in human tumor cells. *Cancer Lett.* **128**, 11-22.

Petronini, P. G., Urbani, S., Alfieri, R., Borghetti, A.F., and Guidotti, G.G. (1996). Cell susceptibility to apoptosis by glutamine deprivation and rescue: survival and apoptotic death in cultured lymphoma-leukemia cell lines. *J. Cell Physiol.* **169**, 175-185.

Prior, C. P., Doyle, K.R., Duffy, S.A., Hope, J.A., Moellering, B.J., Prior, G.M., Scott, R.W., and Tolbert, W.R. (1989). The recovery of highly purified biopharmaceuticals from perfusion cell culture bioreactors. *J. Parent Sci. Tech.* **43**, 15-23.

Pugachev, K. V., and Frey, T.K. (1998). Rubella virus induces apoptosis in cultured cells. *Virology* **250**, 359-370.

Qi, X. M., He, H., Zhong, H., and Distelhorst, C.W. (1997). Baculovirus p35 and Z-VAD-fmk inhibit thapsigargin-induced apoptosis of breast cancer cells. *Oncogene* **15**, 1207-1212.

Qin, S., Minami, Y., Kurosaki, T., and Yamamura, H. (1997). Distinctive functions of Syk and Lyn in mediating osmotic stress- and ultraviolet C irradiation-induced apoptosis in chicken B cells. *J. Biol. Chem.* **272**, 17994-17999.

Rabizadeh, S., LaCount, D.J., Friesen, P.D., and Bredesen, D.E. (1993). Expression of the baculovirus *p35* gene inhibits mammalian neural cell death. *J. Neurochem.* **61**, 2318-2321.

Reddy, S., and Miller, W.M. (1994). Effects of abrupt and gradual osmotic stress on antibody production and content in hybridoma cells that differ in production kinetics. *Biotechnol. Prog.* **10**, 165-173.

Reed, J. C. (1994). Bcl-2 and the regulation of programmed cell death. *J. Cell Biol.* **124**, 1-6.

Rossé, T., Olivier, R., Monney, L., Rager, M., Conus, S., Felley, I., Jansen, B., and Borner, C. (1998). Bcl-2 prolongs cell survival after Bax-induced release of cytochrome c. *Nature* **391**, 496-499.

Sarin, A., Wu, M.L., and Henkart, P.A. (1996). Different interleukin-1 beta converting enzyme (ICE) family protease requirements for the apoptotic death of T lymphocytes triggered by diverse stimuli. *J. Exp. Med.* **184**, 2445-2550.

Scallan, M. F., Allsopp, T.E., and Fazakerley, J.K. (1997). *bcl-2* acts early to restrict Semliki Forest virus replication and delays virus-induced programmed cell death. *J. Virol.* **71**, 1583-1590.

Schafer, R., Karbach, K., and Hoppe, J. (1998). Multiple intracellular pathways interfere with the activation of a CPP32-like protease induced by serum deprivation of AKR-2B cells. *Exp. Cell Res.* **240**, 28-39.

Sheshagiri, S., and Miller, L.K. (1997). Baculovirus inhibitors of apoptosis (IAPs) block activation of Sf-caspase-1. *Proc. Natl. Acad. Sci. USA* **94**, 13606-13611.

Shimizu, S., Eguchi, Y., Kosaka, H., Kamiike, W., Matsuda, H., and Tsujimoto, Y. (1995). Prevention of hypoxia-induced cell death by Bcl-2 and Bcl-x_L. *Nature* **374**, 811-813.

Simpson, N., Milner, A.N., and Al-Rubeai, M. (1997). Preventaion of hybridoma cell death by *bcl-2* during sub-optimal culture conditions. *Biotechnol. Bioeng.* **54**, 1-16.

Simpson, N. H., Singh, R.P., Perani, A., Goldenzon, C., and Al-Rubeai, M. (1998). In hybridoma cultures, deprivation of any single amino acid leads to apoptotic death, which is suppressed by the expression of the *bcl-2* gene. *Biotechnol. Bioeng.* **59**, 90-98.

Singh, R. P., Al-Rubeai, M., Gregory, C.D., and Emery, A.N. (1994). Cell death in bioreactors: A role for apoptosis. *Biotechnol. Bioeng.* **44**, 720-726.

Singh, R. P., Emery, A.N., and Al-Rubeai, M. (1996). Enhancement of survivability of mammalian cells by overexpression of the apoptosis-suppressor gene *bcl-2*. *Biotechnol. Bioeng.* **52**, 166-175.

Singh, R. P., Finka, G., Emery, A.N., and Al-Rubeai, M. (1997). Apoptosis and its control in cell culture systems. *Cytotechnology* **23**, 87-93.

Slater, A. F. G., Kimland, M., Jiang, S.A., and Orrenius, S. (1995). Constitutive nuclear NF-kappa-B/rel DNA-binding activity of rat thymocytes is increased by stimuli that promote apoptosis, but not inhibited by pyrrolidine dithiocarbamate. *Biochem. J.* **312**, 833-838.

Strasser, A., and Anderson, R.L. (1995). Bcl-2 and thermotolerance cooperate in cell survival. *Cell Growth Differ.* **6**, 799-805.

Sugimoto, A., Friesen, P.D., and Rothman, J.H. (1994). Baculovirus *p35* prevents developmentally programmed cell death and rescues a *ced-9* mutant in the nematode *Caenorhabditis elegans*. *EMBO J.* **13**, 2023-2028.

Susin, S. A., Zamzami, N., Castedo, M., Hirsch, T., Marchetti, P., Macho, A., Daugas, E., Geuskens, M., and Kroemer, G. (1996). Bcl-2 inhibits the mitochondrial release of an apoptogenic protease. *J. Exp. Med.* **184**, 1331-1341.

Suzuki, E., Terada, S., Ueda, H., Fujita, T., Komatsu, T., Takayama, S., and Reed, J.C. (1997). Establishing apoptosis resistant cell lines for improving protein productivity of cell culture. *Cytotechnology* **23**, 55-59.

Takayama, S., Cazals-Hatem, D.L., Kitada, S., Tanaka, S., Miyashita, T., and Hovey, L.R. (1994). Evolutionary conservation of function among mammalian, avian, and viral homologs of the Bcl-2 oncoprotein. *DNA Cell Biol.* **13**, 679-692.

Talley, A. K., Dewhurst, S., Perry, S.W., Dollard, S.C., Gummuluru, S., Fine, S.M., New, D., Epstein, L.G., Gendelman, H.E., and Gelbard, H.A. (1995). Tumor necrosis factor alpha-induced apoptosis in human neuronal cells: protection by the antioxidant N-Acetylcysteine and the genes *bcl-2* and *crmA*. *Mol. Cell. Biol.* **15**, 2359-2366.

Tarodi, B., Subramanian, T., and Chinnadurai, G. (1994). Epstein-Barr virus BHRF1 protein protects against cell death induced by DNA-damaging agents and heterologous viral infection. *Virology* **201**, 404-407.

Terada, S., Fukuoka, K., Fujita, T., Komatsu, T., Takayama, S., Reed, J.C., and Suzuki, E. (1997). Anti-apoptotic genes, *bag-1* and *bcl-2*, enabled hybridoma cells to survive under treatment for arresting cell cycle. *Cytotechnology* **25**, 17-23.

Tewari, M., and Dixit, V.M. (1995). Fas- and tumor necrosis factor-induced apoptosis is inhibited by the poxvirus crmA gene product. *J. Biol. Chem.* **270**, 3255-3260.

Tewari, M., Quan, L.T., O'Rourke, K., Desnoyers, S., Zeng, Z., Beidler, D.R., Poirier, G.G., Slavesen., G.S., and Dixit, V.M. (1995). Yama/CPP32 beta, a mammalian homolog of CED-3, is a Crm-A-inhibitable protease that cleaves the death substrate poly(ADP-ribose) polyermase. *Cell* **81**, 801-809.

Thomas, J. N. (1990). Mammalian cell physiology, A. Lubiniecki (ed.), *Bioprocess Technology*, Vol. 10, Marcel Dekker, New York. pp. 93-145.

Thornberry, N. A., and Lazebnik, Y. (1998). Caspases: enemies within. *Science* **281**, 1312-1316.

Tsai, J. C., Jain, M., Hsieh, J.M., Lee, W.S., Yoshizumi, M., Patterson, C., Perrella, M.A.,Cooke, C., Wand, H., Haber, E., Schlegel, R., and Lee, M.E. (1996). Induction of apoptosis by pyrrolidinedithiocarbamate and N-acetylcysteine in vascular smooth muscle cells. *J. Biol. Chem.* **271**, 3667-3670.

Tsujimoto, Y., Finger, L.R., Yunis, J., Nowell, P.C., and Croce, C.M. (1984). Cloning of the chromosome breakpoint of neoplastic B cells with the t(14:18) chromosome translocation. *Science* **226**, 1097-1099.

Vander Heiden, M. G., Chandel, N.S., Williamson, E.K., Schumacker, P.T., and Thompson, C.B. (1997). Bcl-x_L regulates the membrane potential and volume homeostasis of mitochondria. *Cell* **91**, 627-637.
Vaux, D. L., and Weissman, I.L. (1988). *Bcl-2* gene promotes haemopoietic cell survival and cooperates with *c-myc* to immortalize pre-B cells. *Nature* **335**, 440-442.
Vomastek, V., and Franěk, F. (1993). Kinetics of development of spontaneous apoptosis in B cell hybridoma cultures. *Immunol. Lett.* **35**, 19-24.
Wang, L., Miura, M., Bergeron, L., Zhu, H., and Yuan, J. (1994). *Ich-1* and *ICE/ced-3* related gene, encodes both positive and negative regulators of programmed cell death. *Cell* **78**, 739-750.
Waring, P., and Beaver, J. (1996). Cyclosporin A rescues thymocytes from apoptosis induced by very low concentrations of thapsigargin: effects on mitochondrial function. *Exp. Cell Res.* **227**, 264-276.
White, E., Sabbatini, P., Debbas, M, Wold, W.S.M., Kusher, D.I., and Gooding, L.R. (1992). The 19-kilodalton Adenovirus E1B transforming protein inhibits programmed cell death and prevents cytolysis by tumor necrosis factor α. *Mol. Cell. Biol.* **12**, 2570-2580.
Wolfe, J. T., Ross, D., and Cohen, G.M. (1994). A role for metals and free radicals in the induction of apoptosis in thymocytes. *FEBS Lett.* **352**, 58-62.
Wong, G. H. W., Elwell, J.H., Oberley, L.W., and Goeddel, D.V. (1989). Manganous superoxide dismutase is essential for cellular resistance to cytotoxicity of tumor necrosis factor. *Cell* **58**, 923-931.
Yang, J., Liu, X., Bhalla, K., Kim, C.N., Ibrado, A.M., Cai, J., Peng, T-I., Jones, D.P., and Wang, X. (1997). Prevention of apoptosis by Bcl-2: release of cytochrome c from mitochondria blocked. *Science* **275**, 1129-1132.
Yuan, J., Shaham, S., Ledoux, S., Ellis, H.M., and Horvitz, H.R. (1993). The *C. elegans* cell death gene ced-3 encodes a protein similar to mammalian interleukin-1β-converting enzyme. *Cell* **75**, 641-652.
Zamzami, N., Marchetti, P., Castedo, M., Hirsch, T., Susin, S.A., Masse, B., and Kroemer, G. (1996a). Inhibitors of permeability transition interfere with the disruption of the mitochondrial transmembrane potential during apoptosis. *FEBS Lett.* **384**, 53-57.
Zamzami, N., Susin, S.A., Marchetti, P., Hirsch, T., Castedo, M., and Kroemer, G. (1996b). Mitochondrial control of nuclear apoptosis. *J. Exp. Med.* **183**, 1533-1544.
Zamzami, N., Hirsch, T., Dallaporta, B., Petit, P.X., and Kroemer, G. (1997). Mitochondrial implications in accidental and programmed cell death: apoptosis and necrosis. *J. Bioeng. Biomembr.* **29**, 185-193.
Zhivotovsky, B., Burgess, D.H., and Orrenius, S. (1996). Proteases in apoptosis. *Experimentia* **52**, 968-978.
Zhivotovsky, B., Orrenius, S., Brustugun, O.T., and Døskeland, S.O. (1998). Injected cytochrome c induces apoptosis. *Nature* **391**, 449-450.
Zhou, Q., Snipas, S., Orth, K., Muzio, M., Dixit, V.M., and Salvesen, G.S. (1997). Target protease specificty of the viral serpin crmA. *J. Biol. Chem.* **272**, 7797-7800.

CONTROL OF MAMMALIAN CELL PROLIFERATION AS AN IMPORTANT STRATEGY IN CELL CULTURE TECHNOLOGY, CANCER THERAPY AND TISSUE ENGINEERING

M. FUSSENEGGER* AND J. E. BAILEY

Institute of Biotechnology, Swiss Federal Institute of Technology, ETH Zurich, CH-8093 Zurich, Switzerland
** Corresponding author, Tel.: +41 1 633 34 48, Fax: + 41 1 633 10 51, e-mail: fussenegger@ubaclu.unibas.ch, or fussenegger@biotech.biol.ethz.ch*

1. Introduction

In the last decades research on the molecular basis of proliferation and cell-cycle control was mainly motivated by curiosity to understand this complex regulatory network which orchestrates the global changes of cycling cells in a temporal and spatial manner. Based on such mechanistic research, cell-cycle regulatory circuits came to be perceived as a fundamental program which stands at the center of multicellular life, since the cell-cycle machinery integrates several important decisions on a cell's fate such as apoptosis, differentiation and growth. Consequently, cell-cycle control systems are important for cell homeostasis and, when defective, can lead to uncontrolled proliferation and neoplastic growth. Cell-cycle regulatory networks have certainly become a focus in cancer therapy for control of carcinogenic growth and the elimination of neoplastic cells. The bioengineering community remained relatively unaffected by theses key developments in the field of proliferation control and continued to use indefinitely growing, neoplastic cells as their workhorse for biopharmaceutical manufacturing. Only recently, as a more refined picture of cell-cycle regulatory systems emerged from basic research, long-standing critical issues in cell culture technology such as apoptosis, protein production, cell differentiation and growth in serum- and protein-free media have been addressed by metabolic engineering and process development strategies using mammalian cells.

Proliferation management of mammalian cells remains a rapidly developing field which has the potential to cope with current and future challenges of human disease as human therapy moves towards cell-based and genetic solutions in the next millennium. This article summarizes current bioprocess-related aspects of proliferation control in mammalian cells and highlights future applications of controlled proliferation technology in gene and cell therapy.

2. Proliferation control by chemical cell culture additives and medium formulations

Although the development of indefinitely proliferating mammalian cells has enabled cell culture technology for large-scale production of heterologous protein therapeutics, uncontrolled growth constrains batch cultivations to only a short life span. After expansion of the production cell line to the desired cell density, the cell population grows past the optimal production phase and enters a decline phase in which suboptimal nutrient and metabolite conditions result in cell death and rapid termination of the production process.

In most batch processes where heterologous protein production is not growth associated, maximum productivity is limited to a short period in which growth diminishes at high cell density and cell viability. Therefore, in order to prolong the maximum production phase, the design of static cell cultures which maintain high cell viability and cell densities is a current priority in contemporary biopharmaceutical manufacturing and process development. Developments of cytostatic cell culture conditions were further stimulated by pioneering experimental as well as modelling studies by Suzuki and Ollis (1989 and 1990) which demonstrate an inverse correlation between hybridoma cell growth rate and specific productivity. Limiting cell growth is therefore expected to have a positive impact on the production characteristics of bioprocesses for two distinct reasons: first, higher overall product yield due to extension of the production phase, and, second, higher specific productivity of the cells.

Initial methods to control proliferation were based on chemical cell culture additives such as thymidine, hydroxyurea and transforming growth factor β (TGFβ) (Suzuki and Ollis, 1990; Al-Rubeai *et al.*, 1992). All of these substances arrest cells in the G1-phase of the cell cycle and achieve higher specific productivities but suffer problems of cytotoxicity which prevent extended cultivation in a proliferation-arrested state (Marcus *et al.*, 1985; Al-Rubeai *et al.*, 1992). Decreased cell viability was also encountered when production cultures were deprived of serum or other essential medium components such as isoleucine. Nutrient deprivation certainly controls proliferation but most cell lines activate their intrinsic cell death program and die by apoptosis (Mercille and Massie, 1994; Singh *et al.*, 1994; Singh *et al.*, 1998).

Discouraged by the negative effects on cell viability of theses initial attempts to control proliferation, most bioengineers focused on strategies to prevent cell death in cultures to prolong production periods rather than refining strategies to inhibit cell growth by cell culture additives (Fussenegger *et al*, 1998a; Fussenegger *et al.*, 1998b; Al-Rubeai *et al.*, 1998; Dickson, 1998). Also, it has been a major concern that implementation of negative proliferation control on the neoplastic cells usually employed for bioprocesses diverts them into apoptosis (Walker *et al.*, 1997). As contemporary basic research reveals increasing molecular details on cellular proliferation control mechanisms, two main pathways are beginning to emerge: A first one in which cells arrest in response to stress and damage to allow repair and recovery, and a second pathway which initiates or maintains cellular differentiation (Figure 1). While induction of apoptosis is a major strategy for the first pathway (to eliminate cells which suffered

G1-Specific Negative Proliferation Control Pathways

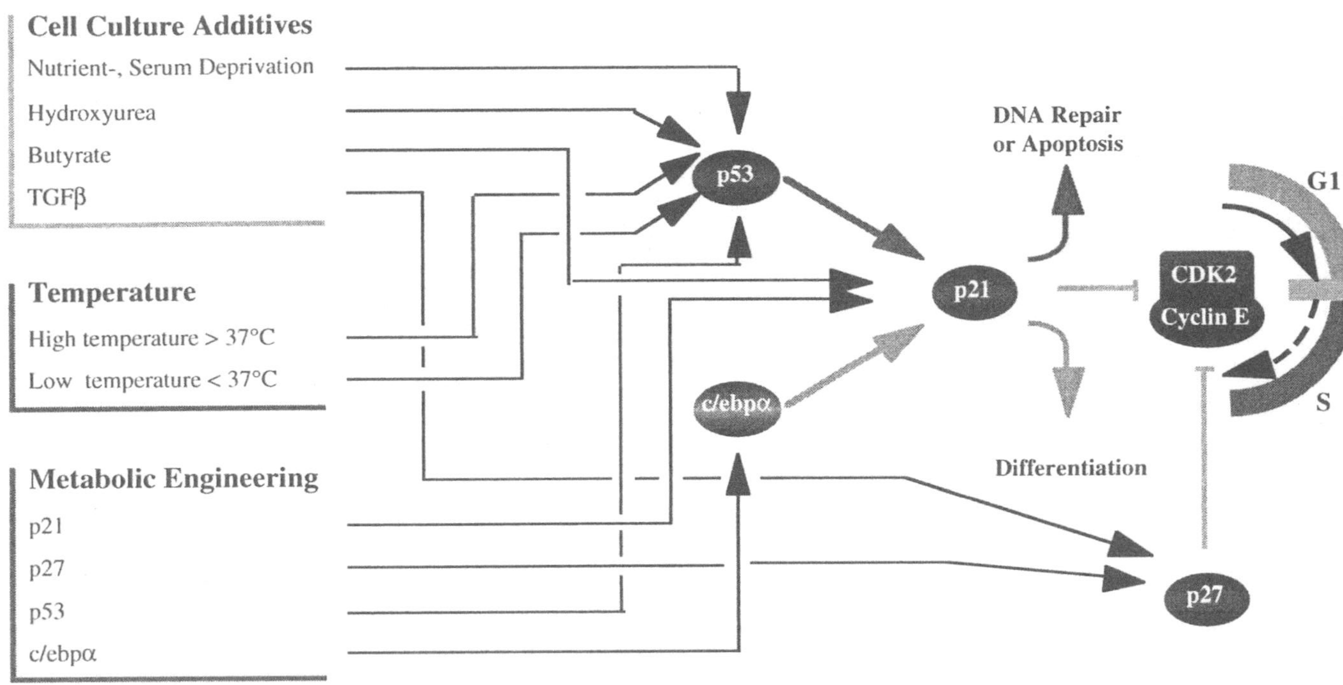

Figure 1. Molecular pathways of negative proliferation control in mammalian cells upon induced by cell culture additives, temperature and metabolic engineering.

irreparable damage), differentiated cells are largely protected from programmed cell death. Therefore, there certainly exist chemical compounds which achieve cell-cycle arrest by induction of differentiation pathways rather than by inducing apoptosis. Proliferation control has always been and still is a major objective of cancer therapy research which reveals an ever growing library of chemical compounds able to suppress growth (Table 1). Certainly, induction of apoptosis concomitant with cell-cycle arrest is perceived as an ideal situation in cancer therapy, but some recent chemotherapeutics rely entirely on growth suppression. It is to be expected that chemical agent-based proliferation technology will see a revival at the intersection of cancer therapy and bioprocess engineering.

The novel nucleotide analogue PMEG (9-(2-phosphonomethoxyethyl)guanine) has recently been reported as a powerful antiproliferative substance when acting on hybridoma cells. As expected for a nucleotide analogue, PMEG interferes with DNA replication and prevents S-phase progression (Franek *et al.*, 1999). Despite its antiproliferative potential PMEG does not induce apoptosis at cell-cycle-inhibitory concentrations. The product yield of the overall production process increases significantly in PMEG-arrested hybridoma cultivations (Franek *et al.*, 1999; Table 1). However, the specific antibody productivity increases only marginally. Several lines of evidence suggest that higher specific productivities of proliferation-inhibited cell populations are exclusively and intimately linked to the G1-phase of the cell cycle (Suzuki and Ollis, 1989 and 1990; Al-Rubeai *et al.*, 1992; Moore *et al.*, 1997; Fussenegger *et al.*, 1997a and 1998c; Mazur *et al.*, 1998; Fussenegger *et al.*, 1999b, see below). Despite its anti-proliferative potential PMEG still awaits studies on its application in cancer therapy.

A paradigm of merged efforts in cancer therapy and cell culture process development is the use of butyrate both to inhibit carcinogenic growth and to enhance bioprocess performance (Archer *et al.*, 1998; Heerdt *et al.*, 1997; Nakano *et al.*, 1997). Numerous epidemiological and experimental studies have identified an association between a high-fiber diet and a decreased incidence of growth of colon cancer. It appears that the protective compound is butyrate generated by bacterial fermentation of the colon flora at concentrations which inhibit growth of colonic carcinomas *in vitro* (Archer *et al.*, 1998; Heerdt *et al.*, 1997). Butyrate has long been known to induce a variety of changes within the nucleus, including hyperacetylation and DNA methylation (Riggs *et al.*, 1977). Core histones contain a high amount of positively charged amino acids (lysine and arginine) which enables their interaction with the negatively charged DNA. In transcriptionally active chromatin, histones appear to be unusually highly acetylated on lysines on the ε-amino groups. The acetyl funcional groups whose addition removes a plus charge and weakens the histone-DNA interaction are constantly added to these histones by the enzyme histone acetylase and removed by histone deacetylase (Grunstein, 1997). Butyrate achieves hyperacetylation by reversible inhibition of the deacetylase, a characteristic which butyrate shares with other hyperacetylating agents such as trichostatin A (TSA) and trapoxin (Kijima *et al*, 1993) (Table 1). Addition of butyrate to the cell culture medium activates chromatin and increases transcription which leads to higher yield in heterologous protein production (Grunstein, 1997). For this reason the addition of butyrate to production cultures has become a widely

TABLE 1. Chemical cell culture additives inducing cell-cycle arrest

Substance	Cell Line	Arrest	Function	Reference
AHPN	Brest/Lung Cancer	G1	Induction of c-Jun, nur77, p21; induces Apoptosis	Li et al., 1998
Aphidicolin	HeLa	G1/G2	nd	Tanaka et al., 1998
Apigenin	Fibroblast	G1	Induction of p21	Lepley and Pelling, 1997
Benzo(a)pyrene	Fibroblast	G1	p53-independent	Vaziri and Faller, 1997
Brefeldin A	Prostate Cancer	G1	Downregulation of CDK2/4, Cyclin D1	Mordente et al., 1998
Butyrate	NIH3T3	G1	Induction of p21	Viziri et al., 1998
Calcium blockers	HUAEC	G1	nd	Zeitler et al., 1997
Concanavalin A	Lymphoma	G1	inhibits cyclin E-Cdk2 and cyclin D-Cdk4	Desrivieres et al., 1997
Curcumin	Muscle cells	G1/G2/S/M	Repression of c-myc, PTK, induces Apoptosis	Chen and Huang, 1998
CVT-313	Several	G1	Inhibitor of CDK2	Brooks et al., 1997
Deferoxamine	Hepatocytes	G1	nd	Chenoufi et al., 1998
Differanisole A	Leukemia cells	G1	nd	Kanatani et al., 1997
EGF blocker	DiFi	G1	Accumulation of p27	Moyer et al., 1997
Estradiol	Breast Cancer	G1	Induction of p21	Wang et al., 1997
GGTI-298	Cancer Cells	G1	Induction of p21	Vogt et al., 1997
Ginsenoside RH2	BALB7c 3T3	G1	Supression of CDK2 Activity	Ota et al., 1997
Gnidimacrin	K562	G1	Activation of Proteinkinase C	Yoshida et al., 1998
Hydroxurea	Hybridoma	G1	nd	Suzuki and Ollis, 1990
IFN-α	Macrophages	G1	Induction of p19 and p21	Matsuoka et al., 1998
L-asparaginase	Leukemia cells	G1	nd; induces Apoptosis	Ueno et al., 1997
Methylseleno-cysteine	Epithelial cells	G1	Inhibition of CDK2	Sinha and Medina, 1997
N-acetylcysteine	NIH3T3	G1	Downregulation of Cyclin D1	Sekharam et al., 1998
NGF	PC12	G1	Induction of p21	Yan and Ziff, 1997
Nitric oxide	Pancreas Cells	G1	Induction of p21	Gansauge et al., 1998
Onconase	U937 lymphoma	G1	Downregulation of CyclinD3; Induction of p16, p21, p27	Ardelt et al., 1998
p21 peptides	Several	G1	inhibits CDK4	Ball et al., 1997
Phorbol ester	Lymphoma	G1	inhibits CyclinE-CDK2	Desrivieres et al., 1997
PMEG	Hybridomas	S/G2	DNA replication	Franek et al., 1999
Prostaglandin A1	A172 glioma	G1	Induction of p21	Tanikawa et al., 1998
Rapamycin	NIH3T3	G1	Degradation of Cyclin D1	Hashemolhosseini et al., 1998
Retinoic acid	neuroblastoma	G1	Induction of p27	Matsuo and Thiele, 1998
Roscovitine	Oocytes	G1/G2	Inhibition of CDK1, CDK2, CDK5	Meijer et al., 1997
Silymarin	Breast Carcinoma	G1	Induction of p21	Zi et al., 1998
Staurosporine	Breast Carcinoma	G1	Induction of p27 and p18	Kwon et al., 1996
Suramin	Neuroblastoma	G1	Inhibition of CDK3	Sullivan et al., 1997
Ribozyme	Glioma Cells	ND	Proteinkinase Cα inhibition	Sioud and Sorensen, 1998
Taxomifen	Prostate Cancer	G1	Induction of p21, Inhibition of Proteinkinase C	Rohlff et al., 1998
TGF-β	Mv1Lu	G1	Induction of p27	Reynisdottir et al., 1995
Trapoxin	NIH3T3	G1	inhibits Histone Deacetylase	Kijima et al., 1993
Vitamine D3 analogues	Breast Cancer	G1	Induction of p27	Koike et al., 1997
Vitamine E	Breast Cancer	G1	Enhances Cyclin A binding to E2F	Turley et al., 1997

standardized practice in biopharmaceutical manufacturing (Chevalot *et al.*, 1995; Lamotte *et al.*, 1998; Oster *et al.*, 1993). However, butyrate also induces G1-phase-specific growth arrest which makes it particularly attractive for cancer therapy (Archer *et al.*, 1998; Heerdt *et al.*, 1997; Nagano *et al.*, 1997; Kijima *et al.*, 1993). Recently, the molecular mechanism of butyrate-mediated growth arrest has been clarified (Nakano *et al.*, 1997; Archer *et al.*, 1998). Butyrate activates the p21 promoter through Sp1 sites (Nakano *et al.*, 1997). Possibly, these Sp1 sites are masked by hypoacetylated histones. Similar induction of p21 is achieved by the other deacetylase inhibitor TSA (Kijima *et al*, 1993). Overexpression of p21 also explains the induction of differentiation which is has recurrently been observed in butyrate-containing cell cultures (Archer *et al.*, 1998). In the light of these new results, increased heterologous protein yields achieved by butyrate addition may not entirely result from chromatin activation but could also be a consequence of p21-mediated G1-phase-specific proliferation control as is suggested by several lines of evidence emerging from metabolic engineering strategies and low temperature cultivations (Fussenegger *et al.*, 1998c; Ohnishi *et al.*, 1998; see below). p21 is a key regulator of the mammalian cell cycle which binds to and inhibits cyclin E-CDK2 kinases, thereby preventing S-phase entry of cells (Fussenegger and Bailey, 1998; Fussenegger *et al.*, 1998b; Fussenegger *et al.*, 1998c; Figure 1). Given the importance of inhibitory interactions between p21 and cyclin E-CDK2 complexes for proliferation control in mammalian cells, several small molecules or p21-derived peptides have been designed which mimic p21 and block the cell cycle by binding to the active center of the cyclin E-CDK2 kinase (Meijer, 1996; Ball *et al.*, 1997; Meijer and Kim, 1997; Table 1).

The potential of culture additives for use in proliferation control of mammalian cell cultures is far from being exhausted. On the contrary, better molecular understanding of central cell-cycle control circuits are expected to stimulate screening for new chemical compounds which, unlike first generation efforts, do not effect growth arrest by induction of global stress situations but exploit cellular pathways evolved to induce and maintain terminal differentiation (Figure 1).

3. Temperature-mediated proliferation control

Adaptation to environmental temperature changes is a constant concern of living organisms, and sophisticated molecular strategies have evolved to extend the tolerable temperature range and allow survival under extreme thermic conditions. In general, extreme temperatures are perceived as a stress situation which is counteracted by growth retardation or proliferation arrest, with concomitant expression of a set of stress proteins which stabilize vital cellular functions: heat shock proteins for elevated temperatures, and cold-shock proteins for low temperatures

Mammalian cells have a relatively narrow permissive temperature range of 36-39°C with an optimum at 37°C (Sisken *et al.*, 1965). It is this optimal temperature of 37°C at which most of today's bioprocesses for the production of protein therapeutics using mammalian cell lines are conducted. Although long accepted as a constant physical parameter, the bioengineering community only recently started to consider alternative temperatures as a low-cost and easy-to-control process variable and to discover the potential of high- or low-temperature cultivations for the improvement of various aspects of mammalian cell culture performance.

3.1. HIGH TEMPERATURE AND HEAT-SHOCK CULTIVATIONS

The response to elevated temperatures has been extensively studied and includes expression of heat-shock proteins, a group of proteins which is highly conserved among all organisms, many of which act as molecular chaperones for protein folding (Craig *et al.*, 1985; Hightower and Hendershot, 1997; Morimoto and Santoro, 1998). One of the most extensively studied heat shock proteins is hsp70, the promoter of which is heat-inducible and remains temperature-regulatable even in heterologous cell types (Voellmy *et al.*, 1985). This finding has stimulated development of heat-inducible mammalian expression systems. However, since the protein yields achieved with such systems are usually relatively low, this technology has never been applied for biotechnological production processes (Pallavicini *et al.*, 1990; Bovenberg *et al.*, 1990; Jenkins and Hovey, 1993; Hovey *et al.*, 1994). In general, physiological studies with mammalian cells grown or heat-shocked at temperatures between 37°C and 43°C did not encourage the use of high-temperature cultivations for industrial applications, since hyperthermia results in growth arrest accompanied by increased glucose consumption, lower specific productivity and product yields as well as in decreased cell viability (Bloemkolk *et al.*, 1992; Jenkins and Hovey, 1993; Hovey *et al.*, 1994; Furukawa and Ohsuye, 1998). Only one study reports increased protein production of heat-shocked mouse hybridomas without affecting the viability of the cells (Passini and Gouchee, 1989).

Despite all of these negative effects of hyperthermic conditions on mammalian cell culture, growth arrest induced under high temperatures attracted much interest since recent cell culture and modelling studies have demonstrated an inverse correlation between cellular growth rate and heterologous protein production and have concluded that exploration of methods to arrest cells while retaining high viability is likely to improve specific productivity (Miller *et al.*, 1988; Suzuki and Ollis, 1989; Suzuki and Ollis, 1990; Al-Rubeai *et al.*, 1992; Coco-Martin *et al.*, 1992). In principle, elevated temperatures induce G1-phase-specific growth arrest; this cell-cycle phase has been shown to be compatible with prolonged cell survival at non-physiological temperatures (Marcus *et al.*, 1985; Bloemkolk *et al.*, 1992; Roti Roti *et al.*, 1992; Jenkins and Hovey, 1993). Although higher temperatures have the potential to induce the desired G1-phase arrest, such conditions are lethal for mammalian cells (Jenkins and Hovey, 1993; Furukawa and Ohsuye, 1998). The G1-arrest as well as the decrease in cell viability are likely the result of stress-induced, p53-dependent accumulation of the cell-cycle inhibitor p21 (Nitta *et al.*, 1997). Therefore, temperature-sensitive mutants were generated which grow like wildtype at 34°C and become G1-phase-arrested at non-permissive temperatures (39°C) following only a brief heat shock (Marcus *et al.*, 1985; Mizuno *et al.*, 1989; Chang *et al.*, 1990; Jenkins and Hovey, 1993; Hovey *et al.*, 1994). In order to maintain the temperature-sensitive cells in an arrested state while retaining high viability, optimized and complex heat-shock regimes using alternating incubations at 34°C and 39°C have been developed, very similar to estrogen regimes elaborated for IRF-1-induced proliferation arrest (see below; Jenkins and Hovey, 1993; Hovey *et al.*, 1994; Müller *et al.*, 1998; Carvalhal *et al.*, 1998). Using this temperature regime technology Jenkins and Hovey (1993; Hovey *et al.*, 1994) achieved 3- to 4-fold increases in specific protein productivity over a process period of 72-130 h using their temperature-sensitive CHO cell-cycle mutants. However, the complex temperature regime required to maintain high cell viability discourages the use of this proliferation control strategy for large-scale industrial applications.

3.2. LOW TEMPERATURE CULTIVATIONS

In contrast to heat-shock responses, the molecular mechanisms governing cellular responses to low temperatures are much less understood. Certainly, low temperatures induce cold-shock proteins (CSP), but, unlike heat-shock proteins, CSPs share no apparent structural or functional characteristics (Thieringer *et al.*, 1998). Apart from the well-known phenomenon of adaptive hypothermia (hibernations) in which mammals lower their body temperatures, even the cells of nonhibernating mammals, which maintain high uniform body temperatures and have a poor tolerance for hypothermia, have a capacity for a wider cold tolerance (Willis, 1987). As a unifying characteristic response of mammals as well as other organisms to low temperatures, cold stress changes the lipid composition, decreases the fluidity of cellular membranes and suppresses the rate of protein synthesis and cell growth, all of which are reasons why hypothermia is now employed in clinics for brain and heart surgeries as well as for the preservation of donor organs (Fuller *et al.*, 1997). The inverse correlation between membrane stability and cultivation temperature represented an attractive strategy to balance shear stress typically encountered during normal bioreactor operation (Peterson *et al.*, 1990; Ludwig *et al.*, 1992; Al-Rubeai *et al.*, 1995). However, one of the most decisive advantages of low temperature bioprocesses resides in the retardation or suppression of cell growth which increases viability and prolongs the cultivation time by delaying the onset of apoptosis, a major cause for premature batch process termination (Franek *et al.*, 1992; Singh *et al.*, 1994; Moore *et al.*, 1995). Therefore, reduction in operating temperature has become a standard practice for commercial production of protein therapeutics and has been applied to various cell lines and cell types (Table 1). However, divergent effects of low temperature cultivations on productivity have been reported for different cell lines (Giard *et al.*, 1982; Sureshkumar and Mutharasan, 1991; Bloemkolk *et al.*, 1992; Reuveny *et al.*, 1993; Weidemann *et al.*, 1994). Hybridoma cell lines exposed to culture temperatures usually become growth-inhibited, maintain high cell viability for a longer period and show suppressed glucose consumption, but cellular productivity of the monoclonal antibody is not enhanced (Bloemkolk *et al.*, 1992; Borth *et al.*, 1992; Reuveny *et al.*, 1986; Sureshkumar and Mutharasan, 1991). In contrast, the yield of interferon induced by certain reagents or viruses could be increased at low temperatures in non-transfected rabbit or human fibroblast cells (Giard and Fleischaker, 1980; Giard *et al.*, 1982;

Kojima and Yoshida, 1974). Furthermore, Takagi and Ueda (1994) reported increased tPA production in human embryo lung cells at low temperatures.

Recent studies on low temperature cultivations also include the biotechnologically relevant CHO and BHK cell lines which behave differently with respect to specific productivity and product yield. Whereas CHO cell lines cultured below 37°C show increased heterologous protein production and higher product yields (Furukawa and Ohsuye, 1998; Kaufmann *et al.*, 1999), the productivity of BHK cells seems not to be affected by temperature reduction (Weidemann *et al.*, 1994). In contrast to CHO cells, BHK show no prominent growth arrest and reduced maximum cell density in response to low temperatures (Weidemann *et al.*, 1994).

Stimulated by several reports of a positive correlation of cellular productivity and G1-phase arrest (Suzuki and Ollis, 1990; Al-Rubeai *et al.*, 1992), research was carried forward especially in CHO cells to understand whether increased productivity at low temperatures correlates with growth inhibition at the G1-phase of the cell cycle. Recently two groups reported a stable G1-phase arrest following cultivation of CHO cells at 30°C (Moore *et al.*, 1997; Kaufmann *et al.*, 1999). Sustained G1-arrest is established within 2-3 days after temperature downshift and covers 80-90% percent of the population, reaching nearly 100% after cultivation at 30°C for several days. Apparently, there exists no molecular exit for cells to escape low temperature-induced growth arrest, since no mutant clones could be observed which resume growth even after extended periods of hypothermia (Kaufmann *et al.*, 1999). Despite this assessment of the cell-cycle dependence of low-temperature cultivation, little is known about the molecular mechanisms governing this response. Recent proteome and phosphotyrosine analyses indicate that cold-shock or cold-tolerance rely on a genetically determined program, adaptation of which to low temperatures requires coordinated regulation of responsive genes as well as cellular signalling cascades (Kaufmann *et al.*, 1999). However, the molecular aspects of these various changes in physiological functions required for low-temperature adaptation are just beginning to be understood (Coles *et al.*, 1996; Denarier *et al.*, 1998).

One molecular key player which connects cold-adaptation with cell-cycle arrest and proliferation control has recently been described as a cold-inducible RNA-binding protein (CIRP; Nishiyama *et al.*, 1997a and b). The Cirp protein consists of an amino-terminal RNA-binding domain and a carboxyl-terminal glycine-rich domain and exhibits overall structural similarity to a class of stress-induced RNA-binding proteins which are found in almost all living organisms from bacteria to plants and humans and which are involved in posttranscriptional regulation of gene expression such as polyadenylation, mRNA stability, splicing and protein export (Burd and Dreyfuss, 1994). CIRP is induced upon a temperature downshift to 32°C in mouse fibroblasts and localizes in the nucleoplasm. Interestingly, overexpression of heterologous Cirp induces impaired growth and results in a prolongation of the G1-phase even at a culture temperature of 37°C (Nishiyama *et al.*, 1997a; Table 2).

While the molecular link of Cirp expression to G1-arrest remains to be investigated, a recent report by Ohnishi *et al.* (1998) shows that cold-shock stress impinges on the central cell-cycle regulatory circuit via p53-dependent induction of the cell-cycle inhibitor p21. p53 exerts its biological functions by inducing downstream mediator

TABLE 2: Temperature-based proliferation control
G: Growth; P: Productivity; V: Viability; C: Medium consumption; I: Release of impurities

Cell type	Product	Temp.	G	P	V	C	I	Reference
BHK	Antithrombin III	< 37°C	-	=	nd	-	nd	Weidemann et al., 1994
CHO	-	30-37°	-	nd	nd	nd	-	Moore et al., 1997
CHO/Mouse L	IGF-I	> 37°C	nd	+	nd	nd	nd	Bovenberg et al., 1990
CHO	tPA	30°C	+	nd	+	-	-	Moore et al., 1997
CHO	c-myc	> 37°C	nd	+	nd	nd	nd	Pallavicini et al., 1990
CHO	TIMP	42°C	-	+	nd	nd	nd	Hovey et al., 1994
CHO	TIMP	34-39°C	-	+	nd	nd	nd	Jenkins and Hovey, 1993
CHO	α-amidase	30-37°C	+	+	+	-	-	Furukawa and Ohsuye, 1998
CHO	α-amidase	37- 43°C	+	-	-	+	nd	Furukawa and Ohsuye, 1998
CHO	SEAP	25-30°C	nd	+	+	nd	nd	Kaufmann et al., 1999
Human Embryonic Lung Cells	tPA	< 37°C	nd	+	nd	nd	nd	Takaki and Ueda, 1994
Human Fibroblast	Interferon	< 37°C	-	+	+	-	nd	Girard and Fleischaker, 1980; Girard et al., 1982
Human Skin	Interferon	32°C	nd	+	nd	nd	nd	Vilcek and Havell, 1973
HeLa	–	43°C	nd	nd	nd	nd	nd	Nitta et al., 1997
Hybridoma	Antibody	33-37°C	=	-	=	-	nd	Barnabé and Butler, 1994
Hybridoma	Antibody	37-39°C	-	+	-	+	nd	Barnabé and Butler, 1994
Hybridoma	Antibody	33-39°C	-	-	+	-	nd	Bloemkolk et al., 1992
Hybridoma	Antibody	< 37°C	-	=	+	-	nd	Borth et al., 1992
Hybridoma	Antibody	34-37°C	-	-	+	-	nd	Reuveny et al., 1986
Hybridoma	Antibody	33-37°C	-	=	+	-	nd	Sureshkumar and Mutharasan, 1991
Hybridoma	Antibody	42°C	nd	+	=	nd	nd	Passini and Gouchee, 1989
Rabbit Cells	Interferon	25°C	-	+	+	-	nd	Kojima and Yoshida, 1994

genes such as p21, *bax* and *gadd45* or through interactions with other regulatory proteins, all of which are involved in cell-cycle control, apoptosis or DNA repair (Ko and Prives, 1997). Thus, p53 represents a central integrator of cellular stress response signals.

Human glioblastoma cell lines were shown to accumulate several-fold more p53 and p21 following a 60 min. cold-shock at 4°C than respective control cell lines constantly grown at 37°C (Ohnishi *et al.*, 1998). Thereby, as an immediate response to cold stress, intracellular p53 concentration is increased via post-transcriptional events (probably including also phosphorylation of p53 as shown to play a major role in UV or radiation-induced stress responses (Woo *et al.*, 1998)) rather than via gene expression since p53 mRNA levels remains constant during and after cold shock. On the contrary, elevated p21 levels result from increased expression of the p21 gene which is not observed in cell lines expressing no or mutated forms of p53. This observation suggests that cold-shock-induced increase in p21 levels is p53-dependent (Ohnishi *et al.*, 1998). Since p21 can act as an inducer of G1-arrest through modification of the retinoblastoma protein and E2F functions, and since p21 was reported to suppress p53-dependent apoptosis (Wang and Walsh, 1996; Polyak *et al.*, 1996; Gorospe *et al.*, 1997; Bissonnette and Hunting, 1998), this cell-cycle inhibitor may be one of the mediators of cells' protective mechanisms against various stresses such as cold shock. This notion is supported by reports that cold shock-induced stress enhances the survival of mouse skin and cultured keratinocytes against genotoxic stress (Ota *et al.*, 1996). p21 induction may also account for the G1-specific growth arrest observed in low temperature cultivations of CHO cells (Moore *et al.*, 1997; Kaufmann *et al.*, 1999), but respective experiments have yet to be performed. However, cold-induced G1-arrest is accompanied in CHO cells by increased resistance to apoptosis and/or significantly delayed onset of programmed cell death (Moore *et al.*, 1997). On the other hand, cold shock activation of the p53 pathway may preferentially lead to apoptosis in some cell types (Kruman *et al.*, 1992; Gregory and Milner, 1994). These findings exemplify again the double-tracked outcome of p53-dependent p21 expression, leading either to stress relief and cell survival or to apoptosis, whereas p53-independent pathways for induction of p21 predominantly lead to cell survival and differentiation (Figure 1; Fussenegger and Bailey, 1998; Fussenegger *et al.*, 1998c).

Despite little knowledge on molecular mechanisms behind the effects of cold shock, it is unlikely that temperature *per se* may directly cause increased productivity, but rather that such increases are secondary effects coupled to primary effects on cell cycle and viability (Ramirez and Mutharasan, 1990). While the contribution of higher viability to increased productivity is evident the positive effect of G1-arrest is not understood. Certainly, several reports indicate a general increase of heterologous protein production in this cell-cycle phase, and mathematical models predict optimal productivity in G1-arrested cells (Coco-Martin *et al.*, 1992; Suzuki and Ollis, 1992). However, the reason for such positive correlation between G1-specific growth inhibition and productivity remains unknown but could in part derive from increased RNA stability (Nishiyama *et al.*, 1997a and 1997b). Two recent reports conclude that the energy charge of G1-arrested cells remains unchanged (Moore *et al.*, 1997; Mazur *et al.*, submitted). However, whether increased productivity can be explained by conservation of intracellular resources and energy in the absence of growth or by catabolite repression is just as a hypothesis as are suggestions that the G1-phase represents an optimal environment for production since the cell replenishes its metabolic precursors and repairs its genetic damage (Ko and Prives, 1996; Linke *et al.*, 1996).

In summary, low temperature cultivation of CHO cells results in (i) G1-phase-specific proliferation control; (ii) enhanced specific productivity; (iii) maintained high cell viability; (iv) suppressed medium consumption; and (v) suppressed release of hydrolytic enzymes and other impurities, all of which are important advantages for pharmaceutical protein manufacturing. Therefore, low temperature cultivation represents an attractive, low-cost and simple strategy for major improvements in industrial process development.

4. Metabolic engineering-based proliferation control of mammalian cells

4.1. NEGATIVE PROLIFERATION CONTROL

The understanding of molecular processes governing cell proliferation has rapidly advanced during the last decade as molecular biology was applied and focused on cancer science. Carcinogenesis represents in fact a cellular process of uncontrolled proliferation. Although the development of cancer can have diverse bases, mutation of a set of genes involved in key regulatory circuits called tumor suppressor genes are one of the major reasons for uncontrolled neoplastic growth (Fussenegger and Bailey, 1998). Much emphasis in current cancer therapy is addressed to use of tumor suppressor genes such as p53 to reprogram carcinogenic tissue and control unlimited proliferation (Bischoff *et al.*, 1996; Pennisi, 1996; Table 3). Reprogramming of cellular proliferation-related control has recently attracted much interest in the bioengineering community, since this metabolic engineering strategy was thought to have the potential to achieve non-growing cell culture states accompanied by the following advantages: (i) higher consistency of the product produced in an identical physiological state of the cells; (ii) lower release of intracellular proteases and glycosidases leading to higher product quality (ii) lower genetic drift of the population due to reduced proliferation; (iii) lower medium consumption; and most importantly (iv) higher cellular productivity as intracellular resources in the absence of growth could be redirected to heterologous protein production.

The use of controlled proliferation technology requires an efficient regulated gene expression system which allows expression of the cytostatic gene exclusively at a desired cell culture state. In principle, cell culture processes based on controlled proliferation technology are biphasic and allow proliferation of the cell population in a non-productive growth phase, in which the cytostatic gene is repressed, followed by a proliferation-inhibited production phase achieved by overexpression of the cytostatic genes. Available mammalian gene regulation systems have recently been reviewed (Fussenegger *et al.*, 1998a; Burcin *et al.*, 1998).

TABLE 3: Cell-cycle inhibitory determinants (applied for bioprocesses in bold); nd not determined

Cytostatic Gene	Cell Line	Growth Arrest	Apoptosis	Production	Reference
IRF-1	BHK-21	Cell-Cycle-Independent	other cell death	Equal	Kirchhoff et al., 1993
p21 (transient)	CHO	G1-Arrest	Resistant	Increased	Fussenegger et al., 1997a
p21 (stable)	CHO	G1-Retardation	nd	Equal	Mazur et al., 1998
p21-c/ebpα	CHO	G1	Resistant	Increased	Fussenegger et al., 1998c
p27	CHO	G1	Resistant	Increased	Fussenegger et al., 1997
p27	CHO	G1	Resistant	Increased	Mazur et al., 1998
p27-bcl-x_L	CHO	G1	Resistant	Increased	Fussenegger et al., 1998c
p53 (transient)	CHO	G1	Resistant	Increased	Fussenegger et al., 1997a
p53 (stable)	CHO	G1	Sensititive	nd	Mazur et al., 1998
c-Jun antisense	F-MEL	G1	Resistant	nd	Kim et al., 1998
APC	NIH 3T3	G1	nd	nd	Baeg et al., 1995
Bcl-2	Fibroblast	G1-Retardation	Resistant	nd	Borner, 1996
BRCA1	HCT116	G1	nd	nd	Somasundaram et al., 1997
C/EBPα	Hep3B2; Saos2; Fibroblast	G1	nd	nd	Hendricks-Taylor and Darlington, 1995
c-Rel	HeLa	G1	nd	nd	Bash et al., 1997
Gadd45	Fibroblasts	G1	nd	nd	Marhin et al., 1997
GATA-6	Fibroblast	G1	nd	nd	Perlman et al., 1998
Gax	Fibroblasts	G1	nd	nd	Smith et al., 1997
HIV-1 Tat	Glial Cells	G1-Retardation	nd	» viral genes	Kundu et al., 1998
MDM2	NIH 3T3	G1	nd	nd	Brown et al., 1998
MyoD	Muscle Cells	G1	nd	nd	Halevy et al., 1995
N-cadherin	3Y1	G1	nd	nd	Wang et al., 1998
Nim1	Fibroblasts	G1	nd	nd	Baldin et al., 1997
p16	Glioma Cells	G1	nd	nd	Uhrbom et al., 1997
p21	BHK-21	G1-Retardation	nd	nd	Sekiguchi and Hunter, 1998
p21	HT29	G1	nd	nd	Archer et al., 1998
p21	SW480	Growth Suppression	nd	nd	El-Deiry et al., 1993
p27	Mv1Lu	G1	nd	nd	Polyak et al., 1994
p27	Saos-2	G1	nd	nd	Toyoshima and Hunter, 1994
p27	HeLa	G1	nd	nd	Kwon and Nordin, 1997
p33ING1	Several	G1	Sensitive	nd	Garkavtsev et al., 1998
p73	SAOS2	G1	Sensitive	nd	Jost et al., 1997
PCK	Epithelial Cells	G1	nd	nd	Frey et al., 1997
PML	HeLa	G1	nd	nd	Chan et al., 1997
PP1	Cancer Cells	G1	nd	nd	Berndt et al., 1997
Raf	Fibroblasts	G1	nd	nd	Sewing et al., 1997
Ras	Primary Cells	G1	nd	nd	Serrano et al., 1997
Ribosomal L7	Jurkat	G1	Sensitive	nd	Neumann and Krawinkel, 1997
TSC2/tuberin	Fibroblast	G1	nd	nd	Soucek et al., 1997
WT1	Leukemia M1 cells	G1	Sensitive	nd	Murata et al., 1997

One of the first metabolic engineering efforts to control proliferation of industrially relevant mammalian cells has been implemented in BHK-21 cells overexpressing the interferon-responsive factor 1 (IRF-1; Kirchhoff *et al.*, 1993; Kirchhoff *et al.*, 1995; Kirchhoff *et al.*, 1996). As a transcription factor IRF-1 is essential for the regulation of the interferon system (a host defence system mounted against viral and bacterial pathogens) inhibits cell growth, and manifests tumor-suppressor activities (Matsuyama *et al.*, 1993; Tanaka *et al.*, 1994; Tamura *et al.*, 1995). IRF-1 has been shown to exert its growth inhibitory effect by cooperating with p53 in upregulating the cyclin-dependent kinase inhibitor p21 in response to DNA damage (Tanaka *et al.*, 1996). However, in contrast to p53, direct interaction of IRF-1 with the p21 promoter has not yet been shown. Furthermore, like p53, IRF-1 has also been implied in inducing apoptosis (Tamura *et al.*, 1995).

In order to use IRF-1 for proliferation control of BHK-21 cells, this transcription factor was fused to the estrogen receptor (ER) which renders IRF-1 activity responsive to the hormone estradiol. BHK-21 stably transfected with the IRF-1-ER fusion could be growth arrested in the presence of 30 nM estradiol, a concentration which is expected to exert no major influence on the cell's intrinsic estradiol-responsive program. Contrary to what was expected from IRF-1's cooperation with p53 for p21 induction, overexpression of IRF-1 did not result in a cell-cycle-specific G1-arrest of the BHK cells which is normally observed in other cell lines at elevated intracellular p21 concentrations. Rather, cell growth in BHK cultures overexpressing IRF-1 slowed at all cell-cycle phases (Müller *et al.*, 1998). The slight accumulation of cells in G1-phase may be a consequence of IRF-1-induced interferon secretion.

Unfortunately, heterologous gene expression from strong viral promoters such as the highly active MPSV promoter was not enhanced in cytostatic BHK cultures following IRF-1 induction. Whereas the productivity of the model protein luciferase was identical to proliferation-competent control cells up to three days following induction of growth arrest, the productivity dropped further to 40% and 20% on days 4 and 5, respectively (Kirchhoff *et al.*, 1996). However, the use of a transcription factor for proliferation control can be combined with heterologous gene expression from promoters activated by the same transcription factor (Kirchhoff *et al.*, 1993; Tamura *et al.*, 1995). This technology enables control of transgene expression concomitant with proliferation arrest and high level expression under IRF-1-mediated growth arrest (Müller *et al.*, 1998). However, expression levels were not enhanced compared to the ones driven by strong constitutive promoters in proliferating cell lines (Müller *et al.*, 1998).

IRF-1-mediated growth arrest of BHK cells results in a drastic decrease in cell viability, reaching values of 50% after 5 days following estradiol addition (Carvalhal *et al.*, 1998). The nature of this estradiol-unrelated cell death remains unknown but might be the result of a continuous anti-pathogenic alert state of the cells resulting from ongoing IRF-1-mediated induction of the interferon system (Matsuyama *et al.*, 1993; Tanaka *et al.*, 1994; Tamura *et al.*, 1995). However, the observed cell death of BHK cells seems to be unrelated to apoptosis (Müller *et al.*, 1998) although a molecular connection between IRF-1 expression and induction of programmed cell death has been suggested (Tamura *et al.*, 1995).

In order to minimize or alleviate deleterious effects of prolonged IRF-1 expression in BHK cells, special bioprocesses can be designed which consist of subsequent rounds of addition and withdrawal of estrogen to and from the cell culture medium. These estrogen regimes allow sufficient IRF-1 induction to effect growth arrest but minimize cell death. Readdition of estrogen prevents growth resumption since growth inhibition is fully reversible (Müller *et al.*, 1998; Carvalhal *et al.*, 1998). The estrogen regime technology is very similar to the temperature regimes developed to extend growth arrested production processes based on temperature-sensitive cell-cycle mutants described above (Jenkins and Hovey, 1993; Hovey *et al.*, 1994).

In a less applied study, Sekiguchi and Hunter (1998) investigated the potential of p21 to control proliferation of BHK cells. Due to its dual growth inhibitory capacity, p21 seems a good candidate for efficient proliferation control. p21 binds and inhibits cyclin-CDK activity necessary for G1-S transition and also binds PCNA thereby inhibiting DNA polymerase (for a review see Fussenegger and Bailey, 1998). Conditional overexpression of p21 was established in BHK cells by using the tetracycline-responsive system (Gossen and Bujard, 1992). Upon withdrawal of tetracycline from the cell culture medium, growth of p21-overexpression BHK cells significantly slowed (by about 50%) with concomitant prolongation of the G1-phase. However, overexpression did not result in complete block of cell proliferation (Sekiguchi and Hunter, 1998). Interestingly, the p21-mediated growth retardation was not reversible, although readdition of tetracycline to slow-growing BHK cells resulted in loss of p21 expression, showing that the tet-regulated promoter could still be inactivated and that the accumulated p21 was degraded. This indicates that overexpression of p21 leads to an irreversible growth retardation in BHK cells. Similarly, 50% growth inhibition has been observed following overexpression of p21 in a human colon cancer cell line and other cell lines (El-Deiry *et al.*, 1993; Archer *et al.*, 1998). In conclusion, p21 overexpression alone is not sufficient to completely block proliferation in BHK cells. Similar results were observed in CHO cells (see below, Fussenegger *et al.*, 1998c; Mazur *et al.*, 1998).

A more complete growth suppression was achieved by overexpression of antisense *c-jun* (Kim *et al.*, 1998a; 1998b). *c-jun* is an immediate early response gene which is rapidly induced in response to external mitogenic stimuli and plays a critical regulatory role in a cell's commitment to proliferate (Ryseck *et al.*, 1988). The nuclear phosphoprotein c-Jun forms together with c-Fos the transcriptional activator AP-1 which promotes transcription of various genes involved in cell-cycle progression (Turner and Tjlan, 1989; Fussenegger and Bailey, 1998 for a review). Consequently, selective inhibition of *c-jun* expression resulted in complete proliferation block in the G1-phase of the cell cycle (Smith and Prochownik, 1992). Furthermore, suppression of *c-jun* by antisense oligonucleotides has also been reported to increase resistance to apoptosis induced by growth factor deprivation or chemical agents (Colotta *et al.*, 1992; Sawai *et al.*, 1995). Therefore, overexpression of *c-jun* antisense for use in controlled proliferation technology would serve a dual purpose: first, to arrest proliferation, and, second, to inhibit apoptosis often encountered when blocking growth of neoplastic cells (Suzuki and Ollis, 1990; Walker *et al.*, 1997).

Friend murine erythroleukemia (F-MEL) cells were constructed which express the *c-jun* antisense gene from a glucocorticoid-inducible MMTV promoter. Addition of dexamethasone to the cell culture medium resulted in sustained and complete growth arrest for 16 days while maintaining the cells in a highly viable state (over 86%; Kim *et al.*, 1998a and b). Additionally, arrested cells were more resistant to apoptosis induced by serum-deprivation than proliferation competent cells in the absence of dexamethasone. Proliferation-inhibited cells resume growth upon withdrawal of dexamethasone from the cell culture medium. The use of *c-jun* antisense technology is an attractive strategy since it combines complete proliferation block with high cell viability. However, the productivity of cells arrested by antisense *c-jun* has not been assessed yet and this technology awaits also transfer into industrially relevant cell lines which express c-Jun at lower levels than the F-MEL carcinoma cell line.

A recent approach in CHO cells uses overexpression of cyclin-dependent kinase inhibitors (CDI) p21 and p27 as well as the tumor suppressor gene p53. A wide variety of signals ranging from genotoxic stress, nutrient depletion, cell-cell contact, differentiation and senescence are integrated by p21, p27 and p53 and impinge on the central proliferation checkpoint of eukaryotic cells, the G1-S-phase transition (Fussenegger and Bailey, 1998). The CDIs effect G1-specific cell-cycle arrest by binding and inhibiting the cyclin E-Cdk2 complex, a kinase which regulates G1-S transition and commits a cell to initiate another round of cell cycle (Fussenegger and Bailey, 1998). Transient tetracycline-repressible overexpression of p21, p27 and p53 induced a strict but reversible G1-arrest in CHO cells with concomitant 4-fold higher specific productivity (Fussenegger *et al.*, 1997a). While overexpression of all three cell-cycle inhibitors resulted in identical levels of G1-arrest and specific productivity, it was already evident from these transient expression experiments that a relatively higher intracellular p21 level must be reached to effect proliferation control than those necessary for p27 and p53 (Fussenegger *et al.*, 1997a; Mazur *et al.*, 1998).

In a stable genetic configuration, these three cell-cycle inhibitors differed in their potential to control CHO proliferation over extended cultivation periods. p53 was expected to be the most attractive solution to impose proliferation control in CHO cells, since p53 is able to induce growth arrest via several routes, one of which is upregulation of p21. Also, since p53 is a transcription activator, heterologous gene expression could have been placed under control of p53-responsive promoter, similar to the IRF-1-based proliferation control system (Kirchhoff *et al.*, 1993). However, even when expression of an apoptosis-deficient p53 mutant (p53175P; Rowan *et al.*, 1996) was induced in CHO cell lines by withdrawal of tetracycline from the cell culture medium, the entire cell population rapidly died, showing morphologies highly suggestive of apoptosis (Mazur *et al.*, 1998). Although the molecular connections between p53 and the cellular cell death program have recently been described, p53 alone in the absence of the stress repertoire normally required for p53-mediated apoptosis induction was not expected to lead to cell death (Fussenegger and Bailey, 1998).

Direct overexpression of the major p53 target gene p21 lead certainly to transient growth retardation, but the cell-cycle inhibitory effect of p21 was not strong enough to inhibit proliferation in stable CHO transfectants, which resumed growth shortly after induction of this CDI (Mazur *et al.*, 1998). Further experiments revealed that the intracellular p21 levels reached by the tetracycline-responsive expression system were not high enough, as indicated by previous transient expression experiments (Fussenegger *et al.*, 1997a). Only when expressing p21 from strong constitutive viral promoters could complete and sustained growth arrest be achieved in transfected cells (Mazur *et al.*, 1998). Similar observations have been reported by Sekiguchi and Hunter (1998) and Archer *et al.*, (1998) where a 10-fold overexpression of p21 by the tetracycline-responsive expression system resulted only in growth retardation and prolongation of the G1-phase without inducing a sustained proliferation block in BHK cells and colon cancer cells, respectively (see above).

Although p27 is very similar to p21 both in its protein structure as well as in its binding to cyclin-dependent kinase complexes, stable expression of this CDI resulted in a complete G1-phase-specific proliferation block in CHO cells (Mazur *et al.*, 1998). G1-arrested cells remain proliferation-inhibited, viable (over 90%), and attached for several weeks and do not show any apoptotic behavior. p27-arrested CHO cells adopt a very characteristic "fried egg-like" morphology, probably as a consequence of reaching out to establish cell-cell contacts (Mazur *et al.*, submitted). Similar morphologies have also been reported for p21-overexpression in BHK cells (Sekiguchi and Hunter, 1998) and are a common phenomenon in fibroblast senescence as a result of residual growth in the absence of DNA replication and cell division (Cristofalo *et al.*, 1989; Uhrbom *et al.*, 1997). The G1-specific proliferation arrest induced by p27 is completely reversible; cells resume growth following addition of tetracycline, thereby reverting their characteristic flattened morphology. Reversibility of the proliferation control system indicates that the tetracycline-repressible promoter can still be inactivated and that the accumulated p27 is degraded, most likely by the ubiquitin pathway by which p27 is eliminated in cycling cells (Elledge and Harper, 1998; Fussenegger and Bailey, 1998). G1-arrested CHO cells based on conditional overexpression of p27 in stable transfectants show on average a 10-fold higher specific productivity compared to proliferation-competent control cells (Mazur *et al.*, 1998). p27 is specially suited for controlled proliferation technology as it is a stress-unrelated and therefore mild inhibitor of mammalian cell growth. In a natural context p27 is upregulated following contact inhibition by transforming growth factor beta (TGFβ) (Polyak *et al.*, 1994a; Reynisdottir *et al.*, 1995) or, like p21, during differentiation of certain cell types (Hauser *et al.*, 1997). Cell interactions are known to protect cells from apoptosis. Cell contact-mediated upregulation of p27 has recently been suggested as a major mediator of adhesion-dependent drug resistance of tumors. Therefore, p27 used in controlled proliferation technology is very likely to increase apoptosis resistance of arrested cells.

Together with other reports and theoretical analyses, results achieved with this G1-specific proliferation arrest strategy suggest that the G1-phase of the cell cycle may be a privileged phase for increased production of heterologous proteins (Suzuki and Ollis 1989 and 1990; Fussenegger *et al.*, 1997a; Mazur *et al.*, 1998; Fussenegger *et al.*, 1998c; Fussenegger *et al.*, 1998b; Furukawa and Ohsuye, 1998; Kaufmann *et al.*, 1999). This contrasts with attempts to control proliferation in a cell-cycle-independent manner which did not result in higher cellular productivities (Müller *et al.*, 1998; Franek *et al.*, 1999). The reason for this phenomenon remains unclear. However, since the G1-phase precedes every round of DNA replication and cell division, a cell is certainly specially prepared for those events by replenishing its energy pools and metabolic precursors and repairs genetic damage (Ko and Prives, 1996; Linke *et al.*, 1997). These logistic genetic programs in G1-phase may therefore enhance the overall physiologic performance of the cells which would coincide with higher production capabilities for heterologous proteins.

p27 is a key regulator for use in cytostatic expression technology since all desired cell culture improvements can be achieved simply by overexpression of this single CDI. Although one-gene metabolic engineering solutions to enhance specific productivity or attain particular cell culture states exist and have been successfully applied for production processes (Holmberg *et al.*, 1997; Renner *et al.*, 1995; Lee *et al.*, 1996), such one-gene metabolic engineering strategies are not likely to cope with future challenges in the field such as gene therapy and tissue engineering which require more complex restructuring of cellular pathways (Papoutsakis, 1998). A good example for limitations in one-gene metabolic engineering is the failure of p21 overexpression to arrest proliferation of CHO cells (Mazur *et al.*, 1988). Based on known mechanisms, p21 should be capable of blocking growth, and this has been shown in transient transfection experiments (Fussenegger *et al.*, 1997a). However, CHO cells somehow bypass such imposed metabolic engineering constraints either by escaping via another pathway or by more rapidly degrading the metabolic engineering effector. In order to target several pathways simultaneously to achieve one specific cell culture state, artificial eukaryotic operons have been developed which allow the simultaneous as well as coordinated and adjustable expression of up to four different genes (Fussenegger *et al.*, 1997b; Fussenegger *et al.*, 1998d; Fussenegger *et al.*, 1999b).

Several cellular signals and pathways such as DNA damage repair, senescence and cell differentiation impinge on p21 and result in cell-cycle arrest (Fussenegger and Bailey, 1998). Using multicistronic expression technology we applied multigene metabolic engineering to achieve p21-based proliferation control of CHO cells by simultaneous overexpression of p21 and *c/ebpα* (Fussenegger *et al.*, 1998c). C/EBPα (CAAT-enhancer-binding protein α) is a differentiation factor which both induces and stabilises p21 to initiate and maintain terminal differentiation in certain cell types (Hendricks-Taylor and Darlington, 1995; Timchenko *et al.*, 1996). By combining the differentiation and the DNA repair pathways of p21, we have been able to achieve proliferation control in CHO cells with nearly identical characteristics with respect to sustenance of G1-arrest, cell morphology, specific productivity (15-fold increase), and reversibility as observed with p27-based cytostatic technology (Fussenegger *et al.*, 1997a; Fussenegger *et al.*, 1998c; Mazur *et al.*, 1998). Also, p21-c/ebpα-arrested cells were highly viable and showed no signs indicative of apoptosis. However, the cell cycle and apoptotic programs are intimately linked at the molecular level (Fussenegger and Bailey, 1998). The role of p21 in at this intersection of both molecular circuits is not clear and depends on the cell lines used as well as on the type of pathway which upregulates p21 (see above). In some cases overexpression of p21 has been shown to protect cells from entering the cell death program (Wang and Walsh, 1996; Polyak *et al.*, 1996; Gorospe *et al.*, 1997; Bissonnette and Hunting, 1998). It is also expected that cells are more resistant to apoptosis in a differentiated state as is simulated by overexpression of C/EBPα (Timchenko *et al.*, 1996). Protection from apoptosis and the G1-phase have also been reported to be intimately linked (Borner, 1996; O'Reilly *et al.*, 1996).

Since blocking growth of neoplastic cells usually diverts them into apoptosis (Walker et al, 1997), cytostatic technology must foresee measures to suppress suicidal cell death although it has not been observed using p21, p27 and *c/ebpα* plus p21 as cytostatic genes. Using multicistronic expression technology we combined expression of p27 and *bcl-x$_L$* in CHO cells. Bcl-x$_L$ belongs to the Bcl-2 family of proteins which suppress apoptosis in higher eukaryotic cells (Fussenegger and Bailey, 1998; Fussenegger *et al.*, 1998b; Fussenegger *et al.*, 1998c; Newton and Strasser, 1998; Al-Rubeai, 1998; Dickson, 1998; Adams and Cory, 1998). Interestingly, CHO cells overexpressing this multicistronic expression units behaved identical to the p27-only strategy with respect to proliferation control (Mazur *et al.*, 1998; Fussenegger *et al.*, 1998). However, specific heterologous glycoprotein production was up to 30 times higher in p27-bcl-x$_L$-expressing cells compared to proliferation-competent control cells, a factor of 3 higher than CHO cells arrested by overexpression of p27 only (Fussenegger *et al.*, 1998c).

The molecular basis for this unexpected bcl-x$_L$-related increase in productivity is not entirely clear but could derive from the tendency of Bcl-x$_L$ to induce polyploidy in mammalian cells (Minn *et al.*, 1996). Apart from sparing cells from programmed cell death, several other apoptosis-unrelated characteristics of the Bcl-2 family have recently been described (Borner, 1996; O'Reilly *et al.*, 1996; Newton and Strasser, 1998; Fussenegger *et al.*, submitted; Al-Rubeai, 1998) one of which is its proliferation-controlling activity (Borner, 1996; O'Reilly *et al.*, 1996). Induction of Bcl-2 expression is consistently associated with retardation of mammalian cell proliferation due to a prolongation of the G1-phase. Whether higher specific productivities observed in hybridoma cells expressing *bcl-2* (Itoh *et al.*, 1992) are a consequence of G1-phase prolongations remains to be investigated. Whereas cells lacking Bcl-2 expression die from any point of the cell cycle in response to apoptotic agents, Bcl-2-overexpressing cells accumulate in the G1-phase and are protected from cell death (Borner, 1996; O'Reilly *et al.*, 1996). Thus, mammalian cells appear to survive better if the G1-phase is prolonged, and Bcl-2 actively contributes to achieve this condition. Indeed, arresting fibroblasts in G1-phase by thymidine block or isoleucine starvation partially rescued cells from induced apoptosis (Belizario and Dinarello 1991; Meikrantz *et al.*, 1994).

Several lines of evidence suggest that the susceptibility of a cell to apoptosis is influenced by its proliferative state. Activation of several growth-related genes (c-Fos, c-Myc, and c-Jun, see above) are associated with apoptosis, and ectopic expression of positive regulators of cell proliferation such as c-Myc or the adenoviral E1A promote programmed cell death (Bissonnette *et al.*, 1992; Rao *et al.*, 1992). Conversely, arresting cells early in G1-phase offers protection from apoptosis (Belizario *et al.*, 1991; Meikrantz *et al.*, 1994; Gorospe *et al.*, 1997; Bissonnette and Hunting, 1998) which is best illustrated by the Abl tyrosine kinase and the retinoblastoma protein, both of which keep cells in G1-phase and act as potent cell survival factors (Sawyers *et al.*, 1994; Haas-Kogan *et al.*, 1995). Similarly, p21 has recently been found to protect mammalian cells from apoptosis (Wang and Walsh, 1996; Polyak *et al.*, 1996; Gorospe *et al.*, 1997; Bissonnette and Hunting, 1998). Recently, arrested cells have also been found to be less susceptible target cells for cytotoxic T lymphocyte-mediated perforin-based killing (De Leon *et al.*, 1998). All these findings may explain why morphologies suggestive of apoptosis could not be found in cytostatic cultures which had been arrested in the G1-phase of the cell cycle.

Therefore, the G1-phase seems to be a privileged state of the mammalian cell cycle which allows increased specific heterologous protein production in an apoptosis-safe environment. Cytostatic expression technology uses these positive cell characteristics by arresting the production cell lines constantly in G1-phase. Besides the cytostatic genes which have already been used for cytostatic expression technology, several other promising candidates remain to be tested. Table 3 gives an overview of genes which have been described for their growth-inhibiting potential.

Despite these multiple advantages, cytostatic expression technology harbors an intrinsic problem: A genetically imposed growth arrest of a permanent cell line establishes an enormous selective pressure on arrested cells, since a single cell which escapes proliferation control and resumes growth will eventually overgrow the entire cytostatic culture and dilute the advantage of the cytostatic expression technology. The situation is much like the development of cancer in a higher organism. In principle, there are several ways for a cell to escape proliferation control including (i) mutations in the regulation system; (ii) increased degradation of the cell-cycle inhibitor; (iii) mutation of the cytostatic gene; or (iv) activation of redundant molecular circuits which bypass proliferation block. We isolated and characterized mutant clones which escaped p27-based proliferation control (Mazur *et al.*, submitted). These mutants resume growth and show growth characteristics identical to the parent host cell line. The tetracycline-regulatory system is still intact in these mutants. However, the dicistronic expression units seems to be absent, since these mutants neither produce detectable amounts of the model glycoprotein product nor of the cytostatic CDI. It is therefore likely that mutation or deletion of the dicistronic expression unit is the reason for recurrent outgrowth of proliferation-competent mutants from growth-inhibited cell populations. This finding is rather encouraging since the proliferation block imposed by overexpression of p27 seems not to be bypassed at the molecular level and since growing mutants do not produce product which could deteriorate product quality. However, spontaneous mutations remain a major challenge for cytostatic cell culture technology.

An important strategy toward total suppression of mutant outgrowth in cytostatic cultures has been developed for the IRF-1 system (Müller *et al.*, 1998). A puromycin resistance gene was genetically linked 3' to the cytostatic gene IRF-1 in a dicistronic configuration. In this autoselective configuration, resistance to puromycin requires intact upstream genetic elements including IRF-1. Individual cells which delete or mutate the cytostatic gene are no longer resistant to puromycin and are counterselected by the superimposed antibiotic selection. Auto-selective systems have recently been developed for the expression of product proteins toxic to the production cell line or for the selection of high producing clones (Fussenegger *et al.*, 1997b; Fussenegger *et al.* 1999a and 1999b, Rees *et al.*, 1996; Gurtu *et al.*, 1996). Other methods to suppress mutant outgrowth include the addition of a cell culture additive which selectively kill dividing cells. As this is the criterion for effective anti-cancer drugs, such substances are usually too expensive for large scale applications. Other substances such as FDU (5-fluoro-2'-deoxyuridine; Nishimoto, 1997) proved not efficient enough to selectively kill growing mutants in the p27-based cytostatic system. However, FDU induces growth arrest in CHO cells and can therefore be used as culture additive to suppress mutant outgrowth in metabolicalic engineering-based cytostatic cultures. Also, FDU seems not to negatively influence the specific productivity of G1-arrested cells (Mazur and Fussenegger, unpublished). A cheap and easy-to-apply method for effective growth suppression is low temperature cultivation which induces a complete G1-block in CHO cells (Moore *et al.*, 1997; Kaufmann *et al.*, 1999, see above). Low temperature has been shown to completely suppress mutant outgrowth in CHO cell cultures grown at 30°C (Kaufmann *et al.*, 1999). Low temperature cultivations in CHO cells were shown to increase specific productivities about two-fold. However, the combination of temperature- and metabolic engineering-based proliferation control systems were not additive with respect to production. Rather, low temperature eliminated productivity increase in p27-arrested CHO cell populations. Consequently, constant low temperature cultivation is not an ideal solution for suppression of mutant outgrowth in p27-based cytostatic expression technology (Mazur *et al.*, submitted). However, the development of a special temperature regimen can be foreseen to maintain high productivities known for the p27-based system while suppressing growth of mutant clones. Similar technology has been successfully applied to conditional cell-cycle mutants (Jenkins and Hovey, 1993; Hovey *et al.*, 1994) and to the IRF-1 system (Carvalhal *et al.*, 1998; Müller *et al.*, 1998).

The combination of the product and the cytostatic genes on the same expression unit in a multicistronic configuration allows concomitant induction of growth arrest and heterologous protein production (Fussenegger *et al.*, 1997a; Fussenegger *et al.*, 1998c; Fussenegger et al., 1998d; Mazur *et al.*, 1998; Fussenegger *et al.*, 1999b). This particular genetic configuration installs a biphasic production process which separates cell growth from production. In this biphasic production process the cells are expanded as rapidly as possible to the desired cell density followed by a proliferation-inhibited production phase. The transition from the non-productive growth phase to the productive non-growth phase is regulated by the withdrawal of the external repressive agent tetracycline (tet_{off}-system; Gossen and Bujard, 1992). The absence of tetracycline in the proliferation-inhibited production phase avoids the presence of the physiological active compound in the process medium.

However, tetracycline may be degraded by at least four possible mechanisms: epimerization, dehydration, hydrolysis and oxidation, of which epimerization is the most important reaction (Day *et al.*, 1978; Miller *et al.*, 1993). Based on this instability of tetracycline, we recently developed a novel autoregulated proliferation-controlled batch culture process for CHO cells which allows the transition from the non-producing growth phase to the non-growing production phase to take place in a self-controlled, "automatic" fashion without any medium change (Mazur *et al.*, submitted). Starting with a defined initial cell population and a chosen initial tetracycline concentration, the transition from growth to production of the self-controlled Tet^{SWITCH} process can be fine-tuned to occur at a particular cell density. This Tet^{SWITCH} process is less dependent on the initial cell population than on the initial tetracycline concentration as the tetracycline degradation rate in cell culture medium is independent of cell density (Mazur *et al.*, submitted).

With the development of metabolic engineering-based controlled proliferation processes, bioprocess engineering enters a new era in which central cellular pathways are redirected and restructured by complex multigene metabolic engineering strategies to achieve major improvements in mammalian cell culture technology (Papoutsakis, 1998). Automatic and self-controlled batch culture processes are expected to play a significant role in future process development strategies.

4.2. POSITIVE PROLIFERATION CONTROL

Apart from negative proliferation control, which arrests cell growth in response to physiologic and genotoxic stress situations or as part of the differentiation program, there exist cellular control circuits for positive proliferation control which play a pivotal role in development and regeneration (Fussenegger and Bailey, 1998). Negative and positive proliferation control pathways are often mutually exclusive as their signals converge at the same central checkpoint of the G1-phase (Fussenegger *et al.*, 1998b; Fussenegger and Bailey, 1998). Proliferation is initiated by external growth factors and other mitogenic signals which are then transduced from the cell membrane to the nucleus by a molecular cascade involving protein binding and modification events including a series of phosphorylations resulting from successive activation of several protein kinases. The extensively studied Ras-MAP kinase pathway exemplifies these signalling cascades (McCormick, 1993; Kerkhoff and Rapp, 1997; Sewing *et al.*, 1997). Activation of growth factor receptors stimulates nucleotide exchange on the Ras low molecular weight GTP binding protein, which then participates in activation of the Raf-1 family of serine/threonine kinases (Avruch *et al.*, 1994). Activated Raf phosphorylates and activates the Mek1 and Mek2 dual specificity kinases, shown to be responsible for phosphorylating the MAP kinases Erk1 and Erk2 on threonine and tyrosine, thus activating them in response to mitogenic stimulation. This pathway is eventually linked with the cell-cycle machinery by giving rise to elevated cyclin D1 levels in G1-phase, concomitant with down-regulation of the cyclin-dependent kinase inhibitor p27 (Liu *et al.*, 1995; Aktas *et al.*, 1997; Greulich and Erikson, 1998). Cyclin D interacts with CDK4 forming a kinase which maintains G1-phase progression and prepares S-phase entry by indirectly inducing cyclin E via phosphorylation-mediated release of E2F from pRB (Fussenegger and Bailey, 1998; Fussenegger *et al.*, 1998b). p27 has the capacity to inhibit both cyclin D-CDK4 and cyclin E-CDK2 and appears to arrest the cell cycle in response to growth factor deprivation and contact inhibition (Polyak *et al.*, 1994a and 1994b; Coats *et al.*, 1996; Rivard *et al.*, 1996). Overexpression of key regulators in this positive proliferation control cascade such as cyclin E, E2F, cyclin D1 and Mek1 have recently been reported to obviate the requirement for Ras function in the induction of cell proliferation in response to mitogenic signalling (Renner *et al.*, 1995; Lee *et al.*, 1996; Aktas *et al.*, 1997; Greulich and Erikson, 1998). Similarly, the suppression of growth-inhibiting regulators such as p27 and *c-jun* lead to growth stimulation in the absence of mitogens (Rivard *et al.*, 1996; Kim *et al.*, 1998; see above).

Short-circuit of growth-stimulatory molecular pathways by metabolic engineering is an attractive strategy which allows operation of mammalian cell cultures in the absence of exogenous growth factors and serum. Elimination of undefined medium components and cell culture additives of animal origin is a current priority in modern pharmaceutical manufacturing. In a pioneering study the overexpression of cyclin E in CHO cells has successfully been shown to obviate the need for serum in the culture medium (Renner *et al.*, 1995). In a wild-type context, overexpression of cyclin E in CHO cells is induced by the basic fibroblast growth factor (bFGF) (Renner *et al.*, 1995). Overexpression of cyclin E results not only in serum- but also in attachment-independent growth. Similar results have also been achieved by suppression of the antagonist of cyclin E-CDK2 p27 (Rivard *et al.*, 1996) as well as by overexpression of the dual specificity kinase Mek1 (Greulich and Erikson, 1998). On the contrary, overexpression of E2F-1 leads to serum-independent growth but attachment-dependent growth is not impaired (Lee *et al.*, 1996). Thus, metabolic engineering of cell-cycle regulators is the basis for a novel strategy for the development of serum-independent production processes. However, cell-cycle regulatory circuits are not the only targets for positive proliferation-modulating engineering efforts in industrially relevant cell lines.

Other strategies rely on overexpression and secretion of growth factors (Hunt *et al.*, 1997; Pak *et al.*, 1996). Insulin and transferrin are the growth factors of choice for serum-free culture of CHO cells. It has been reported that the insulin-like growth factor I (IGF-1) has similar mitogenic effects to insulin and can replace insulin in serum-free medium formulations (Ross and Englesberg, 1993). CHO cells transfected with the gene encoding IGF-1 show autocrine growth stimulation and can be cultivated under fully defined protein-free conditions (Hunt *et al.*, 1997; Pak *et al.*, 1996). Although this technology seems to perform well with CHO cells, it probably requires adaptation for other cell lines each of which may require specific growth factors or combinations thereof. On the contrary, due to the high conservation of cell-cycle regulatory networks in mammalian cells, cell-cycle targeted positive proliferation strategies can be expected to be less restricted to particular host cell types.

Besides the development of serum- and protein-free cell culture systems, positive proliferation control is becoming a core technology for cell and tissue engineering, where expansion of an initial tissue graft is the critical step for such cell-based therapies. Several efforts have been undertaken in recent years to stimulate growth of cells from important organs such as the liver (Michalopoulos and DeFrances, 1997), the brain (McKay, 1997; Studer *et al.*, 1998), the skin (Martin, 1997; Pomahac *et al.*, 1998), the bones (Prockop, 1997; News, 1998) or the hematopoietic system (Ladd *et al.*, 1997; Glaser, 1998). As basic molecular research reveals an increasing number of mitogenic cytokines, great progress has been achieved in expansion of stem cells by applying special regimens of sequential cytokine stimulation (Ladd *et al.*, 1997). Also, substantial progress has been made in the development of culture conditions and the development of artificial biomatrices. A recent breakthrough in cytokine-based cell therapy has been reported for Parkinson's disease (Studer *et al.*, 1998). As established for the rat model, Studer *et al.*, (1998) obtained stem cells (McKay, 1997) from the brain of fetal rats and cultured them in medium containing basic fibroblast growth factor (bFGF) which initiated proliferation of the stem cells. Upon withdrawal of bFGF, the stem cells began to differentiate into different types of neurons including dopaminergic neurons which are lost during Parkinson's pathology. Transplantation of this expanded and differentiated cell population into the brains of parkinsonian rats showed substantial recovery of these animals (Studer *et al.*, 1998).

However, the cultivation and differentiation of stem cells covers only a small part of therapeutic tissue engineering needed to overcome today's life-threatening non-pathogenic diseases. For most tissues or cell populations displaying malignant cell deficiencies, no stem cells are available which can be grown and differentiated and engineered *ex vivo* before reimplantation. Also, gene therapy must target in most cases highly differentiated tissues, cells of which have lost the potential to regenerate and to grow in culture in their differentiated tissue-identical state. Reprogramming cells which are major constituents of organs and tissues to allow proliferation and regeneration will be one of the main challenges of metabolic engineering in the next decade. A major step towards the dream of designing and cultivating artificial tissues and organs has recently been achieved by Thomason et al., (1998) who generated undifferentiated, proliferation competent human embryonic stem cell lines from human blastocysts. These embryo-derived, pluripotent cells maintained the potential to form trophoblast and derivatives of all three embryonic germ layers, including gut epithelium, cartilage, bone, smooth and striated muscle and neural epithelium, embryonic ganglia, and stratified squamous epthelium (Thomasson et al., 1998).

Liver regeneration is a unique natural phenomenon and a paradigm of tissue regeneration and dedifferentiation of fully specialized and differentiated cells: the hepatocytes. The ancient Greeks apparently recognized liver regeneration since it is described in the myth of Prometheus who, after stealing the secret of fire from the gods of Olympus, was condemned to having a portion of his liver eaten daily by an eagle. Transplantation and hepatectomy experiments confirm today the wonderful capacity to regenerate and adapt the size of the organ to the acceptor individual. Several cytokines such as hepatocyte growth factor, epidermal growth factor, transforming growth factor-α, interleukin 6, tumor necrosis factor-α, insulin, and norepinephrine appear to play important roles in this process (Michalopoulos and DeFrances, 1997). However, little is known about the molecular mechanisms governing redifferentiation and growth resumption. One possible key player in the process of liver regeneration is be the transcription and differentiation factor C/EBPα (Hendricks-Taylor and Darlington, 1995; Timchenko *et al.*, 1996). C/EBPα expression is restricted to terminally differentiated non-dividing cells and is most abundant in the liver (Mischoulon *et al.*, 1992). Interestingly, while C/EBPα protein is present at high levels in normal adult liver, its expression is dramatically decreased in regenerating liver. The mechanisms repressing C/EBPα are not clear yet but might rely on intracellular signalling cascades involving IGFBP1 (a plasma protein binding insuline-like growth factor), nuclear factor kappa B (NF-κB), the signal transducer and activator of transcription 3 (STAT3), and the transcription factor AP1 (Michalopoulos and DeFrances, 1997).

Positive proliferation control is only one side of the coin; negative proliferation control is the other. Therefore, mechanisms which stop liver regeneration are currently under investigation. One growth inhibitory mechanism is imposed by C/EBPα which both induces and stabilizes the CDI p21, a characteristic which as been used to implement cytostatic cell culture technology (Fussenegger *et al.*, 1998a; Fussenegger *et al.*, 1998c; see above). Further experiments must be performed to confirm the role of TGF-β in cessation of regeneration. The anti-mitotic potential of TGF-β resides in its capacity to induce the CDI p27 in response to contact inhibition (Polyak *et al.*, 1994a and 1994b; Ciats *et al.*, 1996; Rivard *et al.*, 1996). Cell-cell contacts have long been known to play a major role for negative control of organ expansion. Cell contacts and adhesion within and at the outer surface of tissues is mediated by a family of homologous cell-surface glycoproteins, the cadherins. N-cadherins antagonize Ras-MAP kinase signalling and therefore interrupt and inhibit growth factor induced cell proliferation (Wang *et al.*, 1998). Ectopic overexpression of N-cadherins leads to G1-specific cell-cycle arrest which is effected via upregulation of p27 (Croix and Kerbel, 1997; Wang *et al.*, 1998). Recently, p27 has also been reported to be involved in induction and maintenance of terminal differentiation of keratinocytes (Hauser *et al.*, 1997). Despite these initial molecular interactions involving N-cadherins and p27, the global structure of pathways terminating expansion of regenerating tissues remains largely obscure. Better understanding of these processes is fundamental for the design of metabolic engineering strategies for cell therapies and tissue engineering in the near future.

5. **Concluding Remarks**

Several scientific advances in the understanding of cell-cycle regulatory networks have recently been successfully applied in a variety of cultivation- or metabolic engineering-based strategies to improve the performance of pharmaceutical manufacturing. As a central recurrent consequence of proliferation control, whether achieved by cell culture additives, cultivation at altered temperatures or overexpression of cytostatic tumor suppressor genes, the G1-phase seems to be an optimal cell-cycle phase for heterologous protein production. G1-arrested cells show a higher specific productivity. Provided the proliferation-inhibited state is induced via a differentiation signal rather than by stress conditions, such G1-arrested cells are less threatened by cell death programs. Technology which drives cells into a highly differentiated state is of major importance to control the regeneration of tissues. However, complete solutions for tissue engineering and cell therapies await better understanding of controlling molecular mechanisms. An integrative effort to manipulate several key regulatory pathways by multigene metabolic engineering as well as by process development will be required in order to move proliferation-based metabolic engineering strategies forward into the 21st century.

6. Acknowledgement

This work was supported by the Swiss Bundesamt für Bildung und Wissenschaft (BBW) within the Framework IV Biotechnology Programme of the European Commission.

8. References

Adams, J.M., and Cory, S. (1998) The Bcl-2 protein family: arbiters of cell survival. *Science* **281,** 1322-1326.

Aktas, H., Cai, H., and Cooper, G.M. (1997) Ras links growth factor signaling to the cell cycle machinery via regulation of cyclin D1 and CDK inhibitor p27KIP1. *Mol. Cell Biol.* **17,** 3850-3857.

Al-Rubeai, M. (1998) Apoptosis and cell culture technology. in S. Scheper (ed.) *Advances in Biochemical Engineering/Biotechnology,* Vol. 59, Springer-Verlag Berlin Heidelberg, pp.225-249.

Al-Rubeai, M., Emery. A.N., Chalder, S., and Jan, D.C. (1992) Specific monoclonal antibody productivity and the cell-cycle comparisons of batch, continuous and perfusion cultures. *Cytotechnology* **9,** 85-97.

Al-Rubeai, M., Singh, R.P., Goldman, M., and Emery, A.N. (1995) Death mechanisms of animal cells in conditions of intensive agitation. *Biotechnol. Bioeng.* **45,** 463-472.

Alves, P.M., Carvalhal, A.V., Moreira, J.L., Santos, H., and Carrondo, M.J.T. (1998) Effect of growth arrest in BHK metabolism: on line monitoring by ^{13}C and ^{31}P NMR spectroscopy. in O.W. Merten, P. Perrin and J.B. Griffiths (eds.), *New Developments and New Applications in Animal Cell Technology,* Kluwer Academic Publishers, Dordrecht, pp. 223-225.

Archer, S.Y., Meng, S., Shei, A., and Hodin, R.A. (1998) $p21^{WAF1}$ is required for butyrate-mediated growth inhibition of human colon cancer. *Proc. Natl. Acad. Sci. USA* **95,** 6791-6796.

Avruch, J., Zhang, X., and Kyriakis, J. (1994) Raf meets Ras: completing the framework of a signal transduction pathway. *Trends Biochem. Sci.* **19,** 279-283.

Baeg, G.H., Matsumine, A., Kuroda, T., Bhattacharjee, R.N., Miyashiro, I., Toyoshima, K., and Akiyama, T. (1995) The tumour suppressor gene product APC blocks cell cycle progression from G_0/G to S phase. *EMBO J.* **14,** 5618-5625.

Baldin, V., Cans, C., Watanabe, N., and Ducommun, B. (1997) Evidence for a mammalian Nim1-like kinase pathway acting at the G0/1S transition. *Biochem. Biophys. Res. Commun.* **236,** 130-134.

Ball, K.L., Lain, S., Fahraeus, R., Smythe, C., and Lane, D.P. (1997) Cell-cycle arrest and inhibition of Cdk4 activity by small peptides based on the carboxy-terminal domain of p21WAF1. *Curr. Biol.* **7,** 71-80.

Barnabe, N., and Butler, M. (1994) Effect of temperature on nucleotide pools and monoclonal antibody production in a mouse hybridoma. *Biotechnol. Bioeng.* **44,** 1235-1245.

Bash, J., Zong, W.X., and Gelinas, C. (1997) c-Rel arrests the proliferation of HeLa cells and affects critical regulators of the G1/S-phase transition. *Mol. Cell Biol.* **17,** 6526-6536.

Belizario, J.E., and Dinarello, C.A. (1991) Interleukin 1, interleukin 6, tumor necrosis factor, and transforming growth factor beta increase cell resistance to tumor necrosis factor cytotoxicity by growth arrest in the G1 phase of the cell cycle. *Cancer Res.* **51,** 2379-2385.

Berndt, N., Dohadwala, M., and Liu, C.W. (1997) Constitutively active protein phosphatase 1alspha causes Rb-dependent G1 arrest in human cancer cells. *Curr. Biol.* **7,** 375-386.

Bischoff, J.R., Kirn, D.H., Williams, A., Heise, C., Horn, S., Muna, M., Ng, L., Nye, J.A., Sampson-Johannes, A., Fattaey, A., and McCormi, F. (1996) An adenovirus mutant that replicates selectively in p53-deficient human tumor cells. *Science* **274,** 373-376.

Bissonnette, R.P., Echeverri, F., Mahboubi, A., and Green, D.R. (1992) Apoptotic cell death induced by c-myc is inhibited by bcl-2. *Nature* **359,** 552-554.

Bissonnette, N., and Hunting, D.J. (1998) p21-induced cycle arrest in G1 protects cells from apoptosis induced by UV-irradiation of RNA polymerase II blockage. *Oncogene* **16,** 3461-3469.

Bloemkolk, J.-W., Gray, M.R., Merchant, F., and Mosmann, T.R. (1992) Effect of temperature on hybridoma cell cycle and Mab production. *Biotechnol. Bioeng.* **40,** 427-431.

Borner, C. (1996) Diminished cell proliferation associated with the death-protective activity of Bcl-2. *J. Biol. Chem.* **271,** 12695-12698.

Borth, N., Heider, R., Assadian, A., and Katinger, H. (1992) Growth and production kinetics of human x mouse and mouse hybridoma cells at reduced temperature and serum content. *J. Biotechnol.* **25,** 319-331.

Bovenberg, W.A., Dauwerse, J.G., Pospiech, H.M., Van Buul-Offers, S.C., Van den Brade, J.L., and Sussenbach, J.S. (1990) Expression of recombinant human insulin-like growth factor 1 in mammalian cells. *Mol. Cell Endocrinol.* **74,** 45-59.

Brooks, E.E., Gray, N.S., Joly, A., Kerwar, S.S., Lum, R., Mackman, R.L., Norman, T.C., Rosete, J., Rowe, M., Schow, S.R., Schultz, P.G., Wang, X., Wick, M.M., and Shiffman, D. (1997) CVT-313, a specific and potent inhibtor of CDK2 that prevents neointimal proliferation. *J. Biol. Chem.* **272,** 29207-29211.

Brown, D.R., Thomas, C.A., and Deb, S.P. (1998) The human oncoprotein MDM2 arrests the cell cycle: elimination of tis cell-cycle-inhibory function induces tumorigenesis. *EMBO J.* **17,** 2513-2525.
Burcin, M.M., O'Malley, B.W., and Tsai, S.Y. (1998) A regulatory system for target gene expression. *Frontiers in Bioscience* **3,** 1-7.
Burd, C.G., and Dreyfuss, G. (1994) Conserved structures and diversity of functions of RNA-binding proteins. *Science* **265,** 615-621.
Carvalhal, A.V., Moreira, J.L., Müller, P., Hauser, H., and Carrondo, M.J.T. (1998) Cell growth inhibition by the IRF-1 system. in O.W. Merten, P. Perrin and J.B. Griffiths (eds.), *New Developments and New Applications in Animal Cell Technology*, Kluwer Academic Publishers, Dordrecht, pp. 215-217.
Chan, J.Y., Li, L., Fan, Y.H., Mu, Z.M., Zhang, W.W., and Chang, K.S. (1997) Cell-cycle regulation of DNA damage-induced expression of the suppressor gene PML. *Biochem. Biophys. Res. Commun.* **240,** 640-646.
Chang, P.Y., Gjornstad, K.A., Goncz, K., and Blakely, E.A. (1990) Involvement of protein synthesis in the development of thermotolerance using a CHO temperature-sensitive mutant. *Int. J. Hyperthermia* **6,** 105-116.
Chen, H.W., and Huang, H.C. (1998) Effect of curcumin on cell cycle progression and apoptosis in vascular smooth muscle cells. *Br. J. Pharmacol.* **124,** 1029-1040.
Chenoufi, N., Baffet, G., Drenou, B., Cariou, S., Desille, M., Clement, B., Brissot, P., Lescoat, G., and Loreal, O. (1998) Deferoxamine arrests in vitro the proliferation of porcine hepatocyte in G1 phase of the cell cycle. *Liver* **18,** 60-66.
Chevalot, I., Dardenne, M., Cherlet, M., Engasser, J.M., and Marc, A. (1995) Effect of sodium butyrate on protein production in different culture systems. in E.C. Beuvery, R. Spier and B. Griffiths (eds.), *Developments Towards The 21st Century*, Kluwer Academic Publishers, Dordrecht, pp. 143-147.
Coats, S., Flanagan, W.M., Nourse, J., and Roberts, J.M. (1996) Requirement of p27Kip1 for restriction point control of the fibroblast cell cycle. *Science* **272,** 877-880.
Coco-Martin, J.M., Oberink, J.W., van der Velden-de Groot, T.A.M., and Beuvery, E.C. (1992) The potential of flow cytometric analysis for the characterization of hybridoma cells in suspension cultures. *Cytotechnology* **8,** 65-74.
Coles, L.S., Diamond, P., Occhiodoro, F., Vadas, M.A., and Shannon, M.F. (1996) Cold shock domain proteins repress transcription from the GM-CSF promoter. *Nucleic Acids Res.* **24,** 2311-2317.
Colotta, F., Polentarutti, N., Sironi, M., and Mantovani, A. (1992) Expression and involvement of c-fos and c-jun protooncogenes in programmed cell death induced by growth factor deprivation in lymphoid cell lines. *J. Biol. Chem.* **267:** 18278-18283.
Craig, E.A. (1985) The heat shock response. *CRC Crit. Rev. Biochem.* **18,** 239-280.
Cristofalo, V.J., Doggett, D.L., Frederich-Brooks, K.M., and Phillips, P.D. (1989) Growth factors as probes for cell aging. *Exp. Gerontol.* **24,** 367-374.
Croix, S.T., and Kerbel, R.S. (1997) Cell adhesion and drug resistance in cancer. *Curr. Opin. Oncol.* **9,** 549-556.
Day, S.T., Crouthamel, W.G., Martinelli, L.C., and Ma, J.K.H. (1978) Mechanism of fluorometric analysis of tetracycline involving metal complexation. *J. of Pharmaceutical Sciences* **67,** 1518-1523.
De Leon, M., Jackson, K.M., Cavanaugh, J.R., Mbangkollo, D., and Verret, C.R. (1998) Arrest of cell cycle reduces susceptibility of target cells to perforin-mediated lysis. *J. Cell Biochem.* **69,** 425-435.
Denarier, E., Fourest-Lieuvin, A., Bosc, C., Pirollet, F., Chapel, A., Margolis, R.L., and Job, D. (1998) Nonneural isoforms of STOP protein are responsible for microtubule cold stability in mammalian fibroblasts. *Proc. Natl. Acad. Sci. USA* **95,** 6055-6060.
Desrivieres, S., Volarevic, S., Mercep, L., and Ferrari, S. (1997) Evidence for different mechanisms of growth inhibition of T-cell lymphoma by phorbol esters and concanavalin A. *J. Biol. Chem.* **272,** 2470-2476.
Dickson, A.J. (1998) Apoptosis regulation and its applications to biotechnology. *Trends in Biotechnology* **16,** 339-342.
El-Diery, W.W., Tokino, T., Velculescu, V.E., Levy, D.B., Parsons, R., Trent, J.M., Lin, D., Mercer, W.E., Kinzler, K.W., and Vogelstein, B. (1993) WAF1, a potential mediator of p53 tumor suppression. *Cell* **75,** 817-825.
Elledge, S.J., and Harper, J.W. (1998) The role of protein stability in the cell cycle and cancer. *Biochim. Biophys. Acta* **1377,** 61-70.
Franek, F., Holy, A., Votruba, I., and Eckschlager, T. (1999) Modulation of cell cycle progression and of antibody production in mouse hybridomas by a nucleotide analogue. *Cytotechnology*, in press.
Franek, F., Vomastek, T., and Dolnikova, J. (1992) Fragmented DNA and apoptotic bodies document the programmed way of cell death in hybridoma cultures. *Cytotechnology* **9,** 117-123.
Fry, M.R., Saxon, M.L., Zhao, X., Rollings, A., Evans, S.S., and Black, J.D. (1997) Protein kinase C isozyme-mediated cell cycle arrest involves induction of p21 (waf1/cip1) and p27 (kip1) and hypophosphorylation of the retinoblastoma protein in intestinal epithelial cells. *J. Biol. Chem.* **272,** 9424-9435.
Fuller, B.J. (1987) Storage of cells and tissues at hypothermia for clinical use. *Symp. Soc. Exp. Biol.* **41,** 341-362.
Furukawa, K., and Ohsuye, K. (1998) Effect of culture temperature on a recombinant CHO cell line producing a C-terminal α-amidating enzyme. *Cytotechnology* **26,** 153-164.

Fussenegger, M., Mazur, X., and Bailey, J.E. (1997a) A novel cytostatic process enhances the productivity of Chinese hamster ovary cells. *Biotechnol. Bioeng.* **55,** 927-939.

Fussenegger, M., Moser, S., Mazur, X., and Bailey, J.E. (1997b) Autoregulated multicistronic expression vectors provide one-step cloning of regulated product gene expression in mammalian cells. *Biotechnol. Prog.* **13,** 733-740.

Fussenegger, M., and Bailey, J.E. (1998) Molecular regulation of cell-cycle progression and apoptosis in mammalian cells: implications for Biotechnology. *Biotechnol. Prog.*, in press.

Fussenegger, M., Bailey, J.E., Hauser, H., and Müller, P.P. (1998a) Genetic optimization of recombinant glycoprotein production by mammalian cells. *Trends in Biotechnology*, in press.

Fussenegger, M., Sburlati, A., and Bailey, J.E. (1998b) Metabolic engineering of mammalian cells, in E.T. Papoutsakis and S.Y. (eds.), *Metabolic engineering, Bioprocess Technology*, Marcel Dekker, Inc., in press.

Fussenegger, M., Schlatter, S., Dätwyler, D., Mazur, X., and Bailey, J.E. (1998c) Controlled proliferation by multigene metabolic engineering enhances the productivity of Chinese hamster ovary cells. *Nature Biotechnol.* **16,** 468-472.

Fussenegger, M., Mazur, X., and Bailey, J.E. (1998d) pTRIDENT, a novel vector family for tricistronic gene expression in mammalian cells. *Biotechnol. Bioeng.* **57,** 1-10.

Fussenegger, M., Moser, S., and Bailey, J.E. (1999a) pQuattro vectors allow one-step multigene metabolic engineering and auto-selection of quattrocistronic artificial mammalian operons. *Cytotechnology*, in press.

Fussenegger, M., Moser, S., and Bailey, J.E. (1999b) Regulated multicistronic expression technology for mammalian metabolic engineering. *Cytotechnology*, in press.

Fussenegger, M., Fassnacht, D., Schwartz, R., Zanghi, J., Graf, M., Bailey, J.E., and Pörtner, R. (1999) Overexpression of the survival factor Bcl-2 increases cell density of Chinese hamster ovary (CHO) cell cultures. *Cytotechnology*, submitted.

Gansauge, S., Nussler, A.K., Beger, H.G., and Gansauge, F. (1998) Nitric oxide-induced apoptosis in human pancreatic carcinoma cells lines is associated with a G1-arrest and an increase of the cyclin-dependent kinase inhibitor p21WAF1/CIP1. *Cell Growth Differ* **9,** 611-617.

Garkavtsev, I., Grigorian, I.A., Ossovskaya, V.S., Chernov, M.V., Chumakov, P.M., and Gudkov, A.V. (1998) The candidate tumour suppressor p33ING1 cooperates with p53 in cell growth control. *Nature* **391,** 295-298.

Giard, D.J., and Fleischaker, R.J. (1980) Examination of parameters affecting human interferon production with microcarrier-grown fibroblast cells. *Antimicrob. Agents Chemother.* **18,** 130-136.

Giard, D.J., Fleischaker, R.J., and Fabrikant, M. (1982) Effect of temperature on the production of human fibroblast interferon. *Proc. Soc. Exp. Biol. Med.* **170,** 155-159.

Glaser, V. (1998) Fake blood marked gets hemoglobin transfusion from reticulocytes. *Nat. Biotechnol.* **16,** 709.

Gorospe, M., Cirielli, C., Wang, X., Seth, P., Capogrossi, M.C., and Holbrook, N.J . (1997) p21 (Waf1/Cip1) protects against p53-mediated apoptosis in human melanoma cells. *Oncogene* **14,** 929-935.

Gossen, M. and Bujard, H. (1992) Tight control of gene expression in mammalian cells by tetracycline-responsive promoters. *Proc. Natl. Acad. Sci. USA* **89,** 5547-5551.

Gregory, C.D., and Milner, A.E. (1994) Regulation of cell survival in Burkitt lymphoma: implications from studies of apoptosis following cold-shock treatment. *Int. J. Cancer* **57,** 419-426.

Greulich, H., and Erikson, R.L. (1998) An analysis of Mek1 signaling in cell proliferation and transformation. *J. Biol. Chem.* **273,** 13280-13288.

Grunstein, M. (1997) Histone acetylation in the chromatin structure and transcription. *Nature* **389,** 349-352.

Gurtu, V., Yan, G. and Zhang, G. (1996) IRES bicistronic expression vector for efficient creation of stable mammalian cell lines. *Biochem. Biophys. Res. Commun.* **229,** 295-298.

Haas-Kogan, D.A., Kogan, S.C., Levi, D., Dazin, P., T'Ang, A., Fung, Y.K.T., and Israel, M.A. (1995) Inhibition of apoptosis by the retinoblastoma gene product. *EMBO J.* **14,** 461-472.

Halevy, O., Novitch, B.G., Spicer, D.B., Skapek, S.X., Rhee, J., Hannon, G.J., Beach, D., and Lassar, A.B. (1995) Correlation of terminal cell cycle arrest of skeletal muscle with induction of p21 by MyoD. *Science* **267,** 1018-1021.

Hashemolhosseini, S., Nagamine, Y., Morley, S.J., Desrivieres, S., Mercep, L., and Ferrari, S. (1998) Rapamycin inhibition of the G1 to S transition is mediated by effects on cyclin D1 mRNA and protein stability. *J. Biol. Chem.* **273,** 14424-14429.

Hauser, H. (1997) in H. Hauser and R. Wagner (eds.) *Mammalian cell biotechnology in protein production*, W. DeGruyter, New York. pp. 3-32.

Hauser, P.J., Agrawal, D., Flanagan, M., and Pledger, W.J. (1997) The role of p27kip1 in the in vitro differentiation of murine keratinocytes. *Cell Growth Differ.* **8:** 203-211.

Hendriks-Taylor, L.R., and Darlington, G.J. (1995) Inhibition of cell proliferation by C/EBPα occurs in many cell types, does not require the presence of p53 or Rb, and is not affected by large T-antigen. *Nucleic Acids Res.* **23:** 4726-4733.

Heerdt, B.G., Houston, M.A., and Augenlicht, L.H. (1997) Short-chain fatty acid-initiated cell cycle arrest and apoptosis of colinic epithelial cells is linked to mitochondrial function. *Cell Growth Differ* **8,** 523-532.

Hightower, L.E., and Hendershot, L.M. (1997) Molecular chaperones and the heat shock response at Cold Spring Harbor. *Cell Stress Chaperones* **2,** 1-11.

Holmberg, N., Lilius, G., Bailey, J.E., and Bülow, L. (1997) Transgenic tabacco expressing *Vitreoscilla* hemoglobin exhibits enhanced growth and altered metabolic production. *Nature Biotechnol.* **15,** 244-247.

Hovey, A., Bebbington, C.R., and Jenkins, N. (1994) Control of growth and recombinant protein synthesis by heat-shock in a mutant mammalian cell line. *Biotechnol. Lett.* **16,** 215-220.

Hunt, S.M.N., Pak, S.C.O., Bridges, M.W., Gray, P.P., and Sleigh, M.J. (1997) Chinese hamster ovary cells produce sufficient insulin-like growth factor I to support growth in serum-free medium. *Cytotechnology* **24,** 55-64.

Itoh, Y., Ueda, H., and Suzuki, E. (1995) Overexpression of *bcl-2*, apoptosis suppressing gene: prolonged viable culture period of hybridoma and enhanced antibody production. *Biotechnol. Bioeng.* **48,** 118-122.

Jenkins, N., and Hovey, A. (1993) Temperature control of growth and productivity in mutant Chinese hamster ovary cells synthesizing a recombinant protein. *Biotechnol. Bioeng.* **42,** 1029-1036.

Jost, C.A., Marin, M.C., and Kaelin, Jr. W.G. (1997) p73 is a human p53 related protein that can induce apoptosis. *Nature* **389,** 191-194.

Juan, G., Ardelt, B., Li, Y., Mikulski, S.M., Shogen, K., Ardelt, W., Mittelman, A., and Darzynkiewicz, Z. (1998) G1 arrest of U937 cells by onconase is associated with suppression of cyclin D3 expression, induction of p16INK4A, p21WAF1/CIP1 and p27KIP and decreased pRB phosphorylation. *Leukemia* **12,** 1241-1248.

Kanatani, Y., Makishima, M., Asahi, K., Sakurai, A., Takahashi, N., Motoyoshi, K., Nagata, N., and Honma, Y. (1997) Differanisole A, a novel antitumor antibiotic, enhances growth inhibition and differntiation of human myeloid leukemia cells induced by 9-cis retinoic acid. *Biochim. Biophys. Acta* **1359,** 71-79.

Kaufmann, H., Mazur, X., Fussenegger, M., and Bailey, J.E. (1999) Influence of low temperature on productivity, proteome and protein phosphorylation of CHO cells. *Biotechnol. Bioeng.*, in press.

Kerkhoff, E., and Rapp, U.R. (1997) Induction of cell proliferation in quiescent NIH 3T3 cells by oncogenic c-Raf-1. *Mol. Cell Biol.* **17,** 2576-2586.

Kijima, M., Yoshida, M., Sugita, K., Horinouchi, S., and Beppu, T. (1993) Trapoxin, an antitumor cyclic tetrapeptide, is an irreversible inhibitor of mammalian histom deacetylase. *J. Biol. Chem.* **268,** 22429-22435.

Kim, Y.H., Iida, T., Prochownik, E.V., and Suzuki, E. (1998) Inhibition of c-jun expression in F-MEL cells causes cell cycle arrest and prevention of apoptosis. in O.W. Merten, P. Perrin and J.B. Griffiths (eds.), *New Developments and New Applications in Animal Cell Technology*, Kluwer Academic Publishers, Dordrecht, pp. 247-254.

Kim, Y.H., Iida, T., Fujita, T., Terada, S., Kitayama, A., Ueda, H., Prochownik, E.V., and Suzuki, E. (1998) Establishement of an apoptosis-resistant and growth-controllable cell line by transfecting with inducible antisense c-Jun gene. *Biotechnol. Bioeng.* **58,** 65-72.

Kirchhoff, S., Schaper, F., and Hauser, H. (1993) Interferon regulatory factor 1 (IRF-1) mediates cell growth inhibition by transactivation of downstream target genes. *Nucleic Acids Res.* **21,** 2881-2889.

Kirchhoff, S., Koromilas, A., Schaper, F., Grashoff, M., Sonenberg, N., and Hauser, H. (1995) IRF-1 induced cell growth inhibition and interferon induction requires the activity of the protein kinase PKR. *Oncogene* **11,** 439-445.

Kirchhoff, S., Kröger, A., Cruz, H., Tümmler, M., Schaper, F., and Hauser, H. (1996) Regulation of cell growth by IRF-1 in BHK-21 cells. *Cytotechnology* **22,** 147-156.

Ko, L.J., and Prives, C. (1996). p53: puzzle and paradigm. *Genes Dev.* **10,** 1054-1072.

Koike, M., Elstner, E., Campbell, M.J., Asou, H., Uskokovic, M., Tsuruoka, N., Koeffler, H.P. (1997) 19-nor-hexafluoride analogue of vitamin D3: a novel class of potent inhibitors of proliferation of human breast cell lines. *Cancer Res.* **57,** 4545-4550.

Kojima, Y., and Yoshida, F. (1974) Enhanced production of interferon by temperature shift-down from 37°C to 25°C in rabbit cell cultures stimulated with Newcastle disease virus. Japan. *J. Microbiol.* **18,** 217-222.

Kruman, I.I., Gukovskaya, A.S., Petrunyaka, V.V., Beletsky, I.P., and Trpakova, E.S. (1992) Apoptosis of murine BW5147 thyoma cells induced by cold shock. *J. Cell Physiol.* **153,** 112-117.

Kundu, M., Sharma, S., De Luca, A., Giordano, A., Rappaport, J., Khalili, K., and Amini, S. (1998) HIV-1 tat elongates the G1-phase and indirectly promotes HIV-1 gene expression in cells of glial origin. *J. Biol. Chem.* **273,** 8130-8136.

Kwon, T.K., Buchholz, M.A., Chrest, F.J., and Nordin, A.A. (1996) Staurosporine-induced G1 arrest is associated with the induction and accumulation of cyclin-dependent kinase inhibitors. *Cell Growth Diff.* **7,** 1315-1313.

Kwon, T.K., and Nordin, A.A. (1997) Overexpression of cyclin E and cyclin-dpendent kinase inhibitor (p27Kip1): effect on cell cycle regulation in HeLa cells. *Biochem. Biophys. Res. Commun.* **238,** 534-538.

Laad, A.C., Pyatt, R., Gothot, A., Rice, S., McMahel, J., Traycoff, C.M., and Srour, E.F. (1997) Orderly process of sequential cytokine stimulation is required for activation and maximal proliferation of primitive human bone marrow CD34+ hematopoietic progenitor cells residing in G0. *Blood* **90,** 658-668.

Lamotte, D., Monaco, L., Jenkins, N., and Marc, A. (1998) Influence of Na-butyrate on the production and sialylation of human interferon-γ by α2,6-sialyltransferase engineered CHO cells. in O.W. Merten, P. Perrin and J.B. Griffiths (eds.), *New Developments and New Applications in Animal Cell Technology,* Kluwer Academic Publishers, Dordrecht, pp. 131-133.

Lee, K.H., Sburlati, A., Renner, W.A., and Bailey, J.E. (1996) Deregulated expression of clonded transcription factor E2F-1 in Chinese hamster ovary cells shifts protein patterns and activates growth in protein-free medium. *Biotechnol. Bioeng.* **50,** 273-279.

Lepley, D.M., and Pelling, J.C. (1997) Induction of p21/WAF1 and G1 cell-cycle arrest by the chemopreventive agent apigenin. *Mol. Carcinog.* **19,** 74-82.

Li, Y., Lin, B., Agadir, A., Liu, R., Dawson, M.I., Reed, J.C., Fontana, J.A., Bost, F., Hobbs, P.D., Zheng, Y., Chen, G.Q., Shroot, B., Mercola, D., and Zhang, X.K. (1998) Molecular determinants of AHPN (CD437)-induced growth arrest and apoptosis in human lung cancer cell lines. *Mol. Cell Biol.* **18;** 4719-4731.

Linke, S.P., Clarkin, K.C., Leonardo, A.D., Tsou, A., and Wahl, G.M. (1996) A reversible, p53-dependent G_0/G_1 cell-cycle arrest induced by ribonucleotide depletion in the absence of detectable DNA damage. *Genes Dev.* **10,** 934-947.

Liu, J.J., Chao, J.R., Jiang, M.C., Ng, S.Y., Yen, J.J.Y., and Yang-Yen, H.F. (1995) Ras transformation results in an elevated level of cyclin D1 and acceleration of G1 progression in NIH 3T3 cells. *Mol. Cell Biol.* **15,** 3654-3663.

Ludwig, A., Tomeczkowski, J., and Kretzmer, G. (1992) Influence of the temperature on the shear stress sensibility of adherent BHK-21 cells. *Appl. Microbiol. Biotechnol.* **38,** 323-327.

Marcus, M., Fainsod, A., and Diamond, G. (1985) The genetic analysis of mammalian cell-cycle mutants. *Ann. Rev. Genet.* **19,** 389-421.

Marhin, W.W., Chen, S., Facchini, L.M., Fornace, A.J. Jr., and Penn, L.Z. (1997) Myc represses the growth arrest gene gadd45. *Oncogene* **14,** 2825-2834.

Martin, P. (1997) Wound healing - aiming for perfect skin regeneration. *Science* **276,** 75-81.

Matsuo, T., and Thiele, C.J. (1998) p27Kip1: a key mediator of retinoic acid induced growth arrest in the SMS-KCNR human neuroblastoma cell line. *Oncogene* **16,** 3337-3343.

Matsuoka, M., Tani, K., and Asano, S. (1998) Interferon-alpha-induced G1 phase arrest through up-regulated expression of CDK inhibitors, p19Ind4D and p21Cip1 in mouse macrophages. *Oncogene* **16,** 2075-2086.

Mazur, X., Fussenegger, M., Renner, W.A., and Bailey, J.E. (1998) Higher productivity of growth-arrested Chinese hamster ovary (CHO) cells expressing the cyclin-dependent kinase inhibitor p27. *Biotechnol. Prog.* **14,** 705-713.

Mazur, X., Eppenberger, H., Bailey, J.E., and Fussenegger, M. (1999) A novel autoregulated proliferation-controlled production process using recombinant CHO cells. *Biotechnol. Bioeng.*, submitted.

McCormick, F. (1993) Signal transduction. How receptors turn Ras on. *Nature* **363,** 15-16.

McKay, R. (1997) Stem cells in the nervous system. *Science* **276,** 66-71.

Meijer, L., Borgne, A., Mulner, O., Chong, J.P., Blow, J.J., Inagaki, N., Inagaki, M., Delcros, J.G., and Moulinoux, J.P. (1997) Biochemical and cellular effects of roscovitine, a potent and selective inhibitor of the cyclin-dependent kinases cdc2, cdk2 and cdk5. *Eur. J. Biochem.* **243,** 527-536.

Meikrantz, W., Gisselbrecht, S., Tam, S.W., and Schlegel, R.A. (1994) Activation of cyclin A-dependent protein kinases during apoptosis. *Proc. Nat. Acad. Sci. USA* **91,** 3754-3758.

Meijer, L. (1996) Chemical inhibitors of cyclin-dependent kinases. *Trends in Cell Biol.* **6,** 393-397.

Meijer, L., and Kim, S.H. (1997) Chemical inhibitors of cyclin-dpendent kinases. *Methods in Enzymology* **283,** 113-128.

Mercille, S., and Massie, B. (1994) Induction of apoptosis in nutrient-limited cultures of hybridoma cells. *Biotechnol. Bioeng.* **44,** 1140-1154.

Michalopoulos, G.K., and DeFrances, M.C. (1997) Liver regeneration. *Science* **276,** 60-66.

Miller, R.F., Sokoloski, T.D., Mitscher, L.A., Bonacci, A.C. and Hoener, B. (1973) Use of circular dichroism in analysis of mixtures of tetracycline and 4-epitetracycline and its application to assay of commercial products. *J. of Pharmaceutical Sciences* **62,** 1143-1147.

Miller, W.M., Blanch, H.W., and Wilke, C.R. (1988) A kinetic analysis of hybridoma growth and metabolism in batch and continuous suspension culture: Effect of nutrient concentration, dilution rate, and pH. *Biotechnol. Bioeng.* **32,** 974-965.

Minn, A.J., Boise, L.H., and Thompson, C.B. (1996) Expression of bcl-x_L and loss of p53 can cooperate to overcome a cell cycle checkpoint induced by mitotic spindle damage. *Genes Dev.* **10,** 2621-2631.

Mischoulon, D., Rana, B., Bucher, N.L.R., and Farmer, S.R. (1992) Growth-dependent inhibition of CCAAT enhancer-binding protein (C/EBP alpha) gene expression during hepatocytes proliferation in the regenerating liver and in culture. *Mol. Cell Biol.* **12,** 2553-2560.

Mizuno, S., Ohkawara, A., and Suzuki, K. (1989) Defect in the development of thermotolerance and enhanced heat shock protein synthesis in the mouse temperature-sensitive mutant ts85 cells upon moderate hyperthermia. *Int. J. Hyperthermia* **5,** 105-113.

Moore, A., Donahue, CJ.., Hooley, J., Stocks, D.L., Bauer, K.D., and Mather, J.P. (1995) Apoptosis in CHO cell batch cultures: examination by flow cytometry. *Cytotechnology* **17,** 1-11.

Moore, A., Mercer, J., Dutina, G., Donahue, C.J., Bauer, K.D., Mather, J.P., Etcheverry, T., and Ryll, T. (1997) Effects of temperature shift on cell cycle, apoptosis and nucleotide pools in CHO cell batch cultures. *Cytotechnology* **23,** 47-54.
Mordente, J.A., Konno, S., Chen, Y., Wu, J.M., Tazaki, H., and Mallouh, C. (1998) The effects of brefeldin A (BFA) on cell cycle progression involving the modulatin of the retinoblastoma protein (pRB) in PC-3 prostate cancer cells. *J. Urol.* **159,** 275-279.
Morimoto, R.I., and Santoro, M.G. (1998) Stress-inducible responses and heat shock proteins: New pahrmacologic targets for cytoprotection. *Nature Biotechnol.* **16,** 833-838.
Moyer, J.D., Barbacci, E.G., Iwata, K.K., Arnold, L., Boman, B., Cunningham, A., DiOrio, C., Doty, J., Morin, M.J., Moyer, M.P., Neveu, M., Pollack, V.A., Pustilnik, L.R., Reynolds, M.M., Sloan, D., Theleman, A., and Miller, P. (1997) Induction of apoptosis and cell cycle arrest by CP-358,774 and inhibitor of epidermal growth factor receptor tyrosine kinase. *Cancer Res.* **57,** 4838-4848.
Müller, P.P., Kirchhoff, S., and Hauser, H. (1998) Sustained expression in proliferation controled BHK-21 cells. in O.W. Merten, P. Perrin and J.B. Griffiths (eds.), *New Developments and New Applications in Animal Cell Technology*, Kluwer Academic Publishers, Dordrecht, pp. 209-214.
Murata, Y., Kudoh, T., Sugiyama, H., Toyoshima, K., and Akiyama, T. (1997) The Wilms tumor suppressor gene WT1 induces G1 arrest and apoptosis in myeloblastic leukemia M1 cells. *FEBS Lett.* **409,** 41-45.
Nakano, K., Mizuno, T., Sowa, Y., Orita, T., Yoshino, T., Okuyama, Y., Fujita, T., Ohtani-Fujita, N., Matsukawa, Y., Tokino, T., Yamagishi, H., Oka, T., Nomura, H., and Sakai, T. (1997) Butyrate activates the WAF1/Cip1 gene promoter through Sp1 sites in a p53-negative human colon cancer cell line. *J. Biol. Chem.* **272,** 22199-22206.
Neumann, F., and Krawinkel, U. (1997) Constitutive expression of human ribosomaol protein L7 arrests the cell cycle in G1 and induces apoptosis in Jurkat T-lymphoma cells. *Exp. Cell Res.* **230,** 252-261.
News (1998) Bone morphogenic proteins and tissue engineering. *Nat. Biotechnol.* **14,** 242-244.
Newton, K., and Strasser, A. (1998) The Bcl-2 family and cell death regulation. *Curr. Opin. Gen. Dev.* **8,** 68-75.
Nishimoto, T. (1997) Isolation and characterization of temperature-sensitive mammalian cell cycle mutants. *Methods in Enzymol.* **283,** 292-309.
Nishiyama, H., Itoh, K., Kaneko, Y., Kishishita, M., Yoshida, O., and Fujita, J. (1997a) A glycine-rich RNA-binding protein mediating cold-inducible suppression of mammalian cell growth. *J. Cell Biol.* **137,** 899-908.
Nishiyama, H., Higashitsuji, H., Yokoi, H., Itoh, K., Danno, S., Matsuda, T., and Fujita, J. (1997b) Cloning and characterization of human CIRP (cold-inducible RNA-binding protein) cDNA and chromosomal assignment of the gene. *Gene* **204,** 115-120.
Nitta, M., Okamura, H., Aizawa, S., and Yamaizumi, M. (1997) Heat shock induces transient p53-dependent cell cycle arrest at G1/S. *Oncogene* **15,** 561-568.
Ohnishi, T., Xinjiang, W., Ohnishi, K., and Takahashi, A. (1998) p53-dependent induction of WAF1 by cold shock in human glioblastoma cells. *Ongocene* **16;** 1507-1511.
O'Reilly, L.A., Huang, D.C.S., and Strasser, A. (1996) The cell death inhibitor Bcl-2 and its homologues influence control of cell cycle entry. *EMBO J.* **15,** 6979-6990.
Oster, T., Thioudellet, C., Chevalot, I., Masson, C., Wellman, M., Marc, A., and Siest, G. (1993) Induction of recombinant gamma-glutamyl transferase by sodium butyrate in transfected V79 and CHO chinese hamster cells. *Biochem. Biophys. Res. Commun.* **193,** 406-412.
Ota, T., Hanada, K., and Hashimoto, I. (1996) The effect of cold stress on UVB injury in mouse skin and cultured keratinocytes. *Photochem. Photobiol.* **64,** 984-987.
Oda, T., Maeda, M., Odashima, S., Ninomiya-Tsuji, J., and Tatsuka, M. (1997) G1 phase-specific suppression of the Cdk2 activity by ginsenoside Rh2 in cultured murine cells. *Life Sci.* **60,** 39-44.
Pak, S.C.O., Hunt, S.M.N., Bridges, M.W., Sleigh, M.J., and Gray, P.P. (1996) Super CHO - a cell line capable of autocrine growth under fully defined protein-free conditions. *Cytotechnology* **22,** 139-146.
Pallavicini, M.G., Rosette, C., Reitsma, M., Deteresa, P.S., and Gray, J.W. (1990) Relationship of c-myc gene copy number and gene expression: cellular effects of elevated c-myc protein. *J. Cell Physiol.* **143,** 372-380.
Papoutsakis, E.T. (1998) Express together and conquer. *Nature Biotechnol.* **16,** 416-417.
Passini, C.A., and Gouchee, C.F. (1989) Response of a mouse hybridoma cell line to heat shock, agitation and sparging. *Biotechnol. Prog.* **5,** 175-188.
Pennisi, E. (1996) Will a twist of viral fate lead to a new cancer treatment? *Science* **274,** 342-343.
Perlman, H., Suzuki, E., Simonson, M., Smith, R.C., and Walsh, K. (1998) GATA-6 induces p21(Cip1) expression and G1 cell cycle arrest. *J. Biol. Chem.* **273,** 13713-13718.
Petersen, J.F., McIntire, L.V., and Papoutsakis, E.T. (1990) Shear sensitivity of hybridoma cells in batch, fed-batch and continuous cultures. *Biotechnol. Prog.* **6,** 114-120.
Polyak, K., Kato, J.Y., Solomon, J.M., Sherr, C.J., Massague, J., Roberts, J.M., and Koff, A. (1994a) p27Kip1, a cyclin-Cdk inhibitor, links transforming growth factor-beta and contact inhibition to cell cycle arrest. *Gene Dev.* **8,** 9-22.
Polyak, K., Lee, M.H., Erdjument-Bromage, H., Koff, A., Roberts, J.M., Tempst, P., and Massague, J. (1994b) Cloning of $p27^{Kip1}$, a cyclin-dependent kinase inhibitor and a potential mediator of extracellular antimitogenic signals. *Cell* **78,** 59-66.

Polyak, K., Kato, J.Y., Solomon, M.J., Sherr, C.J., Massagué, J., Roberts, J.M., and Koff, A. (1994) p27Kip1, a cyclin-Cdk inhibitor, links transforming growth factor-β and contact inhibition to cell-cycle arrest. *Gene and Dev.* **8,** 9-22.

Polyak, K., Waldman, T., He, T.-C., Kinzler, K.W., and Vogelstein, B. (1996) Genetic determinants of p53-induced apoptosis and growth arrest. *Genes Dev.* **10,** 1945-1952.

Pomahac, B., Svensjo, T., Yao, F., Brown,, H., and Eriksson, E. (1998) Tissue engineering of skin. *Crit. Rev. Oral Biol. Med.* **9,** 333-344.

Prockop, D.J. (1997) Marrow stromal cells as stem cells for nonhematopoietic tissues. *Science* **276,** 71-74.

Ramirez, O.T., and Mutharasan, R. (1990) Cell cycle- and growth phase-dependent varations in size distribution, antibody productivity and oxygen demand in hybridoma cultures. *Biotechnol. Bioeng.* **36,** 839-848.

Rao, L., Debbas, M., Sabbatini, P., Hockenbery, D., Korsmeyer, S.J., and White, E. (1992) The adenovirus E1A proteins induce apoptosis, which is inhibited by the E1B 19-kDa and Bcl-2 proteins. *Proc. Natl. Acad. Sci. USA* **89,** 7742-7746.

Renner, W.A., Lee, K.H., Hatzimanikatis, V., Bailey, J.E., and Eppenberger, H.M. (1995) Recombinant cyclin E expression activates proliferation and obviates surface attachment of Chinese hamster ovary (CHO) cells in protein-free medium. *Biotechnol. Bioeng.* **47,** 476-482.

Rees, S., Coote, J., Stables, J., Goodson, S., Harris, S., Lee, M.G. (1996) Bicistronic vector for the creation of stable mammalian cell lines that predisposes all antibiotic-resistant cells to express recombinant protein. *BioTechniques* **20,** 48-56.

Reuveny, S., Kim, Y.J., Kemp, C.W., and Shiloach, J. (1993) Effect of temperature and oxygen on cell growth and recombinant protein production in insect cell cultures. *Appl. Microbiol. Biotechnol.* **38,** 619-623.

Reynisdottir, I., Polyak, K., Iavarone, A., and Massagué, J. (1995) Kip/Cip and Ink4 inhibitors cooperate to induce cell-cycle arrest in response to TGF-β. *Genes Dev.* **9,** 1831-1845.

Riggs, M.G., Whittaker, R.G., Neumann, J.R., and Ingram, V.M. (1977) n-Butyrate causes histone modification in HeLa and Friend erythroleukaemia cells. *Nature* **268,** 462-464.

Rivard, N., L'Allemain, G., Bartek, J., and Pouyssegur, J. (1996) *J. Biol. Chem.* **271,** 18337-18341.

Rohlff, C., Blagosklonny, M.V., Kyle, E., Kesar, A., Kim, I.Y., Zelner, D.J., Hakim, F., Trepel, J., and Bergan, R.C. (1998) Prostate cancer cell growth inhibition by tamoxifen is associated with inhibition of protein kinase C and induction of p21. *Prostate* **37,** 51-59.

Ross, S., and Englesberg, E. (1993) The competence progression model in CHO-K1 cells: the relationship between protein kinase C and immediate early gene expression in the insulin mitogenic signal. *Biochim. Biophys. Acta* **1177,** 307-317.

Roti Roti, J.L., Mackey, M.A., and Higashikubo, R. (1992) The effects of heat shock on cell proliferation. *Cell Prolif.* **25,** 89-99.

Rowan, S., Ludwig, R.L., Haupt, Y., Bates, S., Lu, X., Oren, M., and Vousden, K.H. (1996) Specific loss of apoptotic but not cell-cycle arrest function in a human tumor derived p53 mutant. *EMBO J.* **15,** 827-838.

Ryseck, R.P., Hirai, S.I., Yaniv, M., and Bravo, R. (1988) Transcriptional activation of c-jun during the G0/G1 transition in mouse fibroblasts. *Nature* **334,** 535-537.

Sawai, H., Okazaki, T., Yamamoto, H., Okano, H., Takeda, Y., Tashima, M., Sawada, H., Okuma, M., Ishikura, H., Umehara, H., and Domae, N. (1995) Requirement of AP-1 for ceramide-induced apoptosis in human leukemia HL-60 cells. *J. Biol. Chem.* **270,** 27326-27331.

Sawyers, C.L., McLaughlin, J., Goga, A., Havlik, M., and Witte, O. (1994) The nuclear tyrosine kinase c-Abl negatively regulates cell growth. *Cell* **77,** 121-131.

Sekharam, M., Trotti,, A., Cunnick, J.M., and Wu, J. (1998) Suppression of fibroblast cell cycle progression in G1 phase by N-acetylcysteine. *Toxicol. Appl. Pharmacol.* **149,** 210-216.

Sekiguchi, T., and Hunter, T. (1998) Induction of growth arrest and cell death by overexpression of the cyclin-Cdk inhibitor p21 in BHK21 cells. *Oncogene* **16,** 369-380.

Serrano, M., Lin, A.W., McCurrach, M.E., Beach, D., and Lowe, S.W. (1997) Oncogenic ras provokes premature cell senescence associate with accumulation of p53 and p16INK4a. *Cell* **88,** 593-602.

Sewing, A., Wiseman, B., Lloyd, A.C., and Land, H. (1997) High-intensity Raf signal causes cell cycle arrest mediated by p21Cip1. *Mol. Cell Biol.* **17,** 5588-5597.

Singh, R.P., Al-Rubeai, M., Gregory, C.D. and Emery, A.N. (1994) Cell death in bioreactors: a role for apoptosis. *Biotechnol. Bioeng.* **44,** 720-726.

Singh, R.P., Fassnacht, D., Perani, A., Simpson, N.H., Goldenzon, C., Pörtner, R., and Al-Rubeai, M. (1998) Modulation of apoptosis by bcl-2 expression following amino acid deprivation and in high cell density perfusion cultures. in O.W. Merten, P. Perrin and J.B. Griffiths (eds.), *New Developments and New Applications in Animal Cell Technology*, Kluwer Academic Publishers, Dordrecht, pp. 235-240.

Sioud, M., Sorensen, D.R. (1998) A nuclease-resistant protein kinase Cα ribozyme blocks glioma cell growth. *Nature Biotechnol.* **16,** 556-561

Sisken, J.E., Morasca, L., and Kibby, S. (1965) Effects of temperature on the kinetics of the mitotic ceycle of mammalian cells in culture. *Exp. Cell Res.* **39,** 103-116.

Smith, M.J., and Prochownik, E.V. (1992) Inhibition of c-jun causes reversible proliferative arrest and withdrawal from the cell cycle. *Blood* **79;** 2107-2115.

Smith, R.C., Branellec, D., Gorski, D.H., Guo, K., Perlman, H., Dedieu, J.F., Pastore, C., Mahfoudi A., Denefle, P., Isner, J.M., and Walsh, K. (1997) p21CIP+-mediated inhibition of cell proliferation by overexpression of the gax homeodomain gene. *Genes Dev.* **11,** 1674-1689.
Sinha, R., and Medina, D. (1997) Inhibition of cdk2 kinase activity by methylselenocysteine in synchronized mouse mammary epithelial tumor cells. *Carcinogenesis* **18,** 1541-1547.
Somasundaram, K., Zhang, H., Zeng, Y.X., Houvras, Y., Peng, Y., Zhang, H., Wu, G.S., Licht, J.D., Weber, B.L., and El-Deiri, W.S. (1997) Arrest of the cell cycle by the tumour-suppressor BRCA1 requires the CDK-inhibitor $p21^{WAF1/Cip1}$. *Nature* **389,** 187-190.
Soucek, T., Pusch, O., Wienecke, R., DeClue, J.E., and Hengstschlager, M. (1997) Role of the tuberous sclerosis gene-2 product in cell cycle control. Loss of the tuberous sclerosis gen-2 induces quiescent cells to enter S phase. *J. Biol. Chem.* **272,** 29301-29308.
Studer, L., Tabar, V., and McKay, R.D.G. (1998) Transplantation of expanded mesencephalic precursors leads to recovery in parkinsonian rats. *Nature Neurobiology* **1,** 290-295.
Sullivan, K.A., Kim, B., Buzon, M., and Feldman, E.L. (1997) Suramin disrupts insulin-like growth factor-II (IGF-II) mediated autocrine growth in human SH-SY5Y neuroblastom cells. *Brain Res.* **744,** 199-206.
Sureshkumar, G.K., and Muthaeasan, R. (1991) The influence of temperature on a mouse-mouse hybridoma growth and monoclonal antibody production. *Biotechnol. Bioeng.* **37,** 292-295.
Suzuki, E., and Ollis, D.F. (1989) Cell cycle model for antibody production kinetics. *Biotechnol. Bioeng.* **34,** 1398-1402.
Suzuki, E., and Ollis, D.F. (1990) Enhanced antibody production at slowed growth rates: experimental demonstration and a simple structured model. *Biotechnol. Prog.* **6,** 231-236.
Takagi, M., and Ueda, K. (1994) Comparison of the optimal culture conditions for cell growth and tissue plasminogen activator production by human embryo lung cells on microcarriers. *Appl. Microbiol. Biotechnol.* **41,** 565-570.
Tamura, T., Ishihara, M., Lamphie,r M.S., Tanaka, N., Oishi, I., Aizawa, S., Matsuyama, T., Mak, T.W., Taki, S., and Taniguchi, T. (1995) An IRF-1 dependent pathway of DNA-damage-induced apoptosis in mitogen-activated T lymphocytes. *Nature* **376,** 596-599.
Tanaka, H., Takenaka, H., Yamao, F., and Yagura, T. (1998) Aphidicolin induces alterations in Golgi complex and disorganization of microtubules of HeLa cells upon long-term administration. *J. Cell Physiol.* **176,** 602-611.
Tanaka, N., Ishihara, M., Lamphier, M.S., Mak, T.W., Aizawa, S., Tokino, T., Oren, M., and Taniguchi, T. (1996) Cooperation of the tumour suppressors IRF-1 and p53 in response to DNA damage. *Nature* **382,** 816-818.
Tanikawa, M., Yamada, K., Tominaga, K., Morisaki, H., Kaneko, Y., Ikeda, K., Suzuki, M., Kiho, T., Tomokiyo, K., Furuta, K., Noyori, R., and Nakanishi, M. (1998) Potent prostaglandin A1 analogs that suppress tumor cell growth through induction of p21 and reduction of cyclin E. *J. Biol. Chem.* **273,** 18522-18527.
Thieringer, H.A., Jones, P.G., and Inouye, M. (1998) Cold shock and adaptation. *Bioessays* **20,** 49-57.
Thomason, J.A., Itskovitz-Eldor, J., Shapiro, S.S., Waknitz, M.A., Swiergiel, J.J., Marshall, V.S., and Jones, J.M. (1998) Embryonic stem cell lines derived from human blastocysts. *Science* **282,** 1145-1147.
Timchenko, N.A., Wilde, M., Nakanishi, M., Smith, J.R., and Darlington, G.J. (1996) CCAAT/enhancer-binding protein a (C/EBPa) inhibits cell proliferation through the p21 (WAF-1/CIP-1/SDI-1) protein. *Genes Dev.* **10,** 804-815.
Turley, J.M., Ruscetti, F.W., Kim, S.J., Fu, T., Gou, F.V., and Birchenall-Roberts, M.C. (1997) Vitamine E succinate inhibits proliferation of BT-20 human breast cancer cells: increased binding of cyclin A negatively regulates E2F transactivation activity. *Cancer Res.* **57,** 2668-2675.
Turner, R., and Tjlan, R. (1989) Leucine repeats and an adjacent DNA binding domain mediate the formation of functional cFos-cJun heterodimers. *Science* **243,** 1689-1694.
Ueno, T., Ohtawa, K., Mitsui, K., Kodera, Y., Hiroto, M., Matsushima, A., Inada, Y., and Nishimura, H. (1997) Cell cycle arrest and apoptosis of leukemia cells induced by L-asparaginase. *Leukemia* **11,** 1858-1861.
Uhrbom, L., Nister, M., and Westermark, B. (1997) Induction of senescence in human malignant glioma cells by p16INK4A. *Oncogene* **15,** 505-514.
Vazari, C., and Faller, D.V. (1997) A benzo(a)pyrene-induced cell cycle checkpoint resulting in p53-independent G1 arrest in 3T3 fibroblasts. *J. Biol. Chem.* **272,** 2762-2769.
Vaziri, C., Stice, L., and Faller, D.V. (1998) Butyrate-induced G1 arrest results from p21-independent disruption of retinoblastoma protein-mediated signals. *Cell Growth Differ.* **9,** 465-474.
Vilcek, J., and Havell, E.A. (1993) Stabilization of interferon messenger RNA activity by treatment of cells with metabolic inhibitors and lowering the incubation temperature. *Proc. Natl. Acad. Sci. USA* **70,** 3909-3913.
Voellmy, R., Ahmend, A.., Schiller, P., Bromley, P., and Rungger, D. (1995) Isolation and functional analysis of a human 70'000-dalton heat shock protein gene segment. *Proc. Natl. Acad. Sci. USA* **82:** 4949-4953.
Vogt, A., Sun, J., Quan, Y., Hamilton, A.D., Sebti, S.M. (1997) The geranylgeranyltransferase-I inhibitor GGTI-298 arrests human tumor cells in G0/G1 and induces p21 (WAF1/CIP1/SDI1) in a p53-independent manner. *J. Biol. Chem.* **272,** 27224-27229.

Walker, P.R., Smith, B., Carson, C., Leblanc, J., Sikorska, M., Woodhouse, C.S., and Morgan, A.C. (1997) Induction of apoptosis in neoplastic cells by depletion of vitamin B12. *Cell Death Differentiation* **4,** 233-241.

Wang, W., Smith, R., Burghardt, R., and Safe, S.H. (1997) 17 beta-Estradiol-mediated growth inhibition of MDA-MB-468 cells stably transfected with the estrogen receptor: cell cycle effects. *Mol. Cell Endocrinol.* **133,** 49-62.

Wang, J., and Walsh, K. (1996) Resistance to apoptosis conferred by Cdk inhibitors during myocyte differentiation. *Science* **273,** 359-361.

Wang, X., Thant, A.A., Machida, K., Hiraiwa, Y., Iwata, H., Matsuda, S., and Hamaguchi, M. (1998) Suppression of cell growth by ectopic expression of N-cadherin. *Int. J. Oncol.* **12,** 1097-1101.

Weidemann, R., Ludwig, A., and Kretzmer, G. (1994) Low temperature cultivation - A step towards process optimization. *Cytotechnology* **15,** 111-116.

Willis, J.S. (1987) Cold tolerance in mammalian cells. *Symp. Soc. Exp. Biol.* **41,** 285-309.

Woo, R.A., McLure, K.G., Lees-Miller, S.P., Rancourt, D.E., and Lee, P.W.K. (1998) DNA-dependent protein kinase acts upstream of p53 in response to DNA damage. *Nature* **394,** 700-704.

Yan, G.Z., and Ziff, E.B. (1997) Nerve growth factor induces transcription of the p21 WAF1/Cip1 and cyclin D1 genes in PC12 cells by activating the Sp1 transcription factor. *J. Neurosci.* **17,** 6122-6132.

Yoshida, M., Yokokura, H., Hidaka, H., Ikekawa, T., and Saijo, N. (1998) Mechanism of antitumor action of PKC activator, gnidimacrin. *Int. J. Cancer* **77,** 243-250.

Zeitler, H., Ko, Y., Glodny, B., Trotzke, G., Appenheimer, M., Sachinidis, A., and Vetter, H. (1997) Cell-cycle arrest in G0/G1 phase of growth factor-induced endothelial celll proliferation by various calcium channel blockers. *Cancer Detect. Prev.* **21,** 332-339.

Zi, X., Feyes, D.K., and Agarwal, R. (1998) Anticarcinogenic effect of a flavonid antioxidant, silymarin, in human breast cancer cells MDA-MB 468: induction of G1 arrest through an increase in Cip1/p21 concomitant with a decreas in kinase activity of cyclin-dependent kinases and associated cyclins. *Clin. Cancer Res.* **4,** 1055-1064.

DEVELOPMENT OF AN IRF-1 BASED PROLIFERATION CONTROL SYSTEM

P. P. MUELLER[1], A. V. CARVALHAL[2], J. L. MOREIRA[2], C. GESERICK[1,3], K. SCHROEDER[1], M. J. T. CARRONDO[2] and H. HAUSER[1]*

[1] *Department of Gene Regulation and Differentiation, GBF - National Research Center for Biotechnology, Mascheroder Weg 1, D-38124 Braunschweig, Germany,* [2] *ITQB/IBET- Instituto de Tecnologia Quimica e Biologica/Instituto de Biologica Experimental e Tecnologia, Apartado 12, P-2780 Oeiras, Portugal,* [3] *Novo Nordisk, Niels Steensens Vej 1, 2820 Gentofte, Denmark*

* *Corresponding author, fax ++49-531-6181-262, e-mail HHA@GBF.DE*

1. Introduction

1.1. BENEFITS EXPECTED FROM PROLIFERATION CONTROL OF MAMMALIAN PRODUCER CELLS

Mammalian cell cultures are the preferred production system for secreted pharmaceutical proteins (Hauser, 1997). Life of a mammal begins with fertilization of the oocyte that divides and proliferates until the animal reaches its mature size. Further cell growth is tightly controlled. Proliferation and cell death are balanced to keep the total cell mass essentially constant, while the synthesis and secretion of cellular products continues. Despite rapid progress in the understanding of growth regulatory mechanisms, tumorigenic growth due the loss of proliferation control of a single body cell initially is still a leading cause of disease and death.

In contrast to the natural situation, due to the requirement to obtain large numbers of cells needed in production processes, mammalian producer cell lines have been selected for rapid and indefinite proliferation capacity. The cell lines used in industrial production processes are transformed and override natural growth control systems such as contact inhibition, anchorage-dependent growth, naturally limited numbers of cell divisions, organ size control and others. However, unrestricted cultured cell growth is associated with some disadvantages. Uncontrolled cell proliferation is associated with genetic instability that may negatively affect productivity with an increasing number of cell generations. After reaching an optimal cell density, further growth leads to changes in the production conditions associated with decreased quality and consistency of the product. The excess cells lead to nutrient and oxygen depletion, rapid accumulation of toxic products, cell lysis, clogging of cell retention and product purification devices, product contamination with cellular debris and product deterioration due to glycosidases and proteases (Fussenegger et al., 1999).

Regulated cell growth could mimic the natural situation in technical applications, with rapid proliferation initially until they reach an optimal cell density. Then reduced growth would extend the productive period and keep production conditions constant by lowering medium consumption and waste product accumulation. Since fewer resources are consumed for the synthesis of cell mass, production would be more efficient. The reduced cell division rate would also reduce genetic drift, and by that stabilize the productivity. Therefore, growth regulation could increase production, product quality and consistency.

We have genetically engineered BHK-21 cells to express IRF-1, a transcriptional activator of genes which lead to growth inhibition. We investigated several properties that are important to control cell growth in production processes. Primarily, the regulation must be stable despite a strong negative selection pressure. The regulation was stabilized by using a dicistronic construct to couple the transcription of the IRF-1 gene to an antibiotic resistance selection marker gene. Another requirement is high cell viability. Permanent IRF-1 activation for longer than two to four days lead to viability decreases. A schedule was developed of IRF-1 activation and inactivation in several cycles to control cell growth over a period of at least one month while maintaining a high cell viability. An aspect of fundamental importance in production processes is the productivity. Since IRF-1 is a transcriptional activator, promoters containing IRF-1 binding sites can be employed to increase transcription to levels even higher than those achieved with strong viral promoters. Finally, for an improved production process, a high product quality and homogeneity must be maintained throughout the production process. However, most recombinant secreted pharmaceutical proteins are glycoproteins and the glycan synthesis is affected by the cultivation conditions. When IRF-1 is activated, the glycan structure of a relevant pharmaceutical product remained essentially the same and there was no influence on site occupancy or protein integrity. These results demonstrate the promising potential of the IRF-1 proliferation control system for an improved production process.

1.2. APPROACHES TO CONTROL CELL GROWTH

Reduced growth was obtained by starving cells for an essential energy source or by the addition of DNA-synthesis inhibitors such as thymidine, hydroxyurea, TGF-β or genotoxic agents such as adriamycin, or by incubating temperature-sensitive mutant cells at the nonpermissive temperature (Al-Rubeai et al., 1992). All these procedures lowered the growth rate, and in some cases productivity was increased. However, the adaption to applied conditions is hampered by reduced cell viability or low productivity soon after the onset of growth arrest.

Genetic growth control systems have specific advantages. There is no need to change production conditions by the addition of toxic compounds, nutrient limitations or suboptimal temperatures that may affect productivity or product quality. The genetic systems are flexible and allow the stepwise improvement of the recombinant regulatory system, optimization of the producer cell and the production conditions. An additional advantage is that the expression of recombinant genes can be induced during growth arrest by using dedicated promoters. Specific requirements must be met by growth control systems concerning stability, productivity, cell viability, quality and consistency of the product and applicability to industrial fermenter systems. Presently under investigation are a system based on cell cycle inhibitory protein overexpression (Fussenegger et al., 1998; Fussenegger et al., 1997; Mazur et al., 1998, see review by Fussenegger in this volume) and a system using c-jun antisense RNA expression (Kim et al., 1998).

1.3. DEVELOPMENT OF A GENETIC PROLIFERATION CONTROL SYSTEM

Despite recent progress, so far no proliferation control system has met the criteria for an applied process, thus demonstrating the requirement for alternative approaches. Using a genetic approach, we have adapted the growth inhibitory factor interferon regulator factor 1 (IRF-1) to regulate cell growth and increase production in biotechnological applications.

2. Mechanisms of IRF-1 mediated proliferation control

2.1. THE BIOLOGICAL ROLE OF IRF-1

IRF-1 has been identified and characterized during investigations of the cellular response to viruses and interferons by its ability to bind to a consensus sequence present in the IFN-β promoter (Fujita *et al.*, 1988; Harada *et al.*, 1989). Related sequences appear within the promoters of many IFN-stimulated genes (ISGs) and some IFN-α promoters. IRF-1 binding to such promoters may explain why overexpression of IRF-1 to IFN-β gene transcription (Harada *et al.*, 1990; Fujita et al., 1989).

Since IRF-1 overexpression mimics many of the cellular interferon responses, IRF-1 was once thought to play a major role in the expression of IFN-inducible genes (Lengyel, 1993). However, in the mean time it has become clear that a whole family of IRF-1 related factors with potentially overlapping functions exist. For example, even though overexpressed IRF-1 can induce IFN-α expression, the delayed IFN-α induction after virus infection is due to IRF-7 activity (Au et al., 1998), and in mice lacking functional IRF-1 genes virus-induced IFN-β gene expression is normal (Reis *et al.*, 1994). On the other hand, IRF-1 functions independently of interferon to prevent tumorigenic cell growth (Tanaka et al., 1994). Therefore, it is presently believed that IRF-1 contributes to the transcriptional activation of some interferon induced genes and also of genes that are expressed in the absence of interferon signaling, and that the physiological IRF-1 function is to prevent uncontrolled cell proliferation.

2.2. IRF-1 DOMAIN STRUCTURE AND FUNCTION

IRF-1 is a typical modular protein whereby the function of the individual domains is independent of the overall structure of the protein. Indeed, it is possible to functionally transpose different IRF-1 domains of other protein in preserving their activities (Schaper et al., 1999) These functions include DNA-binding, nuclear translocation, heterodimerization with ICSBP and transcriptional activation (Fig 1). IRF-1 is a member of a family of related transcription factors. It is believed that such families develop by gene duplication and further diversification. The N-terminal 125 amino acids encoding the DNA-binding domain are structurally and functionally related among the IRF family of proteins. The C-terminal part of IRF-1 does not share amino acid sequence homology with other IRF-family members.

2.3. INDUCERS OF IRF-1

With the exception of early embryonal cells, IRF-1 mRNA is constitutively expressed at a low level in all cell types examined. IRF-1 is a DNA-binding transcription activator that accumulates in response to interferons, double-stranded RNA, cytokines, and some hormones. Interferon-γ is the strongest known natural inducer of IRF-1. In addition, in murine fibroblasts IRF-1 mRNA levels are cell cycle regulated. In the G1 phase, low mRNA levels are present but increase successively until mitosis (Stevens and Yu-Lee, 1992). Since the IRF-1 protein has a half-life time of 30 min, the IRF-1 protein abundance most probably parallels that of the IRF-1 mRNA (Watanabe et al., 1991).

2.4. IRF-1 INDUCED GENES

IRF-1 recognizes a short sequence domain in the interferon-β promoter that is also present in interferon-stimulated response elements (ISREs) of various interferon-stimulated genes (Reich and Darnell, 1989; Pine et al., 1990; Kirchhoff et al., 1993). IRF-1 directly binds to these promoters and activates transcription (Pine, 1992; Reis *et al.*, 1992; Kirchhoff *et al.*, 1993). Overexpression of IRF-1 leads to the induction of IFN-β and many interferon regulated genes even in the absence of IFN signaling (Fujita et al., 1989; Kirchhoff et al., 1993; Reis et al., 1992). Thereby, IRF-1 induces typical IFN-functions including antiviral state (Pine, 1992; Kimura *et al.*, 1994), induction of histocompatibility antigens (Chang *et al.*, 1992) and growth arrest (Table 1).

IRF-1 is known to induce a number of genes which exert growth inhibitory effects. Among them is the translation inhibitory protein PKR (Kirchhoff et al., 1995), 2'-5' oligo-adenylate synthetase (Benech et al., 1987) that regulates mRNA degradation by RNAse L (Jacobsen et al., 1983, Hassel et al., 1993) and p21 (Tanaka et al., 1996), a cell cycle inhibitor. Other IRF-1 induced enzymes such as the lysyl oxidase (Tan et al., 1996) and indoleamine 2,3-dioxigenase may limit the biosynthetic capacity of the cell by enhancing catabolic degradation of essential precursors. IL-4, IL-5, IL-7-receptor, p53, E-cadherin and MHC-HLA class I genes and some ISGs are induced by IRF-1 (Tanaka et al., 1993) and might contribute to the growth suppressing effects.

IRF-1 deficient mice have shown that IRF-1 is essential for the induction of iNO synthase in response to IFNγ (Matsuyama et al., 1993; Reis et al., 1994; Kamijo et al., 1994), ICE-1 protease gene expression is reduced. These mice are deficient in the differentiation of natural killer cells, in the generation of the TH 1 type of T helper cell and in the DNA damage response (Tanaka et al., 1996). Related factors of the IRF family may substitute for other IRF-1 functions. Whereas several IRF-1 induced genes block of cell cycle progression at the G1 phase (Table 1), this is not the case for IRF-1.

2.5. IRF-1 IS A TUMOR SUPPRESSOR

IRF-1 supresses tumor cell growth in vivo and in vitro (Harada et al., 1993; Tanaka et al., 1994, 1996). In vivo, deletions of the chromosomal IRF-1 locus in humans are associated with myelodysplasia and leukemia, indicating that a loss of IRF-1 function may lead to uncontrolled growth (Boultwood *et al.*, 1993).

TABLE 1. IRF-1 induced antiproliferative genes

Protein	Mechanism of action	Reference
Cell surface proteins, cell cycle regulators	IL-4, IL-5, IL-7-receptor, E-cadherin and MHC-HLA class I genes are cell surface proteins that are involved in trans-membrane signal transduction. p53 and p21 are cell cycle regulators that arrest cells specifically at the G1 phase. These genes contain IRF-1 binding sites in their promoters and are suppressing cell growth and tumorigenesis.	Tanaka et al., 1993
ISGs	Interferon stimulated genes have IRF-recognition motifs within their promoters Some ISGs encode proteins that have antiproliferative and/or tumor-suppressive effects.	Tanaka et al., 1993
PKR	Double-stranded RNA-dependent protein kinase (PKR), is a low level constitutively expressed ISG in a large variety of cells. The PKR is activated by double-stranded RNA (dsRNA) leading to autophosphorylation and by eIF2 phosphorylation indirectly to a decrease of protein biosynthesis. Other substrates such as IκB have been shown to be phosphorylated in vitro. Overexpression of the human kinase in murine cells results in an impairing of protein synthesis. The presence of catalytically inactive dominant negative PKR mutants leads to uncontrolled cell proliferation and oncogenicity. Induction of IRF-1 activity induces PKR expression which correlates with the IRF-1 mediated cell growth inhibitory effect. A catalytically inactive dominant negative PKR mutant abolishes the antiproliferative action of IRF-1 but does not alter the IRF-1-mediated resistance to certain viruses.	Kirchhoff, et al., 1995; Koromilas et al., 1992; Lee et al., 1993; Raveh et al., 1996, Kuhen et al., 1998; Samuel, 1998; Srivastava et al., 1998; Clemens et al., 1998
RNAse L	RNAse L is an ISG which remains latent and is activated by dsRNA. RNAse L levels are increased in growth arrested cells. A dominant negative mutant of the RNAse L suppresses the antiproliferative effects of IFN.	Hassel et al., 1993
OASE	Oligoadenylates are synthesized by 2-5(A)Synthetases. These genes are also ISGs. The 2-5(A)Synthetases are latent unless activated by dsRNA. 2-5(A)synthetase levels are increased in growth arrested cells. Overexpression of 2-5(A)synthetase leads to reduced cell growth.	Benech et al., 1987; Jacobsen et al., 1983
IDO	Indoleamin-2-3-oxygenase might play a role in IFN-mediated growth control. The level of this enzyme, which is responsible for tryptophane degradation is correlated with cell growth activity.	Takikawa et al., 1988

Expression of IRF-1 in aggressive noniummunogenic sarcoma cells inhibits tumor formation in mice (Yim et al., 1997). In vitro, IRF-1 inhibits growth of transformed cells lines, such as NIH3T3 cells transformed by E 1a/b, or by the tumorigenic EGFR mutants HER1 and HER2, (Kirchhoff and Hauser, 1999) as well as MMV infected mouse fibroblasts and a human sarcoma cell line.

IRF-1 can also suppress transformed cell growth induced by different oncogenes, such as myc, fosB (Tanaka et al., 1994) and IRF-2 (Harada et al., 1993). Moreover, embryonic fibroblasts deficient in IRF-1 genes become transformed by over expressing a single c-Ha-ras oncogene (Tanaka et al., 1994). In contrast to wild-type cells, IRF-1$^{-/-}$ cells do not undergo apoptosis upon c-Ha-ras oncogene expression and serum starvation, suggesting that IRF-1 overexpression can induce apoptosis of transformed cells, whereas in nontransformed cells IRF-1 can reduce proliferation without inducing apoptosis (Tanaka, *et al.*, 1994; Kirchhoff and Hauser, 1999).

3. Approaches to regulate ectopic IRF-1 activity

3.1 REQUIREMENT FOR IRF-1 REGULATION

To control mammalian cell growth we have employed a gene that is induced by interferon, the interferon regulatory factor 1 (IRF-1) (Kirchhoff et al., 1996; Koester et al., 1995). Both interferon and IRF-1 induce cell proliferation arrest in mammalian cells, but only a subset of cells is susceptible to the transient interferon-induced growth arrest. However, all mammalian cell lines that we have examined so far are subject to IRF-1-mediated growth arrest. These IRF-1 induced antiproliferative effects last as long as active IRF-1 protein is present in the cell. The inhibitory effect of IRF-1 precludes constitutive overexpression (Kirchhoff et al., 1993), since such cells would not grow. To obtain stable cell clones expressing recombinant IRF-1, either the activity or the expression of the IRF-1 protein must be regulated. Two IRF-1 control systems were established. In one of the systems, the activity of a constitutively expressed IRF-1 fusion protein is regulated, and in the other the transcription of IRF-1 is dependent on a recombinant trans-activator.

3.2. CONTROLLING IRF-1 ACTIVITY USING A FUSION PROTEIN

To allow sufficient cell growth for the formation of stable clones, the activiy of recombinant IRF-1 must be regulated. For this purpose, an IRF-1-estrogen receptor fusion protein IRF-1-hER was constructed to control cell growth in a ligand-dependent manner of the biotechnologically relevant cell line BHK-21 (Kirchhoff et al., 1996). In the absence of the hormone, the fusion protein does not affect cell growth. The addition of β-estradiol or other estrogen receptor ligands to the medium activates IRF-1-hER, leading to transcriptional induction of IRF-1-responsive genes and to a reduced growth rate. The proliferation of IRF-1-hER expressing cells can be gradually regulated and it depends on the β-estradiol concentration in the medium and on the duration of the exposure to the hormone (Kirchhoff et al., 1996; Köster et al., 1995)(Fig. 2A). The ligand concentrations to induce the IRF-1-hER fusion protein are rather high (up to 1 micromolar ß-estradiol for maximal activity) compared to concentrations used to induce native estrogen-receptors. This insensitivity which is due to the use of a receptor mutant in the fusion protein is rather advantageous and avoids potential unwanted interferences with estrogen like activities of media components such as phenol red or estrogen from the serum.

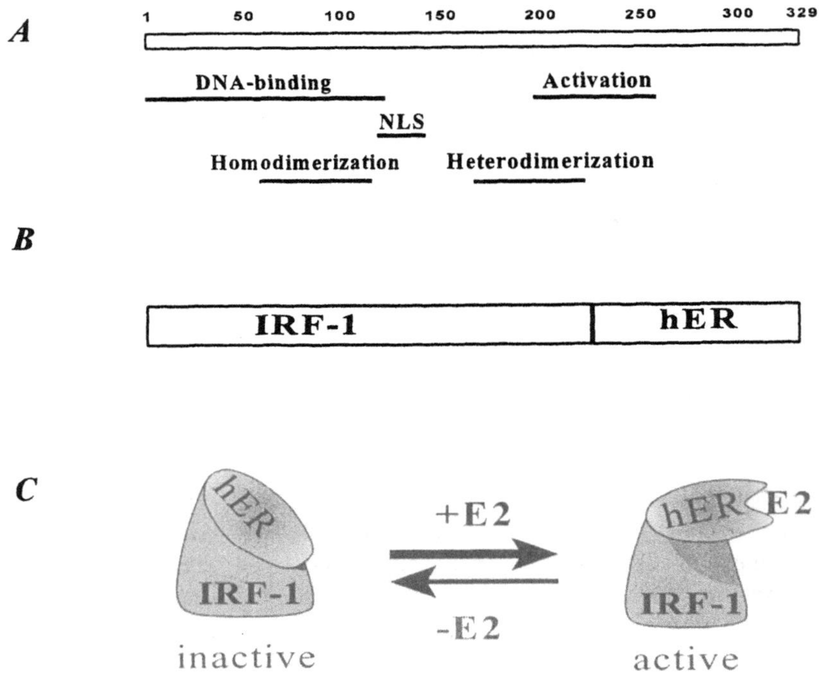

Figure 1. Domain structure of IRF-1 and the construction of the IRF-1hER fusion.
The function of individual IRF-1 domains has been identified by deletion analysis and by fusions of IRF-1 fragments to heterologous DNA-binding domains (A). IRF-1 fusion to the hormone binding domain of the human estrogen receptor (hER) (B). IRF-1 activity is ligand regulated in the IRF-1-hER fusion protein (C). DNA-binding requires the N-terminal 124 amino acids with an arbitrary C-terminal extension, without a specific amino acid sequence requirement. Conserved tryptophane residues in the DNA-binding domain are involved in the formation of the helix-turn-helix-motif(Uegaki et al., 1995) as well as for DNA-binding. For transactivation, an acidic domain in the C-terminal part of IRF-1 was required and sufficient (Lin et al., 1994; Schaper et al., 1999). A short basic sequence or IRF-1 acts as nuclear localization signal and targets heterologous proteins to the nucleus (Schaper et al., 1999). IRF-1 heterodimerization with ICSBP was demonstrated by biochemicaly and confirmed by a two-hybrid-system adapted to mammalian cells (Schaper et al., 1999; Graham and v. der Eb, 1973; Harada et al., 1989; Harada, et al., 1993). The domain required in IRF-1 for ICSBP binding and heterodimer formation is positioned between amino acids 164 and 219. Monomeric IRF- 1 can bind DNA in vitro (Tanaka et al., 1994). Cocrystals of the IRF-1 DNA-binding domain with a cognate DNA sequence element reveal a monomeric IRF-1-DNA complex. In vivo, in a mammalian two-hybrid assay system, DNA-binding of IRF-1 homodimers was demonstrated (Kirchhoff et al., 1999a). The homodimerization domain is located between amino acids 91 and 114 in the N-terminus of IRF-1 that lies within the DNA-binding domain. IRF-1 binding sites contain a direct hexamer repeat, suggesting a tandem array of the DNA-binding moieties of the IRF-1 dimer could interact with opposing sites of the DNA double helix.

3.3. TETRACYCLINE REGULATED IRF-1 EXPRESSION

Gossen and Bujard (1992) have regulated transcription based on the bacterial tetracycline regulatory system. A chimeric transactivator protein (tTA) was constructed by fusing the tetracycline repressor to the activation domain of the Herpes virus transcriptional activator, vp16. The target of this transcription factor is a synthetic promoter composed of tandem repeats of the tetracycline operator and a minimal eukaryotic promoter containing the TATA box and the transcription start site. The presence of tetracycline prevents DNA binding of the transactivator. Removal of tetracycline induces transcription of the gene under the control of the tetracycline responsive promoter. This system was used to establish cell clones for conditional expression of other antiproliferative genes, cyclines D1 and E.

A plasmid expressing the tTA transactivator was stably transfected into mouse C243 fibroblasts and stable transfectant clones were identified by screening with a transcriptional expression assay using a luciferase gene transcribed from a tetO-promoter. A positive clone was then transfected again with a dicistronic construct encoding IRF-1 and secreted alkaline phosphatase (SEAP) as a screening marker. Only a few percent of the clones show a complete absence of expression in the presence of tetracycline, indicating that optimal chromosomal integration of the recombinant gene is required to avoid background expression (Fig. 2B).

Induction of IRF-1-expression has little effect on cell proliferation in the first two days. Only four days after IRF-1 induction cell growth is significantly retarded. Even though there is a substantial decrease in viability at day six, no indications of apoptosis were detected.

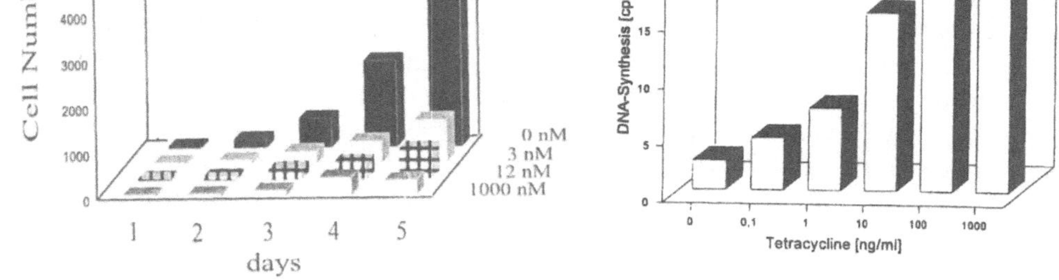

Figure 2. Evaluation of two systems to regulate IRF-1 activity.
C243 mouse cells constitutively expressing the IRF-1-hER fusion protein were grown in the presence of ß-estradiol. The inhibition of cell growth was determined by manually counting the cells using a microscope (A). Cells stably expressing the tetracycline repressible transactivator and wild-type IRF-1 expressed from a tetracycline regulated promoter were grown in the presence of tetracycline. Cell growth was monitored by the determination of the amount ^{3}H-Thymidine labeled DNA (B).

3.4. DICISTRONICALY STABILIZED EXPRESSION

Cells that were stably cotransfected with IRF-1-hER and the puromycin acetyltransferase gene spontaneously lost the responsiveness to ß-estradiol even though the cells remained resistant to high levels of puromycin. Western blot analysis using anti estrogen receptor antibodies showed that the loss of growth control correlated with a drastically reduced amount of fusion protein. To improve the stability of the regulatory system, the IRF-hER gene was linked with a poliovirus internal ribosomal entry site to puromycin acetyltransferase gene expression on a dicistronic construct (Fig. 3). Three out of four growth regulated BHK single clones were growth regulated for over three month of growth in static tissue cultures in the absence of estrogen or for over one month of growth in a fermenter when exposed to ß-estradiol (Carvalhal et al., 1998).

Figure 3. Stabilized expression of IRF-1-hER by dicistronic coupling with a selection marker gene expression.BHK-21 cells were transfected with a dicistronic IRF-1-hER expression construct with a poliovirus derived internal ribosomal entry site (IRES) that allows translation of the downstream puromycin acetylase gene (PAC). The mRNA was transcribd from a constitutive viral (MPSV) promoter (P). In the presence of puromycin, stable expression for several month was achieved.

3.5. COMPARISON OF THE TWO REGULATORY SYSTEMS

All biochemical and morphological effects of IRF-1 studied with the fusion protein approach and the tet-off system are basically identical. The tet-off system is a shut-off system and requires removal of tetracycline to induce IRF-1. Alternatively, the inherent instability of tetracycline in the medium could be used to curb growth at a precalculated time. On the other hand, the tet-on system could be employed to facilitate induction and to avoid medium exchange (Gossen et al., 1995). Both tetracycline regulated systems require repetitive and time consuming screening: for optimal transactivator expression a cell line has to be made and a second round of genetic manipulations is then required to establish efficient tetracycline regulated IRF-1 expression.

Tetracycline derivatives with no antibiotic activity are available and environmentally safe. The same is true for inexpensive ß-estradiol analoga. Both systems can be used in large scale fermentations. Gene transfer with the IRF-1-hER system requires a single transfection to obtain growth-regulated clones and can even be applied to preexisting producer cell lines. We have not observed any effects of ß-estradiol in the absence of IRF-1-hER, since the biotechnologically relevant cell lines currently in use do not express the estrogen receptor. No effects of IRF-1-hER expression on cell growth were detected in the absence of ß-estradiol. For these reasons we have decided to further develop the ß-estradiol inducible IRF-1 system for technological applications.

4. Biotechnology of IRF-1 controlled cell growth

4.1. KINETICS OF CELL GROWTH INHIBITION

An ideal cell culture process would grow cells as rapidly as possible to high cell density, then enter a production phase in which cell proliferation would be inhibited or completely stopped. To a certain extent proliferation of most cell lines is reduced after the cells have reached a high density. However, this is not sufficient for continuous culture, therefore excess cells must be constantly removed. To examine the feasibility of a genetic growth control system, BHK-21 cells have been used to demonstrate proliferation control in a biotechnologically relevant cell line. BHK-21 cells were transfected with IRF-1-hER expression constructs and individual clones were tested for reduced growth in the presence of β-estradiol. Variations in the degree of inhibition between individual clones may be related to the IRF-1-hER expression level. Normal cell growth was observed in all cases in the absence of β-estradiol. Upon estrogen addition to the growth medium, cell proliferation is reduced practically immediately. The degree of growth retardation depends on several parameters, such as the estrogen concentration, the cell density and the duration of exposure to the β-estradiol. The lowest β-estradiol concentration at which effects could be observed was 1 to 3 nM, whereas the response to concentrations more than 100 nM up to 1 mM was essentially the same. After three days in the presence of 30 nM estrogen, inhibition is about five- to tenfold (compare Fig. 2A).

4.2. INDUCERS OF THE IRF-1-hER FUSION PROTEIN

Apart from β-estradiol other compounds are able to activate the IRF-1hER fusion protein. Different steroidal and nonsteroidal compounds act as inducers of the fusion protein. Interestingly, the estrogen receptor antagonist hydroxy-tamoxifen also activates the IRF-1-hER fusion protein (Kirchhoff et al., 1993). This discrepancy may be due to the fact that tamoxifen allows DNA binding of the IRF-1-hER fusion protein, and that, in addition to the estrogen receptor transactivation domain, the IRF-1-hER fusion protein contains an IRF-1 transactivation domain that is not inactivated by tamoxifen, whereas the native estrogen receptor has only a single transactivation site. This mechanism also applies to other antagonists. It thus can be concluded that a great variety of compounds can be used to regulate cell proliferation in IRF-1-hER expressing cells.

4.3. CELL VIABILITY

After extended periods of IRF-1 activation cell viability decreases. The decrease in viability of BHK-21 cells begins at about day 3 after IRF-1 activation without the typical hallmarks of apoptosis. However, at higher cell densities the loss of viability was reduced. To use IRF-1 in technical processes, it cannot remain activated permanently. Either the activity of IRF-1-hER has to be reduced by low doses of estrogen or the induction of IRF-1 may be applied in intervals, alternating with recovery periods.

4.4. SIDE EFFECTS OF IRF-1

IRF-1 is a potent activator for a large number of genes. It is not known whether these might interfere with biotechnological purposes. We have however found that IRF-1 overexpressing cells are still capable of secreting even difficult to express heterologous proteins at practically the same rates as in the absence of recombinant IRF-1 activity (see below).

Thus, the major task is to keep viability high. Since the activities of IRF-1 are mediated by a large number of genes, it may be possible to block the activities of genes that negatively affect viability but allow expression of genes that reduce cell growth. Inhibition of the apoptosis pathway by ectopic expression of Bcl-2 did not have any significant effects. Overexpression of I-κB (Kirchhoff et al., 1999) blocks IRF-1-mediated NF-κB activation and reduces the expression from some IRF-1 induced promoters. However, ectopic I-κB expression leads to spontaneous apoptosis and is thus not applicable. Expression of a dominant negative mutant of PKR transiently blocks IRF-1 induced cell death as well as cell growth arrest, whereas IRF-1 mediated viral resistance is not altered (Kirchhoff et al., 1995). The ongoing studies on the IRF-1 induced genes and their function may resolve the question how cell death can be avoided while maintaining proliferation control.

5. Product formation in proliferation controlled cells

5.1. RELATIONSHIP BETWEEN GROWTH AND PRODUCTIVITY

In recombinant mammalian cells, most frequently productivity increases with the growth rate (Smiley et al., 1989; Cockett et al., 1990; Robinson and Memmert, 1991; Pendse et al., 1992). There are also a few reports of a negative correlation (Mitchell et al., 1991; Bebbington et al., 1992; Tonouchi et al., 1992) or where the specific product formation appears to vary independently of the growth rate (Tsao et al., 1992). This interdependence is influenced by the culture system and the cell line; In suspended aggregate cultures, BHK cells expressing a recombinant antibody show that the cell-specific antibody production rate increased concomitantly with the growth rate; in contrast, for microcarrier culture, the specific production rate of the same BHK cell line increased as the growth rate decreased.

In IRF-1 proliferation controlled BHK-21 cells, productivity of a recombinant immunoglobulin remains constant for up to three days of growth arrest, but decreases thereafter (Kirchhoff et al., 1996). However, recombinant gene expression can be improved by genetic engineering.

5.2. IRF-1 INDUCED RECOMBINANT GENE EXPRESSION

Since IRF-1 acts as a transcriptional activator, promoters with IRF-1 binding sites inserted upstream of a minimal promoter can be employed to obtain expression levels equal to or higher than with commonly used viral promoters (Kirchhoff et al., 1993). However, expression from such promoters is low in the absence of recombinant IRF-1 activity. Except for special circumstances such as the expression of highly cytotoxic products, a maximal productivity independent of IRF-1 would be desirable. By employing two separate expression constructs to express the product, one with a constitutive promoter and a second one with an IRF-1 inducible promoter, production levels are high both in the presence or absence of IRF-1 activity. The constitutive promoter is responsible for expression in the absence of IRF-1 activity, while the inducible promoter compensates any decreases or even enhances expression to higher levels in growth arrested cells (Mueller et al., 1998). The drawback of using two constructs is that both have to be highly active, which may require repeated rounds of transfection and selection procedures.

As an alternative, a promoter with the desired characteristics was constructed by inserting IRF-1 binding sites into a strong constitutive promoter. Whereas IRF-1 binding sites inserted into an MPSV promoter did not lead to significant expression levels and showed little inducibility, a novel composite promoter assembled from three heterologous elements does have the desired characteristics (Fig. 4; Mueller et al., manuscript in preparation).

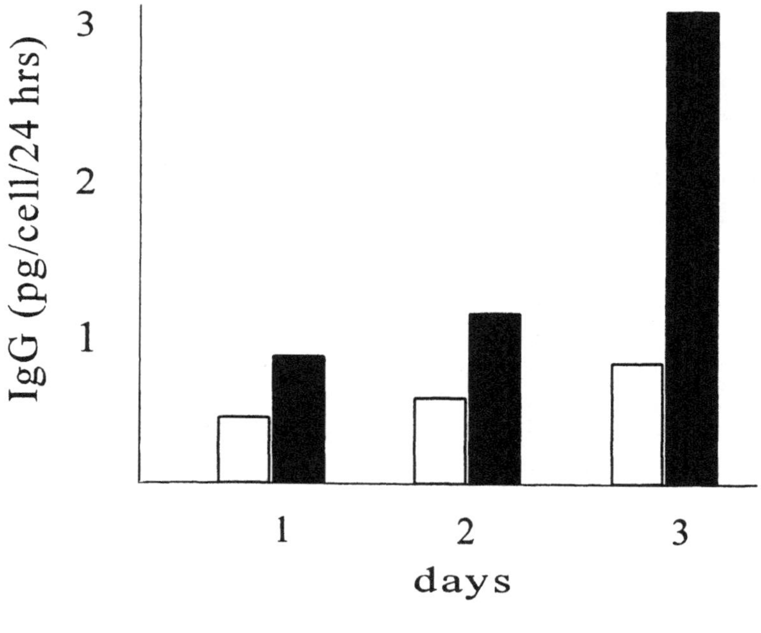

Figure 4. Estrogen stimulated productivity.in IRF-hER expressing BHK cells.
A composite promoter containingconstitutive enhancer elements from a viral promoter and IRF-1 binding sites was used to express IgG in IRF-1-hER growth regulated BHK-21 cells. Open bars; no ß-estradiol added. Black bars; cells grown in the presence of 100 nM ß-estradiol. The amount of IgG secreted into the supernatant wasdetermined by a sandwich ELISA procedure.

6. Product quality in proliferation controlled producer cells

6.1. CONSTANT QUALITY OF A PHARMACEUTICAL GLYCOPROTEIN PRODUCED IN IRF-1 CONTROLLED CELLS

In addition to the expression level, a uniformly high product quality is a principal aim for biotechnological process applications. Protein quality refers to a number of parameters including protein folding, processing, post-translational modifications and protein integrity. Glycosylation is common and important for secreted proteins and shows the highest variability among the posttranslational modifications. Pharmaceutical protein glycosylation influences the biological activity, antigenicity and clearing time from the blood circulation. Glycosylation is influenced by the producer cell and its environment; it can change in the course of cultivation. Toward the end of production processes high cell densities contribute to the ammonia accumulation in the medium that can lead to drastic changes in the glycosylation pattern (Gawlitzek et al., 1998; Gawlitzek et al., 1995; Grammatikos et al., 1998; Jenkins et al., 1996). Similarly, secreted glycosidases and enzymes released by lysed cells can degrade the carbohydrates (Gramer and Goochee, 1990; Gramer et al., 1995; Munzert et al., 1998). Therefore, we have determined human erythropoietin (Epo) protein integrity and glycosylation pattern as sensitive and relevant quality criteria.

Recombinant Epo is used to treat symptoms of anaemic hypoxia, a major complication arising from chronic renal failure. Epo is a highly glycosylated pharmaceutical product that has been characterized in detail in various producer cell lines. Epo glycosylation, in particular terminal sialylation, is essential for the in vivo biological activity (Fibi et al., 1995; Fukuda et al., 1989; Goto et al., 1988). BHK-21 cells have a heterogeneous Epo glycosylation pattern that has been described in great detail (Grabenhorst et al., 1995).

To characterize the impact of the proliferation control on glycoprotein quality, recombinant protein integrity and glycosylation pattern was analyzed of Epo secreted from BHK-21 cells (Mueller et al., submitted). The Epo quality from the proliferation controlled culture was at least equivalent to that from the growing culture. IRF-1 activation did not influence the Epo protein integrity and there were no signs of proteolytic degradation. Therefore, the IRF-1 system can reduce excess cell growth to extend the production phase while keeping the production conditions more constant to yield a consistent and high product quality.

7. IRF-1 mediated proliferation control in different culture systems

7.1. CHARACTERISTICS OF CONTROLLED CELL GROWTH IN BATCH CULTURE

We have investigated the feasibility of IRF-1-hER regulated BHK-21 cell proliferation in technical culture systems. With the addition of ß-estradiol at the beginning of the exponential phase, IRF-1 activation leads to efficient cell growth inhibition within 24 hours. There was a change in cell morphology, the outer membrane became highly irregular and accumulation of intracellular vesicles was observed. Using stirred vessels operated as batch cultures, the cells responded with the same pattern as that obtained in static cultures. This clearly indicates that IRF-1-hER activation effect on the cells is independent of culture system. In stirred suspended cultures the cells could be induced to form aggregates, with a decrease in the aggregate size and a higher number of single cells in suspension after IRF-1 activation. This response is independent of the clone, suggesting that this regulation will be applicable to other BHK-21 producer cells (Carvalhal et al., 1998).

7.2. PROLIFERATION CONTROL IS ESTROGEN INDUCED

In order to define strategies to overcome losses in cell viability after extensive exposure to ß-estradiol, several parameters were evaluated: ß-estradiol concentration, time point of ß-estradiol addition, time span of ß-estradiol exposure and serum concentration. All the studies but the time span of ß-estradiol exposure led to no significant influence upon the cell viability pattern.

Cell growth inhibition is dependent on ß-estradiol. The decrease in cell viability was independent of ß-estradiol concentration, but growth arrest was less drastic below 100 nM. All ß-estradiol concentrations above 100 nM led to similar results. Cell growth inhibition was observed after IRF-1-hER activation at all cell growth phases, but its kinetics is dependent upon the growth phase: at the stationary phase, a reduced response to in the activation of the IRF-1-hER system is apparent. To show that estrogen had no effect on BHK cell growth in the absence of IRF-1-hER, 100 nM ß-estradiol was added at the beginning of the exponential phase of cells without the IRF-1-hER construct (nonregulated clone). There was no difference on cell growth, viability or morphology when compared with the control, confirming that the effect of ß-estradiol is dependent on the presence of IRF-1-hER

The type and concentration of serum were also considered. Estrogens present in the serum may activate IRF-1. Thus, preferentially serum batches with low or no estrogen content have to be used.

7.3. EFFECT OF PULSED ß-ESTRADIOL EXPOSURE

When the time of estrogen exposure is limited cell viability is similar to the untreated control culture. During a recovery phase, the cells passed through two different phases: first a stationary phase where cell growth was still inhibited, followed by an increase of cell concentration. The reduction of the time span of ß-estradiol exposure can be used as a strategy to overcome the cell viability decrease and extend the stationary phase. Thus, the effects of IRF-1 activation are reversible.

The optimum time of contact of the cells with ß-estradiol is in the range of up to 72 hours, ß-estradiol removal after shorter pulses leads to less significant proliferation arrest and cell growth is resumed immediately after ß-estradiol removal. Control experiments revealed that the recovery of viability was due to the removal of ß-estradiol and not to the addition of fresh medium. Further cycles of ß-estradiol addition and removal leads to similar effects, suggesting a possibility to extend cell growth inhibition for longer periods.

7.4. PROLIFERATION CONTROL IN PERFUSION CULTURES

The conditions for a long-term proliferation controlled culture were established in a perfusion system. To extend the growth controlled period in stirred tanks (the most convenient culture system in industry) for ß-estradiol removal the operation of a perfusion reactor is required. The perfusion conditions and the time span of ß-estradiol exposure were optimized to ensure that the ß-estradiol concentration is slowly decreased by gradual dilution with estrogen-free fresh medium.

In Figure 5 the result of two cycles of ß-estradiol addition and removal in a 2 L stirred tank bioreactor operated at a perfusion rate between 0.7 and 1.4 day-1, is presented. ß-estradiol was added in a single step in the middle of the cell exponential phase in the batch phase of the culture. Cell growth arrest could be repeated consecutively several times. This indicates that cells respond to ß-estradiol addition independently of the growth phase and that the IRF-1-hER fusion protein is stably expressed from the dicistronic construct.

These observations led to the definition of strategies to operate IRF-1 regulated cells by pulse ß-estradiol addition (final concentration of 100 nM) followed by periods with ß-estradiol (48 to 72 hours each) and its removal by perfusion (1.4 day^{-1}).

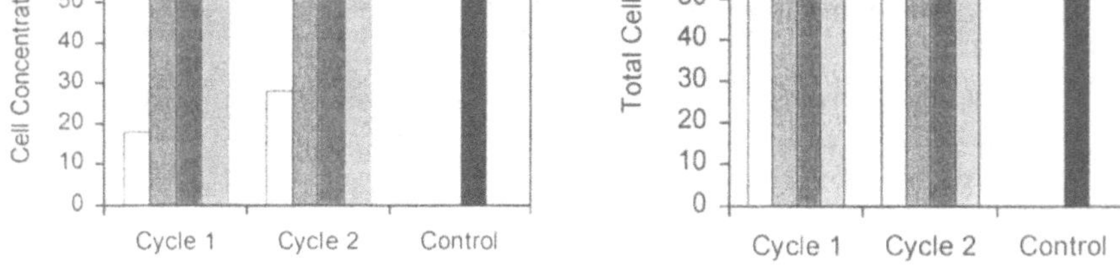

Figure 5: Repeated cycles of cell growth control in a perfusion culture
Effect of cycles of ß-estradiol addition (100 nM) and removal by perfusion with ß-estradiol-free fresh medium at 1.4 and 0.7 day B1, respectively, on cell concentration (A) and cell viability (B). Empty bars, beginning of the experiment; dark grey, cell density and viability at the time point of ß-estradiol addition; hatched bar, beginning of estradiol removal by perfusion; light grey, end of perfusion; black bars, control experiment without ß-estradiol addition.

8. Conclusions and prospect

The applicability of the genetic IRF-1 mediated cell proliferation control system has been demonstrated in the biotechnologically relevant producer cell line BHK-21. Growth can be effectively regulated by adding the inexpensive, stable and nontoxic compound β-estradiol. Cell viability remains high with alternating cycles of β-estradiol addition and dilution. With a dicistronic, self-selecting construct, the regulation was stable for at least one month of controlled growth in a perfusion system. Productivity can be enhanced by employing IRF-1 inducible promoters, while the product remains similar or is perhaps even superior to product quality from uncontrolled growing cells. By avoiding excess cell growth, ammonia production and cell death, the environment of producer cells should remain more constant and is expected to result in improved product quality and consistency in applied production processes. This shows that the IRF-1 proliferation control system has now been developed to an advanced state, ready to test its suitability for an improved applied production process.

9. Acknowledgments

The authors thank Simon Klibisch, Meike Tümmler and Martina Grasshoff for experimental help and for providing results prior to publication, and Rosemary Avram for typing the manuscript.

10. References

Al-Rubeai, M., Emery, A.N., Chalder, S. and Jan, D.C. (1992) Specific monoclonal antibody productivity and the cell cycle-comparisons of batch, continuous and perfusion cultures, *Cytotechnology* **9**, 85-97.

Au, W.C., Moore, P.A., LaFleur, D.W., Tombal, B., Pitha, P.M. (1998) Characterization of the interferon regulatory factor-7 and its potential role in the transcription activation of interferon A genes, *J. Biol. Chem.* **273**, 29210-7.

Bebbington, C.R., Renner, G., Thomson, S., King, D., Abrams, D., Yarranton, G.T., (1992) High-level expression of a recombinant antibody from myeloma cells using a glutamine synthetase gene as an amplifiable selectable marker, *Biotechnology* **10**, 169.

Benech, P., Vigneron, M., Peretz, D., Revel, M., Chebath, J. (1987) Interferon-responsive regulatory elements in the promoter of the human 2',5'-oligo(A) synthetase gene, *Mol. Cell. Biol.* **7**, 4498-504.

Boultwood, J., Fidler, C., Lewis, S., MacCarthy, A., Sheridan, H., Kelly, S, Oscier, D., Buckle, V.J., Wainscoat, J.S. (1993) Allelic loss of IRF1 in myelodysplasia and acute myeloid leukemia: retention of IRF1 on the 5q- chromosome in some patients with the 5q- syndrome, *Blood* **82**, 2611-6.

Carvalhal, A.V., Moreira, J.L., Müller, P.P., Hauser, H., and Carrondo, M.J.T. (1998) Cell growth inhibition by the IRF-1 system, in New developments and new applications in animal cell technology (Merten, O.-W., Perrin, P, and Griffiths, B., eds.) Kluwer Academic Publishers, pp. 215-217.

Chang, C.H., Hammer, J., Loh, J.E., Fodor, W.L., Flavell, R.A. (1992) The activation of major histocompatibility complex class I genes by interferon regulatory factor-1 (IRF-1), *Immunogenetics* **35**, 378-384.

Clemens, M.J. (1997) PKR--a protein kinase regulated by double-stranded RNA, *Int. J. Biochem. Cell. Biol.* **29**, 945-949.

Cockett, M.I., Bebbington, C.R., Yarranton, G.T. (1990) High level expression of tissue inhibitor of metalloproteinases in Chinese hamster ovary cells using glutamine synthetase gene amplification, *Biotechnology* **8**, 662-667.

Fibi, M.R., Hermentin, P., Pauly, J.U., Lauffer, L. and Zettlmeissl, G. (1995) N- and O-glycosylation muteins of recombinant human erythropoietin secreted from BHK-21 cells, *Blood* **85**, 1229-1236.

Fujita, T., Sakakibara, J., Sudo, Y., Miyamoto, M., Kimura, Y. and Taniguchi, T. (1988) Evidence for a nuclear factor(s), IRF-1, mediating induction and silencing properties to human IFN-beta gene regulatory elements *EMBO J.* **7**, 3397-3405.

Fujita, T., Kimura, Y., Miyamoto, M., Barsoumian, E. L. and Taniguchi, T. (1989) Induction of endogenous IFN-α and IFN-ß genes by a regulatory transcription factor, IRF-1, *Nature* **337**, 270-272.

Fukuda, M.N., Sasaki, H., Lopez, L. and Fukuda, M. (1989) Survival of recombinant erythropoietin in the circulation: the role of carbohydrates, *Blood* **73**, 84-89.

Fussenegger, M., Mazur, X. and Bailey, J. E. (1997) A novel cytostatic process enhances the productivity of Chinese hamster ovary cells, *Biotechnol. Bioeng.* **55**, 927-938.

Fussenegger, M., Schlatter, S., Dätwyler, D., Mazur, X. and Bailey, J. E. (1998) Controlled proliferation by multigene metabolic engineering enhances the productivity of Chinese hamster ovary cells, *Nat. Biotechnol.* **16**, 468-472.

Fussenegger, M., Bailey, J., Hauser, H. and Mueller, P.P. (1999) Genetic Optimization of Recombinant Protein Production by Mammalian Cells, *TIBTECH* **17**, 43-50.

Gawlitzek, M., Conradt, H. S. and Wagner, R. (1995) Effect of different cell culture conditions on the polypeptide integrity and N-glycosylation of a recombinant model glycoprotein,*Biotechnol. Bioeng.* **46**, 536-544.

Gawlitzek, M., Valley, U., Wagner, R. (1998) Ammonium ion/glucosamine dependent increase of oligosaccharide complexity in recombinant glycoproteins secreted from cultivated BHK-21 cells, *Biotechnol. Bioeng.* **57**, 518-528.

Gossen, M. and Bujard, H. (1992) Tight control of gene expression in mammalian cells by tetracycline-responsive promoters, *Proc. Natl. Acad. Sci. USA* **89**, 5547-5551.

Gossen, M., Freundlieb, S., Bender, G., Müller, G, Hillen, W. and Bujard, H. (1995) Transcriptional activation by tetracyclines in mammalian cells., Science **268**, 1766-1769.

Goto, M., Akai, K., Murakami, A., Hashimoto, C., Tsuda, E.,Ueda, M., Kawanishi, G., Takahashi, N., Ishimoto A., Chiba, H. and Sasaki, R. (1988). Production of recombinant human erythropoietin in mammalian cells: Host-cell dependency of the biological activity of the cloned glycoprotein, *Bio/Technology* **6**, 67-71.

Grabenhorst, E., Hoffmann, A., Nimtz, M., Zettlmeissl, G. and Conradt, H. S. (1995). Construction of stable BHK-21 cells coexpressing human secretory glycoproteins and human Gal(β1-4)GlcNAc-R α2,6-sialyltransferase. α2,6-linked NeuAc is preferentially attached to the Gal(β1-4)GlcNAc(β1-2)Man(α1-3)-branch of diantennary oligosaccharides from secreted recombinantβ-trace protein, *Eur. J. Biochem.* **232**, 718-725.

Graham, F. and v. der Eb, L. (1973) A new technique for the assay of infectivity of human adenovirus DNA, *Virology* **52**, 456-487.

Gramer, M. J. and Goochee, C. F. (1990) Glycosidase activities in Chinese hamster ovarycell lysate and cell culture supernatant, *Biotechnol. Prog.* **9**, 366-373.

Gramer, M. J., Goochee, C.F., Chock, V.Y., Brousseau, D.T. and Sliwkowsky, M.B. (1995) Removal of sialic acid from a glycoprotein in CHO cell culture supernatant by action of an extracellular CHO cell sialidase, *Bio/Technology* **13**, 692-698.

Grammatikos, S. I., Valley, U., Nimtz, M., Conrad, H. S., Wagner, R. (1998) Intracellular UDP-N-acetyl-hexosamine pool affects N-glycan complexity: A mechanism of ammonium action on protein glycosylation, *Biotechnol. Progr.* **14**, 410-419.

Harada, H., Fujita, T., Miyamoto, M., Kimura, Y., Maruyama, M., Furia, A., Miyata, T. and Taniguchi, T. (1989) Structurally similar but functionally distinct factors, IRF-1 and IRF-2, bind to the same regulatory elements of IFN and IFN-inducible genes, *Cell* **58**, 729-739.

Harada, H., Willison, K., Sakakibara, J., Miyamoto, M., Fujita, T. andTaniguchi, T. (1990) Absence of the type I IFN system in EC cells: Transcriptional activator (IRF-1) and repressor (IRF-2) genes are developmentally regulated, *Cell* **63**, 303-312.

Harada, H., Kitagawa, M., Tanaka, N., Yamamoto, H., Harada, K., Ishihara, M. and Taniguchi, T. (1993) Anti-oncogenic and oncogenic potentials of interferon regulatory factors-1 and -2, *Science* **259**, 971-974.

Hassel, B.A., Zhou, A., Sotomayor, C., Maran, A.,Silverman, R.H. (1993) A dominant negative mutant of 2-5A-dependent RNase suppresses antiproliferative and antiviraleffects of interferon, *EMBO J.* **12**, 3297-304.

Hauser, H.: Heterologous expression of genes in mammalian cells. (1997) In: Mammalian Cell Biotechnology in Protein Production. (H. Hauser and R. Wagner, eds.) De Gruyter, Berlin, pp. 1-32.

Jacobsen, H., Czarniecki, C.W., Krause, D., Friedman, R.M. and Silverman R.H. (1983) Interferon-induced synthesis of 2-5A-dependent RNase in mouse JLS-V9R cells, *Virology* **125**, 496-501.

Jenkins, N., Parekh, R. B. and James, D. C. (1996) Getting the glyosylation right: implications for the biotechnological industry, *Nat. Biotechnol.* **14**, 975-981.

Kim, Y.H., Iida, T., Fujita, T., Terada, S., Kitayama, A., Ueda, H., Prochownik, E.V. and Suzuki, E. (1998) Establishment of an apoptosis-resistant and growth-controllable cell line by transfecting with inducible antisense c-Jun gene, *Biotechnol. Bioeng.* **58**, 65-72.

Kimura, T., Nakayama, K., Penninger, J., Kitagawa, M., Harada, H., Matsuyama, T., Tanaka, N.,Kamijo, R., Vilcek, J., Mak, T.W. et al. (1994) Involvement of the IRF-1 transcription factor in antiviral responses to interferons, Science **264**, 1921-1924.

Kirchhoff, S., Schaper, F. and Hauser, H. (1993) Interferon regulatory factor 1 (IRF-1) mediates cell growth inhibition by transactivation of downstream target genes, Nucleic Acids Res. **21**, 2881- 2889.

Kirchhoff, S., Koromilas, A. E., Schaper, F., Grashoff, M., Sonenberg, N. and Hauser, H. (1995) IRF-1 induced cell growth inhibition and interferon induction requires the activity ofthe protein kinase PKR, *Oncogene* **11**, 439-445.

Kirchhoff, S., Kröger, A., Cruz, H., Tümmler, M., Schaper, F., Köster, M. and Hauser, H. (1996) Regulation of cell growth by IRF-1 in BHK-21 cells, *Cytotechnology* **22**, 147-156.

Kirchhoff, S., Schaper, F., Oumard, A. and Hauser, H. (1998) *In vivo* formation of IRF-1 homodimers, Biochimie, in press.

Kirchhoff, S. and Hauser, H. (1999a) Cooperative activity between HER oncogenes and the tumor suppressor of IRF-1 results in apoptosis, Oncogene, in press.

Kirchhoff, S., Wilhelm, D., Angel. P. and Hauser, H. (1999b) NF-κB activation is required for IRF-1 mediated IFN- induction, Europ. J. of Biochem., in press.

Köster, M., Kirchhoff, S., Schaper, F. and Hauser, H. (1995) Proliferation control of mammalian cells by the tumor suppressor IRF-1, in Animal Cell Technology: Developments towards the 21st Century (Beuvery, Griffiths, Zeijlemaker, eds.) Kluwer Academic Publishers, pp. 33-44.

Koromilas, A.E., Roy, S., Barber, G.N., Katze, M.G. and Sonenberg (1992) Malignant transformation bya mutant of the IFN-inducible dsRNA-dependent protein kinase, *Science* **257**, 1685-1689.

Kuhen, K.L., Vessey J.W. and Samuel, C.E. (1998) Mechanism of interferon action: identification of essential positions within the novel 15-base-pair KCS element required for transcriptional activation of the RNA-dependent protein kinase pkr gene, *J. Virol.***72**, 9934-9939.

Lee, S.B. and Esteban, M. (1993) The interferon-induced double-stranded RNA-activated human p68 protein kinase inhibits the replication of vaccinia virus, Virology **193**, 1037-1041.

Lengyel, P. (1993) Tumor-suppressor genes: news aboutthe interferon connection *Proc. Natl. Acad. Sci. U S A*, **90**, 5893-5895.

Lin, R., Mustafa, A., Nguyen H., Gewert, D. and Hiscott J. (1994) Mutational analysis of interferon (IFN) regulatory factors 1 and 2, J. Biol. Chem. **269**, 17542-17549

Matsuyama, T., Kimura, T., Kitagawa, M., Pfeffer, K., Kawakami, T., Watanabe, N., Kündig, T.M., Amakawa, R., Kishihara, K., Wakeham, A., Potter, J., Furlonger, C.L., Narendran, A., Suzuki, H., Ohashi, P.S., Paige, C.J., Taniguchi, T. and Mak, T.W. (1993) Target disruption of IRF-1 or IRF-2 results in abnormal typel IFN gene induction and aberrant lymphocyte development, *Cell* **75**, 83-97.

Mazur, X., Fussenegger, M., Renner, W.A. and Bailey, J.E. (1998) Higher productivity of growth-arrested Chinese hamster ovary cells expressing the cyclin-dependent kinase inhibitor p27,*Biotechnol. Prog.* **14**, 705-713.

Mitchell, C.A., Beall, J.A., Wells, J.R., Gray, P.P. (1991) Growth and protein production kinetics of a murine myeloma cell line transfected with the human growth hormone gene, *Cytotechnology* **5**, 2232-31.

Miyamoto, M., Fujita, T., Kimura, Y., Maruyama, M., Harada, H., Sudo, Y., Miyata, T. and Taniguchi, T. (1988) Regulated expression of a gene encoding a nuclear factor, IRF-1, that specifically binds to IFN-gene regulatory elements, *Cell* **54**, 903-913.

Mueller, P. P., Kirchhoff, S. and Hauser, H.(1998) in *New Developments and New Applications in Animal Cell Technology* (Merten, O.W., Perrin, P. and Griffiths, J.B., eds) Kluwer Academic Publishers, pp. 209-213.

Mueller, P.P., Schlenke, P., Nimtz, M.,Conradt, H.S. and Hauser, H. (1999).Recombinant glycoprotein product quality in proliferation controlled BHK-21 cells. Submitted.

Munzert, E., Heidemann, R., Büntemeyer, H., Lehmann, J. and Müthing, J. (1998) Production of recombinant human antithrombin III on 20-l bioreactor sale: Correlation of supernatant neuraminidase activity, desialylation, and decrease of biological activity of recombinant glycoprotein, *Biotechnol. and Bioeng.* **56**, 441-447.

Pendse, G.J., Karkare, S. and Bailey, J.E. (1992) Effect of cloned gene dosage on cell growth and hepatitisB surface antigen synthesis and secretion in recombinant CHO cells, *Biotechnol. Bioeng.* **40**, 119-129.

Pine, R., Decker, T., Kessler, D.S., Levy, D.E. and Darnell Jr., J.E. (1990) Purification and cloning of interferon-stimulated gene factor 2 (ISGF2): ISGF2 (IRF-1) can bind to the promoters of both beta interferon- and interferon-stimulated genes but is not a primary transcriptional activator of either, *Mol. Cell. Biol.* **10**, 2448-2457.

Pine, R. (1992) Constitutive expression of an ISGF2/IRF1 transgene leads to interferon independent activation of interferon-inducible genes and resistance to virus infection, *J. Virol.* **66**, 4470-4478.

Raveh, T., Hovanessian, A.G., Meurs, E.F., Sonenberg, N. and Kimchi, A. (1996) Double-stranded RNA-dependent protein kinase mediates c-Myc suppression induced by type I interferons,*J. Biol. Chem.* **271**, 25479-15484.

Reich, N.C., Darnell, J.E. Jr (1989) Differential binding of interferon-induced factors to an oligonucleotide that mediates transcriptional activation, *Nucleic Acids Res.* **17**, 3415-3424.

Reis, L.F.L., Ruffner, H., Stark, G., Aguet, M. and Weissmann, C. (1994) Mice devoid of IRF-1 show normal expression of type I IFN genes, *EMBO J.* **13**, 4798-4806.

Reis, L.R.L., Harada, H., Wolchok, J. D., Taniguchi, T. and Vilcek, J. (1992) Critical role of a common transcription factor, IRF-1, in the regulation of IFN- and IFN-inducible genes, *EMBO J.* **11**, 185-193.

Robinson, D.K., Memmert, K.W. (1991) Kinetics of recombinant immunoglobulin production by mammalian cells in continuous culture, Biotechnol. Bioeng. **38**, 972-976.

Samuel, C.E. (1998) Reoviruses and the interferon system,*Curr. Top. Microbiol. Immunol.* **233**, REOVIR.II, 125-145.

Schaper, F., Kirchhoff, S., Posern, G., Koester, M., Oumard, A., Sharf, R., Levi, B.Z. and Hauser, H. (1998) Functional domains of interferon regulatory factor I (IRF-1), Biochem. J. **335**, 147-157.

Smiley, A.L., Hu, W.S. and Wang, D.I.C. (1989) Production of human immune interferon by recombinant mammalian cells cultivated on microcarriers *Biotechnol. Bioeng.* **33**, 1181- 1190.

Srivastava, S.P., Kumar, K.U. and Kaufman, R.J. (1998) Phosphorylation of eukaryotic translation initiation factor mediates apoptosis in response to activation of the double-stranded RNA-dependent protein kinase, *J. Biol. Chem.* **23**, 273, 2416-2423.

Stevens, A.M., Yu-Lee, L. (1992) The transcription factor interferon regulatory factor-1 is expressed during both early G1 and the G1/S transition in the prolactin-induced lymphocyte cell cycle,*Mol Endocrinol.* **6**, 2236-43.

Suzuki, E. and Ollis, D.E. (1990) Enhanced antibody production at slowed growth rates: experimental demonstration and a simple structured model, *Biotechnol. Prog.* **6**, 231-236.

Tan, R.A., Taniguchi, T. and Harada, H. (1996) Identification of the lysyl oxidase gene as target of the anti oncogenic transcription factor, IRF-1, and its possible role in tumor suppression, *Cancer Res.* **56**, 2417-2421.

Tanaka, N., Kawakami, T., Taniguchi, T. (1993) Recognition DNA sequences of interferon regulatory factor 1 (IRF-1) and IRF-2, regulators of cell growth and the interferon system, *Mol. Cell. Biol.* **13**, 4531-4538.

Tanaka, H. and Samuel, C.E. (1994) Mechanism of interferon action: structure of the mouse PKR gene encoding the interferon-inducible RNA-dependent protein kinase, *Proc. Natl. Sci. USA* **91**, 7995-7999.

Tanaka, N., Ishihara, M., Kitagawa, M., Harada, H., Kimura, T., Matsuyama, T., Lamphier, M. C., Aizawa, S., Mak, T. W. and Taniguchi, T. (1994) Cellular commitment to oncogene-induced transformation or apoptosis is dependent on the transcription factor IRF-1, *Cell* **77**, 829-839.

Tanaka, N., Ishihara, M., Lamphier, M.S., Nozawa, H., Matsuyama, T., Mak, T.W., Aizawa, S., Tokino, T., Oren, M. and Taniguchi, T. (1996) Cooperation of the tumour suppressors IRF-1 and p53 in response to DNA damage. *Nature* **382**, 816-818.

Tonouchi, N., Koyama, N, Miwa, A. (1992) CHO strain producing high-level human IL-6 with the 3' deletion construct, *J. Biotechnol.* **22**, 283-289.

Tsao, E.I., Bohn, M.A., Omstead, D.R., Munster, M.J. (1992) Optimization of a roller bottle process for the production of recombinant erythropoietin, *Ann. N. Y. Acad. Sci.* **665**, 127-136.

Uegaki, K., Shirakawa, M., Harada, H., Taniguchi, T. and Kyogoku, Y. (1995) Secondary structure and folding topology of the DNA binding domain of interferon regulatory factor 2, as revealed by NMR spectroscopy, *FEBS Lett.* **359**, 184-188.

Yim, J.H., Wu, S.J., Casey, M.J., Norton, J.A. and Doherty, G.M. (1997) IFN regulatory factor-1 gene transfer into an aggressive, nonimmunogenic sarcoma suppresses the malignant phenotype and enhances immunogenicity in syngeneic mice. *J. Immunol.* **158**, 1284-1292.

FACTORS INVOLVED IN THE CELL CYCLE ARREST OF ADULT RAT CARDIOMYOCYTES

Possibilities for the reactivation of the cardiomyocyte proliferation in cell culture.

JOSEF P. MAGYAR and HANS M. EPPENBERGER*

Institute of Cell Biology, Swiss Federal Institute of Technology, ETH-Hönggerberg, 8093 Zurich, Switzerland

** Corresponding author; eMail: hans.eppenberger@cell.biol.ethz.ch*

1. Introduction

The mammalian heart consists of two pumps called the left and right ventricles. The left ventricle has to work harder as it must force the blood around the body. Therefore, it is more likely to fail despite the fact that it has more muscle. O_2 carried by the blood is essential to maintain life. If the left ventricle weakens, the body suffers from O_2 deficit. A failing left ventricle will also fail to keep pace with the right ventricle, which is feeding the lungs, thus causing pulmonary blood congestion. If the right ventricle weakens, it will fail to propel enough blood through the lungs to exchange O_2 to meet the needs of the rest of the body. There are four major types of cardiovascular diseases – hypertension, atherosclerosis, ischemic heart disease and thrombosis – which account for most cases where the circulatory system goes wrong. Approximately 400,000 new cases of heart failure are registered in the US each year, summing up to four to five million people (Gheorghiade and Bonow, 1998). Heart attacks may lead to death or debilitation and the risk dying thereafter is considerably increased. (Cowie et al., 1997; Gheorghiade and Bonow, 1998; Sharpe and Doughty, 1998). Survivors suffer severe damage to heart muscle tissue that cannot regenerate. Frequently, the heart reacts to an increase in load caused by disease with a growth of the muscle tissue (hypertrophy) to compensate for pathological haemodynamics. This hypertrophy is solely based on an increase of the cardiac muscle cell size. Despite the facts that during the last 50 years much progress has been made in the recognition of cardiologic risk factors, in diagnosis and therapy of cardiovascular disorders, only little is known about the molecular and cellular fundamentals of these diseases. As mentioned, compensatory heart-hypertrophy is the response to multiple forms of heart damage. Hypertrophy not only causes increase in muscle mass i.e. synthesis of muscle proteins but is accompanied by a reactivation of genes normally only active during early embryogenesis and fetal stages in heart development (Eppenberger-Eberhardt et al., 1990). There are not only quantitative but also qualitative changes. To understand the underlying mechanisms controlling and regulating these events, the use of recombinant DNA-technology and other methods of molecular biology as well as of cell culture technology is essential.

Mammalian heart muscle cells, cardiomyocytes exit the cell cycle shortly after birth. Even under emergency conditions such as tissue damage, they do not re-enter the cell cycle. In the case of cardiovascular diseases, instead of invoking cell division to replace damaged cells, cardiac muscle hypertrophies to maintain heart performance. In humans, mitosis or hyperplasia of cardiomyocytes is lost at about 4-5 months of age and in rats about four days after birth. Normally, from then on only an increase in cell size and mass of existing cardiomyocytes is possible, i.e. cells become larger or hypertrophic. The molecular mechanism responsible for terminal differentiation and in particular for cell cycle arrest, in cardiomyocytes is mostly unknown. Thus, today the most effective therapy for the failing heart is transplantation, an operation that is expensive, risky and hampered by a chronic shortage of transplantable hearts. A feasible alternative to whole or partial organ replacement is the implantation of engineered tissue substitutes. On the one hand, non-dividing cardiomyocytes might be grafted directly into the injured area and might compensate for cell loss (Li et al., 1997); on the other hand, implantation of still proliferating cardiomyocytes (e.g. fetal) could be performed to replace damaged cells (Van Meter et al., 1995).

Expansion of primary cells from mammalian tissues is challenging due to the initial cell population being typically in a differentiated, non-proliferating state. In some cases, mostly involving cell types such as skin or hematopoietic cells which are not highly differentiated and which are known to lose proliferation control and develop cancers, protocols have been developed which permit cells to multiply while also preserving the essential functions of the initial population. Some of these tissue-engineering methods are now in commercial use. Expansion, however, of more differentiated, non-proliferating mammalian cells, such as cardiomyocytes and neurons, is of increasing importance.

Reversal of cell cycle arrest of cardiomyocytes is a prerequisite for the initiation of potential regenerative processes in the heart. An understanding of the factors controlling the cell cycle arrest of adult cardiomyocytes and ability to override such signals to re-induce cardiomyocyte proliferation have important implications in the treatment of heart diseases. For a number of tissues and organs (e.g. skin, cartilage, bone and liver), engineered tissues have been demonstrated to be a feasible alternative to organ transplantation, which in case of autologous expanded samples can additionally circumvent tissue compatibility problems. A precondition for tissue engineering is the ability of the cells to proliferate and to rebuild the original structure and function.

There is also another possible use of cultivated cardiac cells, namely somatic gene therapy. Cardiomyocytes, manipulated *in vitro* to express genes for proteins such as growth factors, could be grafted into the sites of an infarction where they might deliver factors in a paracrinic way to the neighboring cells and to the extracellular environment. However, such a system needs a tight control of proliferation to prevent tumor formation in heart, which is very rare under normal conditions. The use of dividing cardiac cells to replace affected cells must be carefully studied and the vector used to transfer the foreign gene into the cells chosen carefully. The major concern is the continued proliferation of the transplanted cells resulting in tumor formation.

The aim of this article is to review the current knowledge about the processes underlying terminal differentiation of cardiomyocytes and to give an overview of the methods and attempts performed to maintain proliferation in fetal or neonatal or to re-initiate cell cycle progression in adult cardiomyocytes.

2. Primary culture of isolated adult cardiomyocytes.

Currently there is no cardiac cell line, which retain the cardiac phenotype over time. Therefore, in order to study cardiomyocytes *in vitro*, it is necessary to prepare cells from heart tissue and to try to keep them alive as long as possible. The major part of this review will report on attempts to re-induce the cell cycle progression in cardiomyocytes *in vitro* or *in vivo*. Adult rat cardiomyocytes *in vitro* are able to form tissue like networks and establish new cell-cell contacts including intercalated disc-like structures (Eppenberger and Zuppinger, 1998; Hertig et al., 1996).

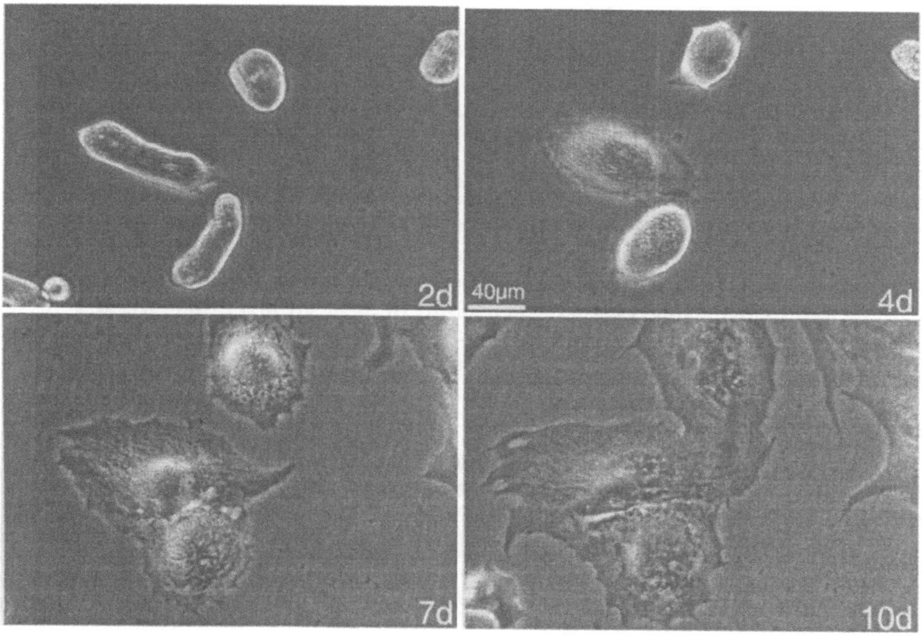

Figure 1
Behavior of adult rat cardiomyocytes in culture. Adult rat cardiomyocytes were isolated and cultured according to (Eppenberger-Eberhardt et al., 1990) for the time indicated in the lower right corner of each picture. Phase contrast images were obtained by video time-lapse microscopy (Rothen Rutishauser et al., 1998). Note the absence of migratory activity of these cells. Image sequence was kindly provided by C. Zuppinger.

These intercellular interactions eventually allow synchronous beating of the cells *in vitro*. These cardiomyocytes achieve contact with each other by a growing in size, thus represent a model for cellular hypertrophy. Figure 1 shows four frames of a time-lapse video microscopy experiment of adult rat cardiomyocytes at different time-points after plating. The re-establishment of cell-cell contacts after 8 days is clearly seen as well as the increase in cell size within 12 days in culture, the time when contacting cells are electrically coupled. Electric coupling is achieved through the formation of gap-junctions, as demonstrated immunohistologically using anti-connexin antibodies (Hertig et al., 1996). These tissue-like cell sheets can beat synchronously demonstrating that the cardiac muscle phenotype has been conserved *in vitro*. We have shown that supplementation of the culture medium with the growth factors IGF-1 and bFGF influenced the cell size and the state of myofibril reformation in culture considerably (Eppenberger-Eberhardt et al., 1997; Harder et al., 1996). Thus, the development of a biocompatible matrix expressing growth factor activities could be an approach to improve cultivation of transplantable cardiomyocytes (L. Polonchuk, personal communication).

3. Cell cycle control and terminal differentiation of cardiomyocytes

Terminal differentiation of cardiomyocytes in adult higher vertebrates is accompanied by an almost complete shut down of DNA synthesis. In rats, cardiomyocyte proliferation can be observed until the fourth postnatal day. DNA synthesis is detectable until the day ten, resulting in binucleated cardiomyocytes (Li et al., 1996). Thereafter, DNA synthesis is detectable in only five out of 10^6 cardiomyocytes in mice (Soonpaa and Field, 1998). Binucleation of ventricular cardiomyocytes in the rat results in tetraploidy of these cells as each nucleus contains a diploid set of chromosomes (Rumyantsev, 1977). In contrast, adult human cardiomyocytes have been shown to have one tetraploid nucleus, whereas the degree of ploidy can be even higher (Brodsky et al., 1993). Although cardiomyocytes constitute the majority of the heart mass, only 20-30% of all cells in the heart are cardiomyocytes. Therefore, one has to be cautious analyzing data obtained from heart tissue and those obtained from freshly isolated cardiomyocytes.

Progression of the cell cycle is controlled by a large number of cell cycle regulating molecules acting in a cascade (figure 2). The periodic expression, activation, inactivation and degradation of these proteins during the cell cycle has been analyzed in other cell systems. Experimental arrest and release of the cell cycle has allowed the assignment of the molecular function for these proteins to a certain phases of the cell cycle. Such cell cycle analyses have revealed a complex interplay between positive and negative regulators. An arrest of the cell cycle at one of the so-called cell cycle checkpoints can be evoked by the enhanced expression or activation of a negative regulator of the cell cycle such as $p21^{Cip1}$, $p27^{Kip1}$ or $p16^{Ink4a}$. Conversely, an arrest of the cell cycle can be overridden by the overexpression of an appropriate positive regulator of the cell cycle like the members of the E2F, cyclin or cyclin dependent kinase (CDK) gene families. It is assumed that a detailed knowledge of the developmental regulation of the expression of such cell cycle regulating molecules will help to identify genes suitable for the induction of cardiomyocyte proliferation.

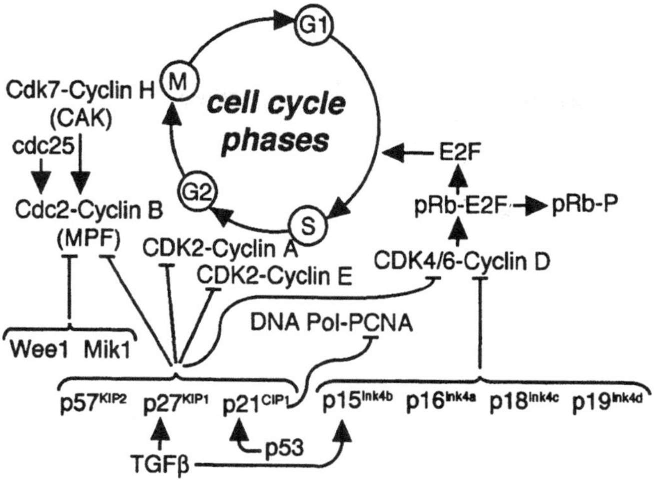

Figure 2
Schematic representation of a number of factors involved in the regulation of the cell cycle. Cyclin-CDK complexes are positioned around the cell cycle according to their role in cell cycle regulation.

Expression and activity of CDK and their regulatory subunits, the cyclins, are confined to certain phases of the cell cycle. The developmental regulation of cyclins and CDKs has been extensively studied with the intention to identify the cell cycle phase in which adult cardiomyocytes are arrested (Brooks et al., 1997; Flink et al., 1998; Kang and Koh, 1997; Poolman and Brooks, 1998; Yoshizumi et al., 1995). Activity and expression of all of these genes are, irrespectively of the differences in the exact time-course, downregulated in the adult (CDK7 and cyclin G and H have not been analyzed so far). Cyclin A is the first of the molecules downregulated during cardiomyocyte development at postnatal day 2 (Kang and Koh, 1997; Poolman and Brooks, 1998) followed by cyclin B, which becomes undetectable at postnatal day 6 (Kang and Koh, 1997). Cyclin B, together with cdc2, forms the so-called mitosis-promoting factor (MPF; for reviews see King et al., 1994 and Russell, 1998). The developmental regulation of cdc2 expression strictly follows that of cyclin B, and both, the amount of detectable protein and kinase activity become minimal by postnatal day 5 (Poolman and Brooks, 1998). However, a decrease in the cdc2 and Cdk2 kinase activities is already detectable at neonatal stages (Brooks et al., 1997). Activity of the MPF has been shown to be essential for the passage of dividing cells through the M-phase and cytokinesis. Inhibition of the MPF activation cascade can result in morphologically distinguishable figures of M phase arrest, but it is not yet clear to what extent these differences might be species specific. The effects can range from mitotic figures with condensed chromosomes upon deletion of protein phosphatase 1 (PP1) in D. melanogaster (Axton et al., 1990) to binucleation with de-condensed chromosomes upon inhibition of PPA1 and PPA2 in S. pombe (Kinoshita et al., 1993). The morphological coincidence of binucleation in adult rat cardiomyocytes together with the disappearance of MPF activity, suggest that the establishment of cell cycle arrest in cardiomyocytes might be initiated by an arrest in the M-phase of the cell cycle. So far, no data are available addressing the expression or activities of the regulators of MPF activity in cardiomyocytes. A decrease in the expression or activity of cyclin activating kinase

(CAK consisting of Cdk7-cyclin H) or of cdc25 or an increase in the expression of Mik1 or Wee1 could explain the observed binucleation and might therefore, be part of the regulatory mechanisms involved in the cell cycle arrest in adult cardiomyocytes.

The downregulation of the expression of the D-type cyclins and it's associated kinase, CDK4, occurs much later, namely after postnatal day 21 (Kang and Koh, 1997), at a time when the cardiomyocytes have already obtained their adult phenotype. By day 5, the cyclins D2 and D3 are expressed at levels comparable to that observed at birth (Poolman and Brooks, 1998) but both cyclins are undetectable in the adult (Brooks et al., 1997). Expression of the other cyclin D associated kinase, Cdk6, is already low in fetal cardiomyocytes and is undetectable in neonatal cells (Brooks et al., 1997). Cyclin E is strongly expressed in neonatal cardiomyocytes but the exact time-course of it's downregulation to undetectable levels in the adult remains to be analyzed (Brooks et al., 1997).

There is one possible combination of the available data for cyclin and CDK expression with the observations made of cellular processes during cardiomyocyte development. The ongoing DNA synthesis until the tenth postnatal day could likely reflect the sustained expression of the D-type cyclin-CDK complexes, whereas binucleation at the same time could well be the result of the absence of MPF activity. Therefore, triggering of M phase progression in neonatal cells might answer the question of whether the first step in establishing cell cycle arrest in cardiomyocytes is an inhibition of M phase progression. Whether the cell cycle arrest in adult cardiomyocytes could be reversed, remains to be shown by similar experiments in adult cardiomyocytes.

4. Cell cycle inhibitors

Assuming that the default setting in life is proliferation, cell cycle arrest should be an active process, similarly to apoptosis. It has been shown for example, that following DNA damage or non-advantageous environmental conditions, progression through the cell cycle can actively be blocked, by the expression of inhibitors. Therefore, developmental up-regulation of cell cycle inhibitor genes could be a direct indication of a potential involvement of that particular gene in the cell cycle arrest of adult cardiomyocytes. Negative regulators of the cell cycle progression, interacting directly with the cyclin-CDK complexes can be divided into three families: the Cip/Kip family of cyclin dependent kinase inhibitors (CDKI) comprising $p21^{Cip1}$ (El-Deiry et al., 1993; Harper et al., 1993; Noda et al., 1994), $p27^{Kip1}$ (Polyak et al., 1994; Toyoshima and Hunter, 1994) and $p57^{Kip2}$ (Lee et al., 1995), the Ink-family of CDKI comprising $p15^{Ink4b}$, $p16^{Ink4a}$ (Quelle et al., 1995), $p18^{Ink4c}$ and $p19^{Ink4d}$ (Chan et al., 1995; Hirai et al., 1995), and finally the retinoblastoma gene family comprising pRb (Harlow et al., 1986), $p107^{RBL1}$ (Ewen et al., 1991) and $p130^{RBL2}$ (Mayol et al., 1993).

The CDKI $p21^{Cip1}$, as a mediator of p53 induced cell cycle arrest (El-Deiry et al., 1994), binds to and inhibits the activity of complexes associated with Cdk2, Cdk3, Cdk4 and Cdk6. Furthermore, it binds to the proliferating cell nuclear antigen (PCNA) and can thereby directly inhibit DNA replication (Flore-Rozas et al., 1994; Li et al., 1994; Waga et al., 1994). In the course of postnatal cardiac development, cardiac expression of the $p21^{Cip1}$ protein is downregulated, similarly to $p57^{Kip2}$ (Koh et al., 1998). The CDKI $p27^{Kip1}$ can bind to and inhibit the activity of cyclin D-Cdk4, cyclin

A-Cdk2, cyclin B-Cdc2 and cyclin E-Cdk2 complexes (Polyak et al., 1994; Toyoshima and Hunter, 1994). We analyzed the developmental regulation of p27^{Kip1} expression by comparing protein levels in neonatal and adult cardiac cells (figure 3). We found that p27^{Kip1} is downregulated in freshly isolated postmitotic adult rat cardiomyocytes, but it is detectable in whole heart lysates throughout postnatal cardiac development, probably because being expressed by other cell types. The sustained expression of p27^{Kip1} in whole heart lysates has also been observed by others (Koh et al., 1998).

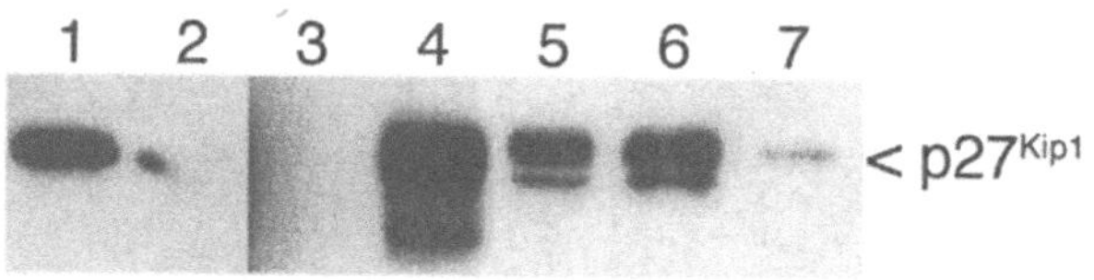

Figure 3
Western blot analysis of the developmental regulation of p27^{Kip1} expression. The amount of detectable protein expression was determined using a monoclonal antibody directed against p27^{Kip1}. Expression of p27^{Kip1} is clearly detectable in lysates of freshly isolated neonatal rat cardiomyocytes (lane 1) and is downregulated in the adult, as shown by the analysis of lysates of freshly isolated adult rat cardiomyocytes (lane 2). To demonstrate the specificity of the antibody used, lysates of CHO cells, engineered to inducibly express p27^{Kip1} (Mazur et al., 1998) were analyzed before (lane 3) and after induction of p27^{Kip1} expression (lane 4). Lysates from total neonatal (lane 6) and adult hearts (lane 5) showed a decrease of p27^{Kip1} during cardiac development, however, the sustained presence of some p27^{Kip1} is likely a result of non-cardiomyocytic expression. HeLa cell lysate was loaded as positive control in lane 7. The weak signal in lane 2 is spillover from lane 1.

The members of the retinoblastoma gene family, if hypo-phosphorylated, can inhibit cell cycle progression by direct binding to the transactivation domain of the members of the E2F family of transcription factors. Expression of pRb and of p107^{RBL1} is downregulated in ventricular cardiomyocytes of adult rats (Tam et al., 1995). Furthermore, E2F proteins in fetal cardiomyocytes are mainly attached to p107^{RbL1} while at postnatal day two p130^{RBL2}–associated E2F complexes are prominent (Flink et al., 1998). These findings suggest, that in adult cardiomyocytes E2F-4 or E2F-5 might be functional, since p130^{RBL2} do not bind to E2F1, E2F-2 or E2F-3. Because E2F-1, 2 and 3 are most effective in promoting S phase entry (DeGregori et al., 1997; Lukas et al., 1996), the indicated absence of these transcription factors is one of the possible reasons for the proliferation stop of adult cardiomyocytes.

We analyzed the tumor suppressor gene p53, to complete our knowledge on the developmental regulation of the expression of major tumor suppressor genes in cardiomyocytes. p53 can both negatively influence cell cycle progression of proliferating cells and induce apoptotic processes via the induction of a number of downstream genes (Polyak et al., 1997). In adult cardiomyocytes, the p53 protein level is elevated two to three fold in comparison to neonatal rat cardiomyocytes (Fig. 4, lanes 1 and 2), which is in contrast to the reported decrease of the p53 mRNA level in adult rat cardiomyocytes (Kim et al., 1994). These observations indicate a possible change in the stability of p53 during cardiomyocyte development. Analyses of the intracellular distribution of p53 in adult cardiomyocytes revealed that p53 is localized within different structures of the cell. Specific immunoreactivity could be seen associated with cytoskeleton and plasma membrane, and ER structures, suggesting that most of the p53 protein might be inactive in these cells (P. David, personal communication).

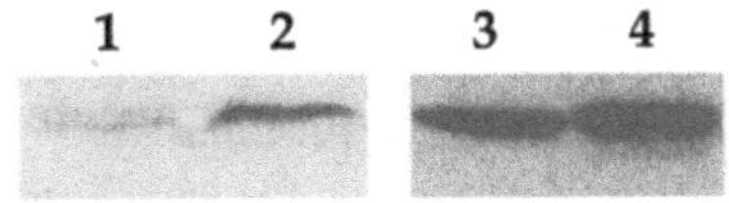

Figure 4
Developmental regulation of p53 protein content in cardiomyocytes. Lysates of freshly isolated neonatal (lanes 1 and 3) and adult rat cardiomyocytes (lanes 2 and 4) were probed with a polyclonal anti p53 antibody (lanes 1 and 2) or with the monoclonal antibody PAb421 (lanes 3 and 4). Because the epitope of the antibody PAb421 is phosphorylation sensitive, lanes 3 and 4 were de-phosphorylated on blot with calf intestinal phosphatase. An about two-fold increase in the amount of detectable p53 protein in the adult is evident from these two blots.

5. Mouse mutants

No proliferation of adult cardiomyocytes could be detected in the hearts of mutant mice deficient for different tumor suppressor genes. To date, "knock out" mice for p53 (Donehower et al., 1992), $p21^{Cip1}$ (Deng et al., 1995), $p27^{Kip1}$ (Fero et al., 1996; Kiyokawa et al., 1996; Nakayama et al., 1996), $p57^{Kip2}$ (Zhang et al., 1997), $p16^{Ink4a}$ (Serrano et al., 1996), pRb (Clarke et al., 1992; Jacks et al., 1992; Lee et al., 1992), $p107^{RBL1}$ and $p130^{RBL2}$ (Cobrinik et al., 1996; Lee et al., 1996) have been generated. Although in the case of the $p27^{Kip1}$ mouse mutant a hyperplasia in fetal and perinatal stages of heart development has been observed, the mutation did not interfere with the establishment and maintenance of adult cardiomyocyte cell cycle arrest. The absence of cardiomyocyte proliferation in these mice suggests that due to the redundancy of these genes, none of them alone would be responsible for the cell cycle arrest. Due to the observed downregulation of $p21^{Cip1}$, $p27^{Kip1}$ and $p57^{Kip2}$ in the adult, the members of the Cip/Kip family of CDKI seem unlikely to play a major role in the maintenance of the cell cycle arrest. There is a growing number of regulatory, downstream effector molecules involved in the regulation of cell cycle. These include cdc25, Wee1, Mik1, PP1a and PP2A. A closer look at these molecules may help to reveal the molecular cues leading to the cardiomyocyte cell cycle arrest. The existence, on the other hand, of unknown, eventually cardiac-specific, proliferation suppressor genes can not be excluded.

The question remains, in which phase of the cell cycle are adult cardiomyocytes arrested, and how could cell cycle progression in adult cardiomyocytes be re-initiated? Mitotically quiescent cells are usually arrested in the G1/G0 phase of the cell cycle, which has led to various attempts targeting this phase to abrogate cell cycle arrest in cardiomyocytes. The expression of viral oncogenes offers the possibility to immortalize different cell types. The ability of oncogenes to induce cardiomyocyte proliferation was tested by a number of groups. Introduction of oncogene sequences into cardiomyocytes was achieved either by the generation of transgenic mice, expressing the particular oncogene under a tissue specific promoter or by the use of adeno or retroviral systems.

The first report on the generation of cardiac tumors by ectopic expression of the viral oncogene SV40 T-antigen was published a decade ago (Behringer et al., 1988). Although the transgene expressed resulted in the loss of the cardiomyocyte phenotype, the observed increase in DNA synthesis initiated further experiments on the generation of transgenic mice expressing the same gene in a strictly cardiomyocyte restricted manner. Mice either expressing the SV40 T-antigen under the regulation of the heart-specific, α-myosin heavy chain promoter (Katz et al., 1992) or under the atrium-specific, atrial natriuretic peptide (ANF) promoter showed proliferation of atrial muscle cells, allowed maintenance of some characteristics of differentiated cardiomyocytes for a limited time in culture and gave rise to the generation of the atrial cardiomyocyte derived AT-1 cell line. AT-1 cells can be propagated subcutaneously in mice for some period of time (Steinhelper et al., 1990). The question, whether a regulated expression of the SV40 T antigen gene would allow to regain the cardiac phenotype and function of normal, differentiated cardiomyocytes was addressed by the infection of cardiomyocytes with MLV retrovirus expressing a temperature sensitive mutant of the SV40 T-antigen (Miller et al., 1994). Using this system, after chasing of the proliferation induction by a shift to the non-permissive temperature, the cells failed to return to the differentiated cardiomyocyte phenotype. This finding might partly be due to pleiotropic effects of temperature shifts, inducing the expression of a number of other endogenous genes (Rossi and Blau, 1998). The transforming activity of SV40 T-antigen is based upon it's ability to inhibit the growth suppressor activity of p53, pRb and $p107^{RBL1}$ by forming of multimeric complexes with these molecules (Ludlow, 1996). In the above experiments, the SV40 T-antigen was active in fetal or neonatal cardiomyocytes. Whether the SV40 T antigen gene is also able to reactivate the cell cycle in postmitotic adult cardiomyocytes remains to be shown. The polyvalent binding capacity of SV40 T-antigen makes this assumption likely, because it can both activate the E2F transcription factors by competitive binding to the members of the retinoblastoma gene family, and also inactivate p53 which otherwise could transduce the expression of the CDKI $p21^{Cip1}$ and the successive expression of apoptotic genes (Madden et al., 1997; Polyak et al., 1997). Furthermore, overexpression of the SV40 T-antigen in diploid cells leads to an increase in the expression of cyclin A, cyclin B, cdc2 and cdc25 activities (Chang et al., 1997). Since these cell cycle regulating genes are downregulated as first ones during the postnatal development of cardiomyocytes (Kang and Koh, 1997; Poolman and Brooks, 1998; Yoshizumi et al., 1995), the molecular effects of SV40 T-antigen overexpression appear to directly counterbalance the processes underlying the establishment of the cell cycle arrest in adult cardiomyocytes.

The necessity for a multivalent effect is also demonstrated by the expression of the adenoviral oncogene E1A (Kirshenbaum and Schneider, 1995; Liu and Kitsis, 1996) or of its cellular counterpart, E2F-1 (Kirshenbaum et al., 1996) in cardiomyocytes. Adenoviral gene transfer mediated expression of these genes resulted in the induction of apoptosis in cardiomyocytes. The E1A or E2F-1 induced apoptotic processes could be inhibited by the co-expression of the apoptosis inhibitory gene E1B, whereby an accumulation of cardiomyocytes in the M phase of the cell cycle was seen (Kirshenbaum et al., 1996). Besides these results, the demonstration of a suppression of cardiomyocyte specific genes by the overexpression of the E1A oncoprotein (Bishopric et al., 1997) has, in addition, indicated that this particular oncogene may not be compatible with the cardiomyocytes phenotype. E2F-1 is the only member of the E2F gene family shown to induce apoptosis (Pan et al., 1998; Tsai et al., 1998), and it well may be possible that other members of the E2F-family could induce cardiomyocyte proliferation without apoptosis. Since the different members of the E2F gene family can induce the transcription of different subsets of genes, they are likely to have distinct roles in cell growth control and apoptosis (DeGregori et al., 1997). An overexpression of such genes in cardiomyocytes might give results that differ in their proliferation inducing efficiency and in the preservation of the cardiomyocyte phenotype.

Retrovirus mediated overexpression of the v-myc oncogene in chicken hearts, if injected before completion of heart looping into the embryo, has been shown to induce proliferation of cardiomyocytes (Saule et al., 1987). Similarly, infection of cultured fetal rat cardiomyocytes with retroviruses expressing the c-myc oncogene resulted in cells that expressed some heart specific genes, but did not beat and displayed several features atypical for cardiac myocytes (Engelmann et al., 1993). Lastly, transgenic mice, constitutively expressing c-myc display cardiomyocyte proliferation only during fetal development (Jackson et al., 1990).

Overexpression of cyclin D1 using the α-cardiac myosin heavy chain promoter resulted in enhanced DNA synthesis and forming of multiple nuclei in the adult cardiomyocytes, but no cytokinesis of could be seen (Soonpaa et al., 1997). The marked increase in the number of cardiomyocyte nuclei in these mice indicates the probable existence of a strong arrest of the M-phase or of the septation in these cells as described earlier by the lack of MPF activity in these cells.

The main morphologic feature of cytokinesis is the formation and the tightening of the cleavage furrow, a process called septation. Several genes were shown to regulate septation (Balasubramanian et al., 1998; Jwa and Song, 1998), the initiation of which is normally linked to the cell cycle. In the fission yeast S. pombe, septation can be uncoupled from the cell cycle by the overexpression of positive or negative regulators of septation. Mammalian homologues of some of these genes have been identified, like the human or mouse homologue of the yeast CDC7 gene (Hess et al., 1998; Jiang and Hunter, 1997; Kim et al., 1998) or $p55^{CDC}$ (Kallio et al., 1998). Since targeted regulation of the activities of cdc7 or Spg1p in S. pombe for example can induce the initiation of septation independently of the cell cycle state (Sohrmann et al., 1998), alteration of such gene activities might provide a tool to induce septation in binucleated adult rat cardiomyocytes.

6. Conclusion

Although the experiments reviewed here have revealed a number of aspects concerning the establishment of the cell cycle arrest in adult cardiomyocytes, the process as a whole is still not clearly understood. The possibility that this process would only rely on an inhibition of the cell cycle progression in the G1-phase has to be rethought, and future work will have to be directed also towards the other phases of the cell cycle. Recent and future progress elucidating the cell cycle regulation mechanisms in simpler cellular systems might provide hints on possible players which could have the ability to re-initiate cell cycle progression also in cardiomyocytes. A parallel development of cardiomyocyte cell culture technologies will allow the production of tissue equivalents and the adaptation of methods for the introduction of *in vitro* produced cells or tissues into the heart will help to the establish successful cardiac tissue replacement as a tool for the treatment of cardiovascular diseases. The increasing number of cardiovascular derived diseases in industrialized countries (Cowie et al., 1997; Sharpe and Doughty, 1998) as well as in developing countries (Cheng, 1988) urges research in this field.

7. Acknowledgement

We thank to Jean-Claude Perriard, Frank Stockdale and Verdon Taylor for careful reading of the manuscript.

8. References

Axton, J. M., Dombràdi, V., Cohen, P. T. W., and Glover, D. M. (1990). One of the protein phosphatase 1 isoenzymes in Drosphila is essential for mitosis. *Cell* ***63***, 33-46.

Balasubramanian, M. K., McCollum, D., Chang, L., Wong, K. C., Naqvi, N. I., He, X., Sazer, S., and Gould, K. L. (1998). Isolation and characterization of new fission yeast cytokinesis mutants. *Genetics* ***149***, 1265-1275.

Behringer, R. R., Peschon, J. J., Messing, A., Gartside, C. L., Hauschka, S. D., Palmiter, R. D., and Brinster, R. L. (1988). Heart and bone tumors in transgenic mice. *Proc. Natl. Acad. Sci. USA* ***85***, 2648-2652.

Bishopric, N. H., Zeng, G. Q., Sato, B., and Webster, K. A. (1997). Adenovirus E1A inhibits cardiac myocyte-specific gene expression through its amino terminus. *J. Biol. Chem.* ***272***, 20584-20594.

Brodsky, V., Sarkisov, D. S., Arefyeva, A. M., and Panova, N. W. (1993). DNA and protein relations in cardiomyocytes. Growth reserve in cardiac muscle. *Eur. J. Histochem.* ***37***, 199-206.

Brooks, G., Poolman, R. A., McGill, C. J., and Li, J.-M. (1997). Expression and activities of cyclins and cylin-dependent kinases in developing rat ventricular cardiomyocytes. *J. Mol. Cell. Cardiol.* ***29***, 2261-2271.

Chan, F. K., Zhang, J., Cheng, L., Shapiro, D. N., and Winoto, A. (1995). Identification of human and mouse p19, a novel CDK4 and CDK6 inhibitor with homology to p16ink4. *Mol. Cell. Biol.* ***15***, 2682-2688.

Chang, T. H., Ray, F. A., Thompson, D. A., and Schlegel, R. (1997). Disregulation of mitotic checkpoints and regulatory proteins following acute expression of SV40 large T antigen in diploid human cells. *Oncogene* ***14***, 2383-93.

Cheng, T. O. (1988). Cardiovascular disease in China. *Nature Med.* ***4***, 1209-1210.

Clarke, A. R., Maandag, E. R., van Roon, M., van der Lugt, N. M., van der Valk, M., Hooper, M. L., Berns, A., and te Riele, H. (1992). Requirement for a functional Rb-1 gene in murine development. *Nature* ***359***, 328-330.

Cobrinik, D., Lee, M.-H., Hannon, G., Mulligan, G., Bronson, R. T., Dyson, N., Harlow, E., Beach, D., Weinberg, R. A., and Jacks, T. (1996). Shared role of the pRB-related p130 and p107 proteins in limb development. *Genes Dev.* ***10***, 1633-1644.

Cowie, M. R., Mosterd, A., Wood, D. A., Deckers, J. W., Poole Wilson, P. A., Sutton, G. C., and Grobbee, D. E. (1997). The epidemiology of heart failure. *Eur. Heart. J.* ***18***, 208-225.

DeGregori, J., Leone, G., Miron, A., Jakoi, L., and Nevins, J. R. (1997). Distinct roles for E2F proteins in cell growth control and apoptosis. *Proc. Natl. Acad. Sci. USA* ***94***, 7245-7250.

Deng, C., Zhang, P., Harper, J. W., Elledge, S. J., and Leder, P. (1995). Mice lacking $p21^{CIP1/WAF1}$ undergo normal development, but are defective in G1 checkpoint control. *Cell* ***82***, 675-684.

Donehower, L. A., Harvey, M., Slagle, B. L., McArthur, M. J., Montgomery, C. A., Butel, J. S., and Bradley, A. (1992). Mice deficient for p53 are developmentally normal but susceptible to spontaneous tumours. *Nature* ***356***, 215-221.

El-Deiry, W. S., Harper, J. W., O'Connor, P. M., Velculescu, V. E., Canman, C. E., Jackman, J., Pietenpol, J. A., Burrell, M., Hill, D. E., and Wang, Y. (1994). WAF1/CIP1 is induced in p53-mediated G1 arrest and apoptosis. *Cancer Res.* ***54***, 1169-1174.

El-Deiry, W. S., Tokino, T., Velculescu, V. E., Levy, D. B., Parsons, R., Trent, J. M., Lin, D., Mercer, W. E., Kinzler, K. W., and Vogelstein, B. (1993). WAF1, a potential mediator of p53 tumor suppression. *Cell* ***75***, 817-825.

Engelmann, G. L., Birchenall-Roberts, M. C., Ruscetti, F. W., and Samarel, A. M. (1993). Formation of fetal rat cardiac cell clones by retroviral transformation: retention of select myocyte characteristics. *J. Mol. Cell. Cardiol.* ***25***, 197-213.

Eppenberger, H. M., and Zuppinger, C. (1998). In vitro reestablishment of cell-cell contacts in adult rat cardiomyocytes. Functional role of transmembrane components in the formation of new intercalated disc-like contacts. ***FASEB J. in press.***

Eppenberger-Eberhardt, M., Aigner, S., Donath, M. Y., Kurer, V., Walther, P., Zuppinger, C., Schaub, M. C., and Eppenberger, H. M. (1997). IGF-I and bFGF differentially influence atrial natriuretic factor and alpha-smooth muscle actin expression in cultured atrial compared to ventricular adult rat cardiomyocytes. *J. Mol. Cell. Cardiol.* ***29***, 2027-2039.

Eppenberger-Eberhardt, M., Flamme, I., Kurer, V., and Eppenberger, H. M. (1990). Reexpression of alpha-smooth muscle actin isoform in cultured adult rat cardiomyocytes. *Dev. Biol.* ***139***, 269-278.

Ewen, M. E., Xing, Y., Lawrence, J. B., and Livingston, D. M. (1991). Molecular cloning, chromosomal mapping, and expression of the cDNA for p107, a retinoblastoma gene product-related protein. *Cell* ***66***, 1155-1164.

Fero, M. L., Rivkin, M., Tasch, M., Porter, P., Carow, C. E., Firpo, E., Polyak, K., Tsai, L. H., Broudy, V., Perlmutter, R. M., Kaushansky, K., and Roberts, J. M. (1996). A syndrome of multiorgan hyperplasia with features of gigantism, tumorigenesis, and female sterility in p27(Kip1)-deficient mice. *Cell* ***85***, 733-44.

Flink, I. L., Oana, S., Maitra, N., Bahl, J. J., and Morkin, E. (1998). Changes in E2F complexes containing retinoblastoma protein family members and increased cyclin-dependent kinase inhibitor activities during terminal differentiation of cardiomyocytes. *J. Mol. Cell. Cardiol.* ***30***, 563-578.

Flore-Rozas, H., Kelman, Z., Dean, F. B., Pan, Z. Q., Harper, J. W., Elledge, S. J., O'Donnel, M. O., and Hurwitz, J. (1994). Cdk-interacting protein 1 directly binds with proliferating cell nuclear antigen and inhibits DNA replications catalysed by the DNA polymerase delta holoenzyme. *Proc. Natl. Acad. Sci. USA* ***91***, 8655-8659.

Gheorghiade, M., and Bonow, R. O. (1998). Chronic heart failure in the United States: a manifestation of coronary artery disease. *Circulation* ***97***, 282-289.

Harder, B. A., Schaub, M. C., Eppenberger, H. M., and Eppenberger-Eberhardt, M. (1996). Influence of fibroblast growth factor (bFGF) and insulin-like growth factor (IGF-I) on cytoskeletal and contractile structures and on atrial natriuretic factor (ANF) expression in adult rat ventricular cardiomyocytes in culture. *J. Mol. Cell. Cardiol.* ***28***, 19-31.

Harlow, E., Whyte, P., Franza, B. R. J., and Schley, C. (1986). Association of adenovirus early-region 1A proteins with cellular polypeptides. *Mol. Cell. Biol.* ***6***, 1579-1589.

Harper, J. W., Adami, G. R., Wei, N., Keyomarsi, K., and Elledge, S. J. (1993). The p21 Cdk-ineracting protein Cip1 is a potent inhibitor of G1 cyclin-dependent kinases. *Cell* ***75***, 805-816.

Hertig, C. M., Eppenberger-Eberhardt, M., Koch, S., and Eppenberger, H. M. (1996). N-cadherin in adult rat cardiomyocytes in culture. I. Functional role of N-cadherin and impairment of cell-cell contact by a truncated N-cadherin mutant. *J. Cell. Sci.* ***109***, 1-10.

Hess, G. F., Drong, R. F., Weiland, K. L., Slightom, J. L., Sclafani, R. A., and Hollingsworth, R. E. (1998). A human homolog of the yeast CDC7 gene is overexpressed in some tumors and transformed cell lines. *Gene* ***211***, 133-140.

Hirai, H., Roussel, M. F., Kato, J. Y., Ashmun, R. A., and Sherr, C. J. (1995). Novel INK4 proteins, p19 and p18, are specific inhibitors of the cyclin D-dependent kinases CDK4 and CDK6. *Mol. Cell. Biol.* ***15***, 2672-2681.

Jacks, T., Fazeli, A., Schmitt, E. M., Bronson, R. T., Goodell, M. A., and Weinberg, R. A. (1992). Effects of an Rb mutation in the mouse. *Nature* ***359***, 295-300.

Jackson, T., Allard, M. F., Sreenan, C. M., Doss, L. K., Bishop, S. P., and Swain, J. L. (1990). The c-myc proto-oncogene regulates cardiac development in transgenic mice. *Mol. Cell. Biol.* ***10***, 3709-3716.

Jiang, W., and Hunter, T. (1997). Identification and characterization of a human protein kinase related to budding yeast Cdc7p. *Proc. Natl. Acad. Sci. USA.* ***94***, 14320-14325.

Jwa, M., and Song, K. (1998). Byr4, a dosage-dependent regulator of cytokinesis in S. pombe, interacts with a possible small GTPase pathway including Spg1 and Cdc16. *Mol. Cells* ***8***, 240-245.

Kallio, M., Weinstein, J., Daum, J. R., Burke, D. J., and Gorbsky, G. J. (1998). Mammalian p55CDC mediates association of the spindle checkpoint protein Mad2 with the cyclosome/anaphase-promoting complex, and is involved in regulating anaphase onset and late mitotic events. *J. Cell. Biol.* ***141***, 1393-1406.

Kang, M. J., and Koh, G. Y. (1997). Differential and dramatic changes of the cyclin-dependent kinase activities in cardiomyocytes during the neonatal period. *J. Mol. Cell. Cardiol.* ***29***, 1767-1777.

Katz, E. B., Steinhelper, M. E., Delcarpio, J. B., Daud, A. I., Claycomb, W. C., and Field, L. J. (1992). Cardiomyocyte proliferation in mice expressing alpha-cardiac myosin heavy chain-SV40 T-antigen transgenes. *Am. J. Physiol.* ***262***, H1867-H1876.

Kim, J. M., Sato, N., Yamada, M., Arai, K., and Masai, H. (1998). Growth regulation of the expression of mouse cDNA and gene encoding a serine/threonine kinase related to Saccharomyces cerevisiae CDC7 essential for G1/S transition. Structure, chromosomal localization, and expression of mouse gene for s. cerevisiae Cdc7-related kinase. *J. Biol. Chem.* ***273***, 23248-23257.

Kim, K. K., Soonpaa, M. H., Daud, A. I., Koh, G. Y., Kim, J. S., and Field, L. J. (1994). Tumor suppressor gene expression during normal and pathologic myocardial growth. *J. Biol. Chem.* ***269***, 22607-22613.

King, R. W., Jackson, P. K., and Kirschner, M. W. (1994). Mitosis in transition. *Cell* ***79***, 563-571.

Kinoshita, N., Yamano, H., Niwa, H., Yoshida, T., and Yanagida, M. (1993). Negative regulation of mitosis by the fission yeast protein phosphatase ppa2. *Genes Dev.* ***7***, 1059-1071.

Kirshenbaum, L. A., Abdellatif, M., Chakraborty, S., and Schneider, M. D. (1996). Human E2F-1 reactivates cell cycle progression in ventricular myocytes and represses cardiac gene transcription. *Dev. Biol.* ***179***, 402-411.

Kirshenbaum, L. A., and Schneider, M. D. (1995). Adenovirus E1A represses cardiac gene transcription and reactivates DNA synthesis in ventricular myocytes, via alternative pocket protein- and p300-binding domains. *J. Biol. Chem.* ***270***, 7791-7794.

Kiyokawa, H., Kineman, R. D., Manova, T. K., Soares, V. C., Hoffman, E. S., Ono, M., Khanam, D., Hayday, A. C., Frohman, L. A., and Koff, A. (1996). Enhanced growth of mice lacking the cyclin-dependent kinase inhibitor function of p27(Kip1). *Cell* ***85***, 721-32.

Koh, K. N., Kang, M. J., Frith-Terhume, A., Park, S. K., Kim, I., Lee, C. O., and Koh, G. Y. (1998). Persistent and heterogenous expression of the cyclin-dependent kinase inhibitor $p27^{KIP1}$, in rat hearts during development. *J. Mol. Cell. Cardiol.* ***30***, 463-474.

Lee, E. Y., Chang, C. Y., Hu, N., Wang, Y. C., Lai, C. C., Herrup, K., Lee, W. H., and Bradley, A. (1992). Mice deficient for Rb are nonviable and show defects in neurogenesis and haematopoiesis. *Nature* ***359***, 288-294.

Lee, M.-H., Williams, B. O., Mulligan, G., Mukai, S., Bronson, R. T., Dyson, N., Harlow, E., and Jacks, T. (1996). Targeted disruption of p107: functional overlap between p107 and Rb. *Genes Dev.* ***10***, 1621-1632.

Lee, M. H., Reynisdottir, I., and Massague, J. (1995). Cloning of $p57^{KIP2}$, a cyclin-dependent kinase inhibitor with unique domain structure and tissue distribution. *Genes Dev.* ***9***, 639-649.

Li, F., Wang, X., Capasso, J. M., and Gerdes, A. M. (1996). Rapid transition of cardiac myocytes from hyperplasia to hypertrophy during postnatal development. *J. Mol. Cell. Cardiol.* ***28***, 1737-1746.

Li, R., Waga, S., Hannon, G., Beach, D., and Stillman, B. (1994). Differential effects by the p21 CDK inhibitor on PCNA dependent DNA replication and repair. *Nature* ***371***, 534-537.

Li, R. K., Mickle, D. A., Weissel, R. D., Mohabeer, M. K., Zhang, J., Rao, V., Li, G., Merante, F., and Jia, Z. Q. (1997). Natural history of fetal rat cardiomyocytes transplanted into adult rat myocardial scar tissue. *Circulation* ***96***, 179-186.

Liu, Y., and Kitsis, R. N. (1996). Induction of DNA synthesis and apoptosis in cardiac myocytes by E1A oncoprotein. *J. Cell Biol.* ***133***, 325-334.

Ludlow, J. W. (1996). Use of sequential immunoprecipitation to reveal discrete, separable populations of SV40 T-antigen binding to host cellular proteins. *J. Virol. Methods* ***59***, 105-112.

Lukas, J., Petersen, B. O., Holm, K., Bartek, J., and Helin, K. (1996). Deregulated expression of E2F family members induces S-phase entry and overcomes $p16^{INK4A}$-mediated growth suppression. *Mol. Cell. Biol.* ***16***, 1047-1057.

Madden, S. L., Galella, E. A., Zhu, J., Bertelsen, A. H., and Beaudry, G. A. (1997). SAGE transcript profiles for p53-dependent growth regulation. *Oncogene* ***15***, 1079-1085.

Mayol, X., Grana, X., Baldi, A., Sang, N., Hu, Q., and Giordano, A. (1993). Cloning of a new member of the retinoblastoma gene family (pRb2) which binds to the E1A transforming domain. *Oncogene* ***8***, 2561-2566.

Mazur, X., Eppenberger, H. M., Bailey, J. E., and Fussenegger, M. (1998). A novel autoregulated proliferation-controlled production process using recombinant CHO cells. *Biotechnol. Bioeng.* ***submitted***.

Miller, C., Rilfs, J., Jaspers, S. R., Buckholt, M., and Miller Jr., T. B. (1994). Transformation of adult ventricular myocytes with the temperature sensitive A58 (tsA58) mutant of the SV40 large T antigen. *Mol. Cell. Biochem.* ***136***, 29-34.

Nakayama, K., Ishida, N., Shirane, M., Inomata, A., Inoue, T., Shishido, N., Horii, I., Loh, D. Y., and Nakayama, K. (1996). Mice lacking p27(Kip1) display increased body size, multiple organ hyperplasia, retinal dysplasia, and pituitary tumors. *Cell* ***85***, 707-20.

Noda, A., Ning, Y., Venable, S. F., Pereira-Smith, O. M., and Smith, J. R. (1994). Cloning of senescent cell-derived inhibitors of DNA synthesis using an expression system. *Exp. Cell Res.* ***211***, 90-98.

Pan, H., Yin, C., Dyson, N. J., Harlow, E., Yamasaki, L., and Van Dyke, T. (1998). Key roles for E2F1 in signaling p53-dependent apoptosis and in cell division within development. *Mol. Cell* **2**, 283-292.

Polyak, K., Lee, M.-H., Erdjument-Bromage, H., Koff, A., Roberts, J. M., Tempst, P., and Massagué, J. (1994). Cloning of p27^{Kip1}, a cyclin-dependent kinase inhibitor and a potential mediator of extracellular antimorphogenic signals. *Cell* **78**, 59-66.

Polyak, K., Xia, Y., Zweier, J. L., Kinzler, K. W., and Vogelstein, B. (1997). A model for p53 induced apoptosis. *Nature* **389**, 300-305.

Poolman, R. A., and Brooks, G. (1998). Expression and activities of cell cycle regulatory molecules during the transition from myocyte hyperplasia to hypertrophy. *J. Mol. Cell. Cardiol.* **30**, 2121-2135.

Quelle, D. E., Ashmun, R. A., Hannon, G. J., Rehberger, P. A., Trono, D., Richter, K. H., Walker, C., Beach, D., Sherr, C. J., and Serrano, M. (1995). Cloning and characterization of murine p16^{INK4a} and p15^{INK4b} genes. *Oncogene* **11**, 635-645.

Rossi, F. M. V., and Blau, H. M. (1998). Recent advances in inducible gene expression systems. *Curr. Op. Biotech.* **9**, 451-456.

Rothen Rutishauser, B. M., Ehler, E., Perriard, E., Messerli, J. M., and Perriard, J. C. (1998). Different behaviour of the non-sarcomeric cytoskeleton in neonatal and adult rat cardiomyocytes. *J. Mol. Cell. Cardiol.* **30**, 19-31.

Rumyantsev, P. P. (1977). Interrelations of the proliferation and differentiation process during cardiac myogenesis and regeneration. *Int. Rev. Cytol.* **51**, 187-273.

Russell, P. (1998). Checkpoints on the road to mitosis. *TIBS* **23**, 399-402.

Saule, S., Mérigaud, J. P., Al-Moustafa, A.-E. M., Ferré, F., Rong, P. M., Amouyel, P., Quatannens, B., Stéhelin, D., and Dieterlen-Lièvre, F. (1987). Heart tumors specifically induced in young avian embryos by the v-myc oncogene. *Proc. Natl. Acad. Sci. USA* **84**, 7982-7986.

Serrano, M., Lee, H., Chin, L., Cordon Cardo, C., Beach, D., and DePinho, R. A. (1996). Role of the INK4a locus in tumor suppression and cell mortality. *Cell* **85**, 27-37.

Sharpe, N., and Doughty, R. (1998). Epidemiology of heart failure and ventricular dysfunction. *Lancet* **352**, SI3-7.

Sohrmann, M., Schmidt, S., Hagan, I., and Simanis, V. (1998). Asymmetric segregation on spindle poles of the Schizosaccharomyces pombe septum-inducing protein kinase Cdc7p. *Genes Dev.* **12**, 84-94.

Soonpaa, M. H., and Field, L. J. (1998). Survey of studies examining mammalian cardiomyocyte DNA synthesis. *Circ. Res.* **83**, 15-26.

Soonpaa, M. H., Koh, G. Y., Pajak, L., Jing, S., Wang, H., Franklin, M. T., Kim, K. K., and Field, L. J. (1997). Cyclin D1 overexpression promotes cardiomyocyte DNA synthesis and multinucleation in transgenic mice. *J. Clin. Invest.* **99**, 2644-2654.

Steinhelper, M. E., Lanson, N. A., Jr., Dresdner, K. P., Delcarpio, J. B., Wit, A. L., Claycomb, W. C., and Field, L. J. (1990). Proliferation in vivo and in culture of differentiated adult atrial cardiomyocytes from transgenic mice. *Am. J. Physiol.* **259**, H1826-H1834.

Tam, S. K., Gu, W., Mahdavi, V., and Nadal Ginard, B. (1995). Cardiac myocyte terminal differentiation. Potential for cardiac regeneration. *Ann. NY Acad. Sci.* **752**, 72-79.

Toyoshima, H., and Hunter, T. (1994). p27, a novel inhibitor of G1 cyclin-Cdk protein kinase activity, is related to p21. *Cell* **78**, 67-74.

Tsai, K. Y., Hu, Y., Macleod, K. F., Crowley, D., Yamasaki, L., and Jacks, T. (1998). Mutation of E2f-1 suppresses apoptosis and inappropriate S phase entry and extends survival of Rb-deficient mouse embryos. *Mol. Cell* **2**, 293-304.

Van Meter, C. H., Claycomb, W. C., Delcarpio, J. B., Smith, D. M., deGruiter, H., Smart, F., and Ochsner, J. L. (1995). Myoblast transplantation in the porcine model: a potential technique for myocardial repair. *J. Thorac. Cardiovasc. Surg.* ***110***, 1442-1448.

Waga, S., Hannon, G., Beach, D., and Stillman, B. (1994). The p21 inhibitor of cyclin-dependent kinase controls DNA replication by interaction with PCNA. *Nature* ***369***, 574-578.

Yoshizumi, M., Lee, W.-S., Hsieh, C.-M., Tsai, J.-C., Li, J., Perrella, M. A., Patterson, C., Endege, W. O., Schlegel, R., and Lee, M.-E. (1995). Disappearance of cyclin A correlates with permanent withdrawal of cardiomyocytes from the cell cycle in human and rat hearts. *J. Clin. Invest.* ***95***, 2275-2280.

Zhang, P., Liégeois, N. J., Wong, C., Finegold, M., Hou, H., Thompson, J. C., Silverman, A., Harper, J. W., DePinho, R. A., and Elledge, S. J. (1997). Altered cell differentiation and proliferation in mice lacking $p57^{KIP2}$ indicates a role in Beckwith-Wiedemann syndrome. *Nature* ***387***, 151-158.

IMMORTALIZATION OF HEPATOCYTES THROUGH TARGETED DEREGULATION OF THE CELL CYCLE.

G. S. JENNINGS[1]* and M. STRAUSS[2]
[1]HepaVec AG & [2]Department of Cell Biology, Humboldt University, Max-Delbrück-Center for Molecular Medicine
Robert-Rössle-Strasse 10, D-13125 Berlin, Germany.
**Corresponding author, fax +49-30-9489-2281*
e-mail gjennings@hepavec.com

1. Introduction

The generation of continuously proliferating hepatocytes expressing differentiated functions has been a goal long sought after by workers in the fields of liver toxicology and pathology, in recombinant protein technology and in gene therapy and medicine. The requirements of these disciplines will define the parameters for selecting an immortalization strategy and will ultimately pass judgement on the success of newly created cell lines (Fig. 1).

(i) *In vitro* toxicology: The chemical and phamaceutical industries release thousands of new products yearly, all of which require testing for potential cyto- and genotoxicity. The liver is the primary organ of xenobiotic detoxication and hepatocytes are abundent in cytochromes P-450, glutathione-*S*-transferases and UDP-glucuronyltransferases, the enzymes crucially involved in the activation of promutagens and protoxins (Meyer, 1996). Many of the subfamilies of cytochrome P-450 are induced by their substrates, an essential property contributing to the toxicological profile of a xenobiotic. A human hepatocyte cell line expressing activities which mirror those *in vivo* would be a boon to toxicologists providing highly relevant information on the effect of human exposure to new chemicals as well as providing opportunity for reduction in animal testing (Guillouzo, 1998).

(ii) Recombinant protein production: Hepatocytes offer high levels of post-translational modification activities and secrete large amounts of protein, ideal for the mass production of new recombinant proteins in so-called 'cell factories'. Furthermore, as hepatocytes contain enzymes for the synthesis of essential amino-acids, fatty acids and complex molecules (Arias *et al.*, 1994), growth media can be simplified if these fuctions are maintained, contributing to cost effectiveness. The removal of ammonia from blood plasma is also the duty of hepatocytes (Arias *et al.*, 1994) and this toxic metabolite continues to limit the life of cells in culture (Schneider *et al.*, 1996) and in productive bioreactors. Ideally, a hepatocyte derived line would allow for longer term mass cultures by incorporation of medium ammonia into urea and glutamine. For these applications it may be sufficient to develop cell lines with the desired characteristics whilst forgoing other non-essential functions which are more difficult to retain.

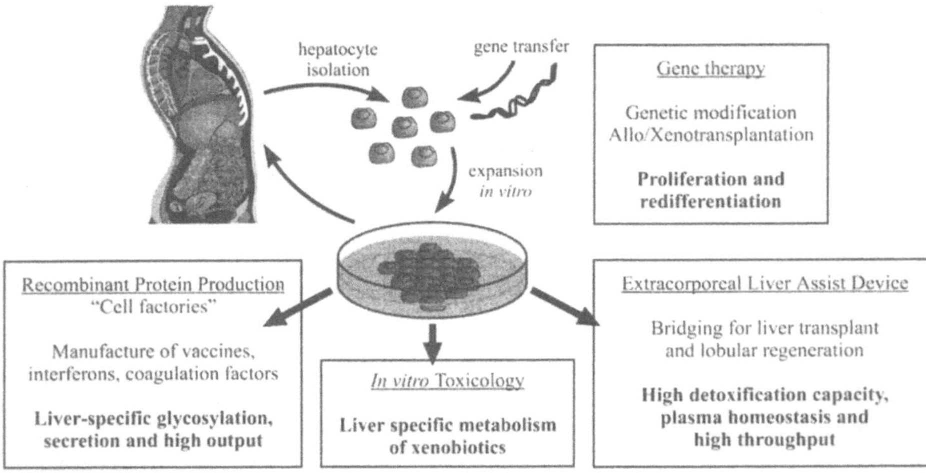

Figure 1. Potential applications for functional immortalized human hepatocytes. The demands on the cell are listed in bold text.

(iii) Clinical liver support: One of the applications in medicine for immortalized hepatocytes takes the form of an extracorporeal liver assist device (ELAD) for patients awaiting a suitable donor liver for transplantation or suffering acute liver failure. Isolated liver cells are seeded into a bioreator where they may reorganize themselves and reassume their tissue specific functions. Patient plasma is fed into the bioreactor and toxins removed or neutralized and solute and electrolyte levels regulated before being returned to the patient's blood. Such devices containing primary porcine hepatocytes have already been tested successfully (Chen *et al.*, 1997; Gerlach, 1996) but the substitution of these with an immortalized cell type would provide significant logistical and financial advantages. Unlike cells which may serve for recombinant protein production, the requirements placed on the cell for liver support would be of the highest order as the ELAD is responsible for maintaining systemic homeostasis (Sussman, 1997). Detoxication activities, synthesis and secretion of clotting factors and plasma transport proteins and maintenance of electrolyte exchange are amongst the most essential of functions, however, each patient presents a different spectrum of complications to which the cell should be capable of responding.

(iv) Gene therapy: Gene therapy for diseases of the liver (Horwich, 1991) was originally envisaged to involve manipulation of hepatocytes *in vitro* and introducton of functional cells into the liver of patients (Grossman, 1994). However, essential to the technique is the expansion of hepatocytes to generate sufficient cells for therapy, yet primary cultured hepatocytes undergo only a few rounds of replication before senescing. One answer to the problem would be to extend the proliferative phase of the cultured cells. Here, of prime concern to the safety of the patient is the method used to achieve immortalization and this will be discussed. In recent years, gene therapy for the liver has shifted its focus towards *in vivo* intervention (Li *et al.*, 1993) thus bypassing the need for cell isolation, culture, expansion and readministration. Nevertheless, the possibilities brought to the fore by developments in allo- and xenotransplantation (Rhim *et al.*, 1994; Gunsalus *et al.*, 1997), the transplantation of healthy cells to diseased organs, keeps the chapter open on the use of proliferating hepatocytes in gene therapy.

2. The Hepatocyte

Studies of the liver in its normal and diseased states and of cultured liver cells have allowed us to gain important insights into the nature of cell division control in hepatocytes. With the advent of transgenic and knockout mouse models, the molecular mechanisms guiding hepatocyte entry into the cell cycle are also becoming clearer. Moreover, experiments *in vivo* have addressed the question of how proliferating cells maintain differentiated functions. However, the establishment of continuously proliferating functional hepatocytes still presents a major challenge to somatic cell engineering.

2.1 THE HEPATOCYTE IN THE LIVER

The hepatocyte carries out hundreds of specific activities essential for systemic homeostasis (Arias *et al.*, 1994). Amongst the vital funtions uniquely or mostly provided by hepatocytes are: the synthesis of blood plasma proteins including the complement factors and α_1-antitrypsin; synthesis of plasma carrier proteins such as albumin, α_2-macroglobulin and transferrin; synthesis of blood clotting factors such as fibrinogen; detoxication of ingested xenobiotics and conjugation of hormones and bile acids; regulation of levels of circulating lipids, cholesterol and glucose; and regulation of the levels of ions and trace elements. Reduction or loss of any of these activities leads to disease states, most of which are seriously debilitating or life threatening if left untreated.

Liver function may be acutely compromised, for example in acetaminophen poisoning or by toxicosis, as with ingestion of large quantities of aflatoxin B_1 contaminated foods (Farrell, 1997; Beaune and Lecoeur, 1997). Chronic liver damage manifested by cirrhosis results from alcoholism, viral infection eg. hepatitis B (HBV) and C (HCV) viruses, and during the progression of hereditary illnesses affecting hepatocyte function. Monogenetic diseases, characterized by a defect at a single genetic locus result in damage to liver cells through accumulation of toxic metabolites; Wilson's disease, for example, results from a deficiency in copper export out of the cell (Bull *et al.*, 1993); familial hypercholesteroleamia, a disease typified by accumulation of cholesterol in the blood vessels and liver is caused by a dysfunctional low-density-lipoprotein receptor on the surface of hepatocytes (Goldstein and Brown, 1973); hereditary tyrosinemia type I arises due to a defect in the enzyme fumarylacetoacetate hydrolase in the pathway of tyrosine catabolism (Lindblad *et al.*, 1977). The outcome in all of these diseases is the accumulation of intermediates to toxic levels causing reduced efficiency of function and eventually the necrotic death of hepatocytes.

2.2 COMPENSATORY HYPERPLASIA

The liver possesses a property rare amongst mammalian tissues in that it is capable of regrowth after loss of functional mass such as occurs in necrotic disease. A rodent model for the study of liver growth and function is induced hyperplasia after partial hepatectomy (PH) (Higgins and Anderson, 1931). Removal of two-thirds of the liver mass, corresponding to excision of right anterior and median lobes, stimulates the remaining cells to enter highly coordinated proliferative and remodelling phases involving all cell types in the liver (Leffert *et al.*, 1988; Fausto and Webber, 1994). Organ mass and architecture are restored within ten to fourteen days after PH and the liver is once again fully functional. Although often referred to as 'regeneration', the process is not strictly regenerative as the excised lobes are not reproduced and the term 'compensatory hyperplasia' best summarizes the growth phase. The loss of functional hepatocytes during liver damage leads to an increased metabolic burden on the remaining cells and this is transmitted to other organs via the changed blood composition and factors secreted by the damaged liver (Leffert *et al.*, 1988; Fausto *et al.*, 1995). The body responds by inducing the transition of hepatocytes from quiescence into the cell cycle through the activation of signal cascades. Several mitogens have been identified and posited to play a role in initiation and progression of compensatory hyperplasia, such as hepatocyte growth factor (HGF), epidermal growth factor (EGF), transforming growth factor β (TGFβ), tumour necrosis factor (TNF) and insulin (Fausto *et al.*, 1995; Michalopoulos and DeFrances, 1997). These are produced in both a juxtracrine and paracrine fashion by neighbouring liver cells and in an exocrine fashion by distal organs implying that the growth programme is steered systemically. While compensatory hyperplasia is sometimes sufficient for recovery from an acute hepatic insult, in the case of chronic liver damage hepatocytes continue to be eliminated and in the long term sustained, compensatory cell division may lead to the development of hepatocellular carcinoma (HCC).

For most cell types the differentiated and proliferative phenotypes are to a great extent mutually exclusive. However, although expression of hepatocyte-specific proteins varies during compensatory hyperplasia, the activities essential to systemic homeostasis must be maintained (Friedmann, 1984). This is not due to subpopulations of cells, some of which maintain differentiated status while others undergo mitosis but that dividing hepatocytes can also express specific proteins. For example, Greenbaum *et al.* (1995) demonstrated the expression of the differentiation specific transcription factors C/EBPα and C/EBPβ in hepatocytes undergoing DNA replication after PH. In attempting to generate immortalized hepatocytes which express differentiated functions we should pay close attention to the complements of second messanger molecules and transcription factors that are present in hyperplastic hepatocytes as control over the expression of these may be the key to reproducing the phenotypic duality *in vitro*.

2.3 THE HEPATOCYTE IN PRIMARY CULTURE

Despite the phenomenal regeneration capacity of the liver, isolated hepatocytes have only a very limited proliferative potential *in vitro* and a correspondingly short life-span in primary culture (Leffert *et al.*, 1988). Current understanding maintains that loss of tissue-specific activities is a result of the failure to reproduce the hepatic microenvironment in culture. Dissociation of the tissue during cell isolation destroys the three-dimensional organisation of the liver lobule including homo- and heterotypic cell-cell contacts and bidirectional flow of blood and bile through the sinusoids and bile ductules respectively. The lobular architecture generates positional cues for hepatocytes within the acinus and directs apico-basal membrane polarization and cell-specific gene expression. Although freshly seeded hepatocytes demonstrate significant activities for some functions, the phenotype is unstable and the level and activities of enzymes such as cytochromes P-450, and the secretion of albumin and other plasma proteins fall to a few percent within 10-14 days (Sirica and Pitot, 1980; Guguen-Guillouzo *et al.*, 1983). Improvements in medium composition, by addition of hormones, fatty acids, soluble metal complexes and growth factors as well as the seeding of cells onto collagen and fibronectin, and co-culture with liver non-parenchymal cells has enabled the stabilization of some enzyme activities albeit at levels much lower than *in vivo* and only for around eight weeks at best (eg. Hoffmann *et al.*, 1989; Waxman *et al.*, 1990). Once again we are made aware of the complex interactions of liver cells and their systemic connexions to other tissues contribute to the differentiated hepatocyte phenotype which we would like to reproduce in culture. Isolation of primary hepatocytes is also a time consuming and expensive business and is subject to the variation inherent in primary cell culture. This, coupled with the limitation of short-term studies has forced liver biologists to make use of cell lines derived from liver neoplasms and *in vitro* transformed hepatocytes.

2.4 HEPATOCYTE CELL LINES

There are numerous cell lines which have been established from hepatoma or HCC explant material (Table 1) and several express hepatocyte-specific proteins. While these cells provide a suitable genetic background for studies of hepatocyte gene expression they present major problems when considered for use in ELAD, gene therapy, recombinant protein production and toxicology.

Growth control pathways are radically altered during tumour evolution such that a semi-autonomous to autonomous state is achieved. Selection of mutations or deletions in growth suppressing genes and activating mutations or amplification of growth promoting genes renders the process irreversible. One of the regulatory circuits to be broken is the responsiveness to circulating factors which coordinate growth and regeneration of tissues with the requirements of the organism. TGFβ for example, is a potent inhibitor of mitosis for hepatocytes *in vivo* and *in vitro* (Wollenberg, 1987) and operates in multiple pleiotropic pathways. Signal transduction occurs through the plasma-membrane bound TGFβ–receptor and secondary cytoplasmic signalling molecules. Evolving HCC can become partially or fully independent of the effects of TGFβ by accumulating mutations in one or more of these critical messengers and cell lines subsequently derived from the HCC will have a similar deficiency in TGFβ responsiveness. A further example of a loss of critical function is the development of insulin independence in cell culture. Insulin is involved in the regulation of blood glucose homeostasis, glycogenesis, fatty acid synthesis and is a growth factor for hepatocytes (Fausto *et al.*, 1995; Michalopoulos and DeFrances, 1997). In medium containing insulin, cultured cells may become insulin independent if sufficient glucose and other substrates are present, and thereby gain a growth advantage over their neighbours. However, if alterations at the level of the genome have occurred then these cells will no longer be able to subsist in simpler media where, on the addition of insulin, metabolic pathways should be stimulated to synthesize essential intermediates. Here we see that the choice of medium plays an important role in selecting the phenotype and even the genotype of continuously proliferating cells. In trying to generate novel immortalized hepatocytes for application in the clinic, it is our aim to maintain sensitivity to hepatotropic factors which may be present in the plasma of patients and through which we might access control over cell proliferation in a bioreactor setting.

TABLE 1. A comparison of G_1 to S phase checkpoint regulators in hepatocyte cell lines.

Cell line	Reference	*RB1* status	*p53* status
PLC/PRF/5	McNab *et al.*, 1976	Normal	Mutation in codon 249
Hep3B	Aden *et al.*, 1979	mRNA absent	Large deletion in gene
HepG2	Aden *et al.*, 1979	Normal	Normal
Huh-7	Nakabayashi *et al.*, 1982	Normal	Mutant
FOCUS	He *et al.*, 1984	Rapid turnover	Not expressed
Mahlavu	Alexander, 1984	Normal	Truncated protein
THLE-2	Pfeiffer *et al.*, 1993	SV40 T-antigen	SV40 T-antigen

None of 31 hepatocyte derived cell lines tested was able to express the full spectrum of liver-specific activities or to reproduce the high levels of activity characteristic of hepatocytes *in vivo* or in primary cuture (Enosawa *et al.*, 1996). Furthermore, loss of specificity in gene expression is common amongst these cells with embryonic genes and non-cell type specific alleles being expressed. Deregulation of growth control is part and parcel of the tumorigenic process and in achieving this the transformed cell must sacrifice some of its claims on tissue specificity. From an evolutionary standpoint, the selection pressure for rapid growth in tumour cells diverts available resources to this end and thus activities which are not directly disadvantagous to cell proliferation will also be suppressed. Because of the difficulty in obtaining high levels of hepatocyte specific gene expression it is our opinion that liver tumour-derived cell lines do not provide the most promising starting material for development of ELAD or bioreactors in general and in addition the transformed nature of the cell is questionable on medical ethical grounds. For this reason we have investigated the possibility of directly modulating elements of cell cycle regulation to achieve continuous proliferation in hepatocytes.

3. Immortalization and Transformation

In vitro, normal cells divide a limited number of times before succumbing to senescence. Proliferative lifespan is thought to be determined by the length of telomeres, specialized repeat structures at the ends of chromosomes which serve to maintain genome integrity. Telomeres shorten with each generation and are ultimately eroded causing karyotypic instability and senescence. However, cells may overcome this 'crisis' in culture by reactivating telomerase, the enzyme responsible for repair of telomeres and which is repressed in most post-embryonic somatic cells. Cells which survive crisis continue to proliferate and are considered immortalized (Fig. 2). Ectopic expression of telomerase in primary human cells was shown to block senescence and allow indefinite proliferation (Bodner *et al.*, 1998). While reactivation of telomerase may allow continuous division in some cells, it has recently been reported that this is insufficient to overcome senescence in two epithelial cell types (Kiyono *et al.*, 1998).

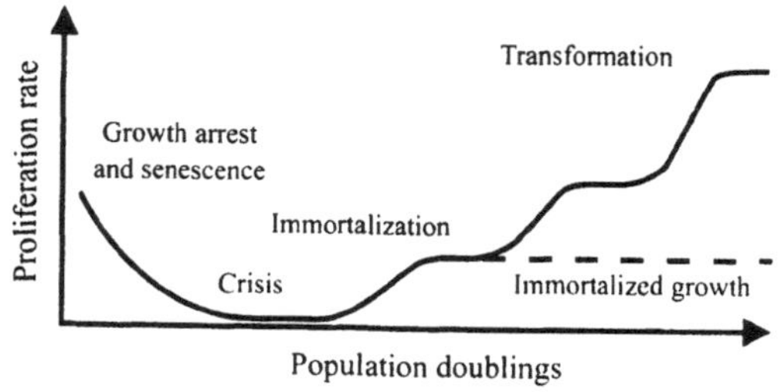

Figure 2. Changes in proliferation rate of normal cells explanted into culture. Cells are supplied with optimized survival media such that changes in proliferative phenotype are a consequence of genetic alterations occurring in culture.

Once the cell gains the ability to proliferate indefinitely, spontaneous mutation becomes a significant factor. The selective pressure to overcome the limits to cell division in culture such as growth factor requirement and contact inhibition results in accumulation of mutations in genes regulating these parameters. Ultimately, a transformed phenotype may arise defined by rapid growth rate (Fig. 2), the ability to form foci in monolyer culture or to grow in soft-agar. Transformed cells may also be tumorigenic when injected into nude mice, indeed cells transformed in culture resemble those established from nepolasms both in phenotype and genotype and for the reasons described (section 2.4) are not suitable for our purposes. There are many requirements which have to be satisfied before a cell can enter into the mitotic cycle and these are coordinated and stringent. To gain control over these cell cycle 'checkpoints' we first have to determine which genes are responsible for growth regulation and what would be the consequence of the deregulation of checkpoints as a means to overcoming growth suppression.

3.1 IDENTIFYING THE TARGETS FOR CELL CYCLE MODULATION

3.1.1. *Cell Cycle Entry*

Regulation of somatic cell division follows a similar pattern in the majority of cell types. Differentiated cells in tissues exist in a 'G0' state where they execute specific functions and do not proliferate (Fig. 3a). Exit of G0 and entry into the G1 phase of the cell division cycle is stimulated by various extracellular stimuli, eg. activation of cell surface receptors by growth factors, internalization hormones, tissue disaggregation or a combination of these. The cell also makes an assessesment of its nutrient supply, nucleotide pools and microenvironment in order to ensure that conditions are amenable to support growth. These diverse signals are integrated at the molecular level in the decision to move from G1 to S phase.

The motors driving mammalian cell division are the cyclin-dependent kinases (cdk) (Hunter and Pines, 1994). During the cell cycle, distinct windows of activation open for each of the cdks and this is achieved by transcriptional induction of the respective cyclin subunit (Fig. 3b). After executing its function, the kinase is inactivated by rapid destruction of the cyclin polypeptide. The product of the retinoblastoma susceptibility gene (*RB1*), pRB is a substrate for phosphorylation by cdks. pRB, lies at the core of a G1 checkpoint (Fig. 4) (Weinberg, 1995; Herwig and Strauss, 1998). Among its many functions, pRB sequesters members of a family of transcription factors, the E2F proteins (Helin *et al.*, 1992; Kaelin *et al.*, 1992). E2F has been shown to regulate the expression of several genes essential for S phase, including DNA polymerase α, dihydrofolate reductase, thymidine kinase and the cell cycle genes cyclin D, cyclin E, c-myc and E2F-1 itself (Slansky and Farnham, 1996). As a complex with pRB, E2F is not only unable to activate expression of these S phase genes but actively takes part in the repression of these and other genes (Brehm *et al.*, 1998; Magnaghi-Jaulin *et al.*, 1998) and thus a block to cell cycle progression is established in late G1 phase. In proliferating cells, the block is relieved by the phosphorylation and inactivation of the pRB protein (Hinds *et al.*, 1992; Kato *et al.*, 1993; Draetta *et al.*, 1994) thereby liberating E2F. Phosphorylation is carried out by cdks in association with regulatory cyclin subunits and the first of these to be activated in the G1 phase is cdk4-cyclin D1 (Fig. 4) (Weinberg, 1995). Various mitogens are known to cause transcriptional induction of cyclin D1 (Lukas *et al.*, 1996) and so a link is forged between growth factor stimulus and cell cycle progression in G1 through the inactivation of pRB.

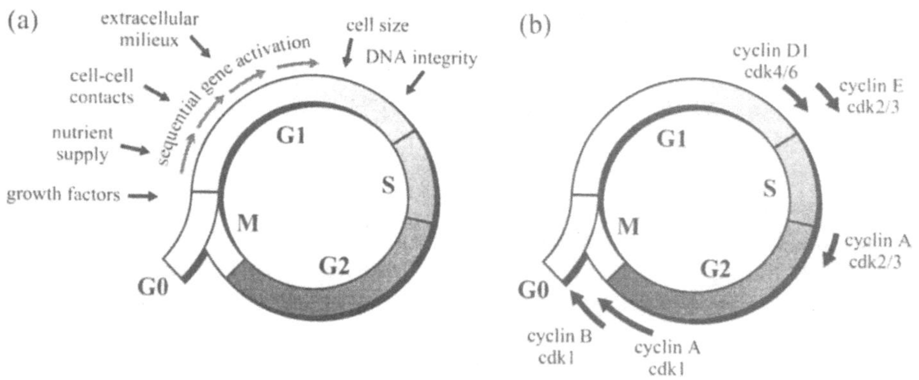

Figure 3. (a) extra- and intracellular events regulating entry of hepatocytes into the cell cycle. Extracellular factors exert their influence in G1. (b) windows of activation of cyclin-cdk complexes regulating cell cycle phase transitions.

Several levels of fine tuning are superimposed upon this basic machinary, each having different mechanisms and kinetics, allowing for transient delays in G1 progression to the complete arrest of cell division and activation of programmed cell death (PCD). One important molecule in this respect is the product of the *TSP53* gene, p53 protein. Although cell division can proceed in the absence of p53, this protein is crucial to correct maintenance of genome integrity and to this effect p53 coordinates repair and replication of damaged DNA (Huang *et al.*, 1996). Depending upon the extent and timing of the lesion, cell cycle progession can be transiently arrested or PCD may be initiated (Liebermann *et al.*, 1995). p53 accomplishes G1 arrest in its role as a transcription factor to induce expression of the *CIP-1/WAF-1* gene whose product, $p21^{CIP-1}$, can bind to and inhibit all cdks (Fig. 4) (Xiong *et al.*, 1993). In this fashion p53 can cause arrest at several points in the cell cycle. *CIP-1* is a cyclin-dependent kinase inhibitor (CKI) gene, a recently identified class of genes composed of two families (Sherr and Roberts, 1995). As described, $p21^{CIP-1}$ binds to cdks promiscuously and exemplifies one of these families; $p16^{INK4a}$ exemplifies the other family which specifically binds to D-type cyclin kinases (Serrano *et al.*, 1993). By virtue of their specificity, *INK4* gene products are integrally involved in the pRB checkpoint. For example, TGFβ is a growth inhibitor for epithelial cell types and one of its effects is to induce transcription of *INK4b* whose gene product, $p15^{INK4b}$, can bring about G1 arrest by binding to and blocking activation of cdk4 (Hannon and Beach, 1994).

3.1.2 *Deregulation of the G1 Checkpoint*

We can uncover clues as to how we might gain control over cell proliferation by analysing the changes most commonly seen during tumorigenesis and in the mechanisms employed by the oncogenic viruses. As has already been mentioned, extracellular influences on cell cycle progression are exerted primarily in G1 phase. For the evolving tumour independence from growth repression is an important step towards cell autonomy. By nullifying pathways signalling to the cell cycle machinary at a common point, the cell can become resistent to several negative growth regulators simultaneously. pRB is the ultimate target of many growth regulatory circuits and consequently loss of a functional pRB protein is observed in many tumour types (Friend *et al.*, 1987; Horowitz *et al.*, 1990). The *INK4a* gene product, which is known to mediate the effects of growth supressors by inhibiting the phosphorylation of pRB, is also frequently inactivated in tumours of different origin (Kamb *et al.*, 1994; Nobori *et al.*, 1994). In fact it is estimated that the majority of all tumours contain inactivating changes in either *RB1* or *INK4a* and most of the rest demonstrate overexpression of the D- or E-type cyclins, all of which result in a weakening or loss of the G1 checkpoint (Strauss *et al.*, 1995; Weinberg, 1995). Consequently, growth factor requirement is lowered and mitogenic inhibition is derailed.

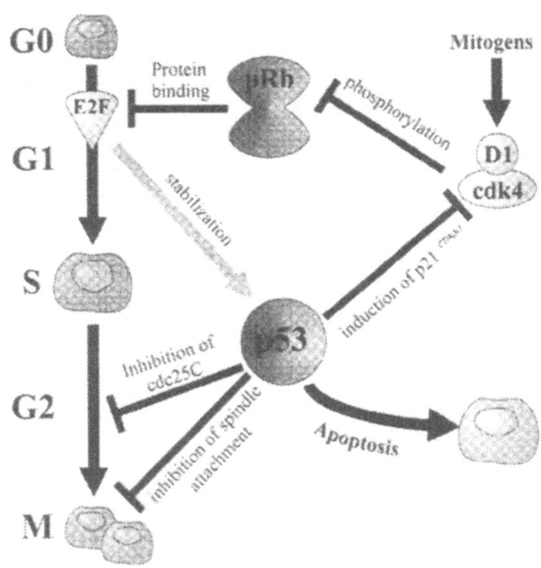

Figure 4. The G1 checkpoint regulated by the retinoblastoma protein (pRB) and the involvement of the p53 protein in G1 and G2 checkpoints.

3.1.3. *Suppression of Apoptosis*

With the overriding of the pRB checkpoint, the transcription factor E2F is released promoting entry into S phase. However, in cells undergoing a normal cell cycle, G1 phase also allows for the synthesis of new organelles and macromolecules and for cell expansion in preparation for DNA synthesis and cytokinesis. Checkpoints must also exist to ensure that transcription of the genes involved is complete before G1 to S transition and further checkpoints are involved in coordinating other events which must take place before DNA replication begins. This we know because usurping of the pRB checkpoint and unscheduled entry into S phase in quiescent cells leads to apoptosis (Morganbesser, *et al.*, 1984). It was recently shown that ectopic expression of E2F-1 in primary fibroblast cultures indirectly activates the p53 protein thereby causing G1 arrest or apoptosis, and that the mechanism depends upon the transcription product of the alternative reading frame (ARF) of the *INK4a* gene, $p14^{ARF}$ ($p19^{ARF}$ in mice) (De Stanchina *et al.*, 1998; Zindy *et al.*, 1998) (Fig. 5). Thus a fail-safe mechanism exists to prevent aberrant cell proliferation after stimulation by abnormal growth signals.

As might be expected from the above discussion, evolving neoplasms are placed under a selection pressure for loss of p53 function and indeed up to half of all tumours have mutated or lost the *TSP53* gene (Nigro *et al.*, 1989; Lane and Benchimol, 1990). Furthermore, the reported inactivation of the *INK4a* gene in tumours (Kamb *et al.*, 1994; Nobori *et al.*, 1994) may also neutralize $p14^{ARF}$ (Zindy *et al.*, 1998) thereby breaking the link between loss of pRB and p53 activation. Taken together this means that the majority of tumours in which the pRB checkpoint is deregulated will have compromised p53 function and this is borne out by analysis of these checkpoints in tumour derived cell lines (Table 1 and Hainaut *et al.*, 1997).

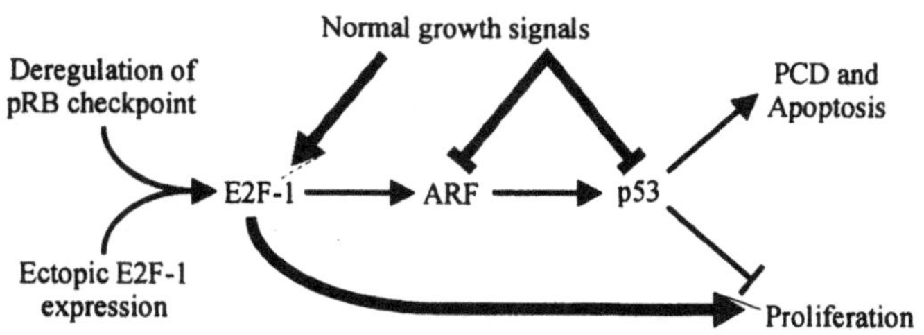

Figure 5. The link between pRB checkpoint deregulation and induction of p53-dependent G1 arrest and programmed cell death (PCD). Deregulation of the pRB checkpoint frees E2F which participates in activation of ARF. p53 becomes stabilized through $p14^{ARF}$ allowing induction of p53 responsive genes involved in growth arrest or PCD. Normally replicating cells must somehow suppress the $p14^{ARF}$-p53 axis to allow proliferation while immortalized cells select for mutations in p53 or $p14^{ARF}$.

4. Immortalization Strategies Utilizing Viral Proteins

pRB is a target for sequestration and inactivation by DNA virus transforming proteins such as the SV40 large T-antigen (svT-ag), polyoma virus large T-antigen (pT-ag), the adenovirus E1A and the human papillomavirus virus E7 proteins (DeCaprio *et al.*, 1988; Whyte *et al.*, 1988; Dyson *et al.*, 1989). In a similar fashion, these DNA tumour virus oncoproteins neutralize p53 through binding; the svT-ag sequesters both pRB and p53 while the adenovirus E1B 55 kDa and papillomavirus E6 proteins also bind p53 (Werness *et al.*, 1990). Thus these viruses are able to rapidly attenuate the growth inhibitory effect of two cooperating tumour suppressor proteins. In both the tumour and virus systems rapid proliferation and escape from PCD is achieved at the cost of reduced sensitivity to extracellular growth signals and a curtailment of maintenance of DNA integrity.

Recombinant viral transforming proteins have been used to great effect in the immortalization of a variety of primary cultured cell types including hepatocytes (Höhne *et al.*, 1987; Woodworth *et al.*, 1987; Strauss *et al.*, 1990). Cells in human liver explants could also be induced to proliferate by transfection with a plasmid containing pT-ag and cell lines expressing pT-ag were established from several different lineages (Hering *et al.*, 1991). However, viral oncoproteins are able to transform as well as immortalize cells which may result in tumorigenicity. This level of deregulation of growth control in established cells would be disadvantageous for their use in ELAD, gene therapy, recombinant protein production or toxicology for the same reasons discussed for tumour derived cell lines (see section 2.4). Furthermore, viral oncoproteins possess other activities aside from sequestration of pRB and p53, which purturb cell differentiation programmes (Bulera *et al.*, 1997). It should be mentioned that the immortalizing and transforming functions of some viral proteins are separable and reside either on different polypeptides or within different domains of the same protein. However, it has as yet not been possible to immortalize primary hepatocytes through the introduction of elements such as adenovirus E1A or polyoma virus middle T-antigen alone without activation of PCD.

Finally, an important ethical consideration arises as to the nature of the immortalizing strategy for applications of established cells in medicine and therapeutic protein production. A suitable cell line would be non-tumorigenic and would not express cancerogenic proteins, such as those produced by the DNA viruses mentioned.

4.1 SHORT-TERM SV40 LARGE T-ANTIGEN EXPRESSION

We devised strategies for bringing the svT-ag into cells transiently with the belief that the mitogenic stimulus provided would extend the life of primary cultures and might lead to continuous proliferation by accumulation of mutations in growth regulatory genes other than pRB and p53. Two methods were employed to achieve this; non-replicating plasmids and Cre-*lox* mediated excision.

4.1.1 *Non-replicating Plasmids*

The plasmid pCMV-Tagori (Fig. 6a) was constructed in which the svT-ag is driven by the strong cytomegalovirus promoter (CMV) and the origin of replication (ori), which stems from the primate SV40 virus, is not compatible with replication in murine cells. Transfection of this plasmid into primary mouse hepatocytes results in expression of svT-ag and the cells are induced to proliferate. The plasmid is not replicated with the genome and eventually, the svT-ag protein is diluted out by cell division and destroyed by protein turnover. Despite this we were able to observe the continuing growth of several colonies and these were isolated and clonal lines (mHCT) established. Spontaneous integration of plasmid DNA into the genome of replicating cells occurs at a rate of 10^{-6} to 10^{-7} and we expected that some lines would express svT-ag.

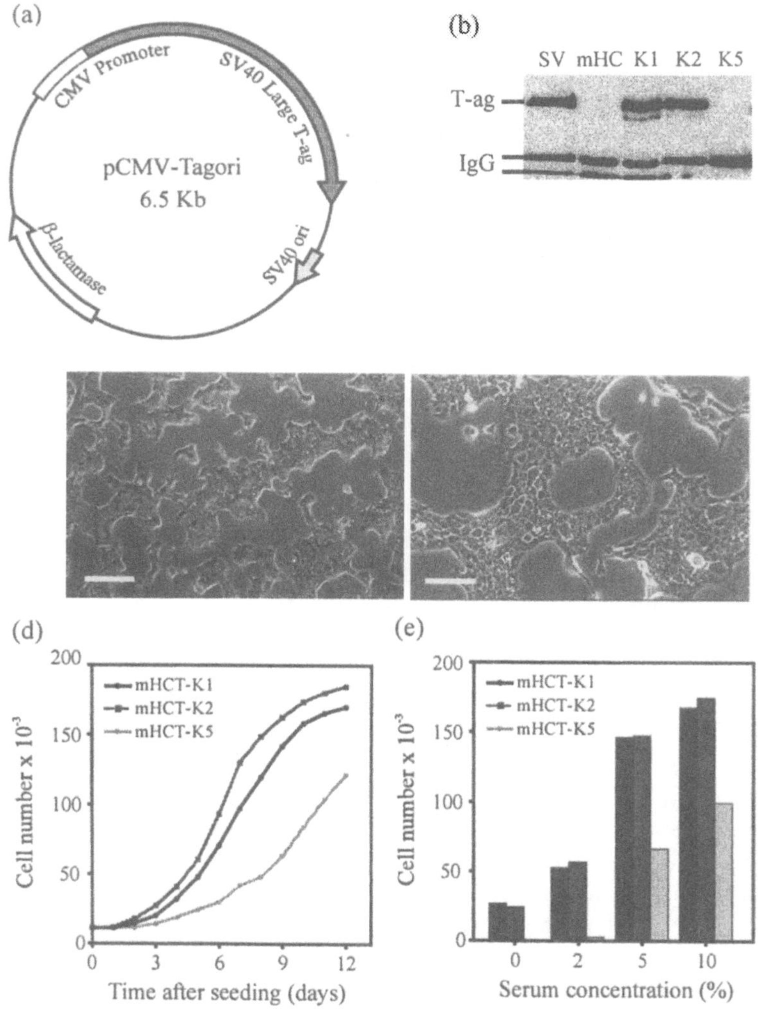

Figure 6. (a) the plasmid pCMV-Tagori. (b) immunoprecipitation with antibody PAb121 (PharMingen) of svT-ag from lysates of mHCT-K1, -K2 and -K5 cells; SV, HepSV40; mHC, primary mouse hepatocytes. IgG indicates position of immunoprecipitating antibody. (c) mHCT-K1 (left) and -K5 (right) cells in monolayer culture Scale bar = 125 μm. (d) growth curves and (e) serum-dependence of growth of mHCT-K1, -K2 and -K5 lines. Cells were plated at 20% confluence and cell number determined after 4 days.

Immunoblotting of cell lysates revealed that two of the cell lines, mHCT-K1 and mHCT-K2 did indeed contain svT-ag and this was present at levels equivalent to that of a cell line established from hepatocytes of an svT-ag transgenic mouse (Paul *et al.*, 1988) (Fig. 6b). Four other cell lines, typified by mHCT-K5, were negative for the svT-ag protein. The morphology of mHCT-K5 in monolayer culture is more organized than that shown by the svT-ag expressing lines, showing a trabecular arrangement of cells and intercellular canaliculi more reminiscent of primary hepatocytes (Fig. 6c). As might be expected, the svT-ag expressing cells grew more rapidly (Fig. 6d) and were less dependedent on serum (Fig. 6e) for cell division than mHCT-K5, however both cell types arrested their growth at confluence. Clearly, changes in the regulation of the mitogenic cycle have been effected in svT-ag negative cells which are different to those resulting from continual expression of this transforming protein. We analysed numerous cell cycle proteins for changes in levels of expression and for mutations, including pRB, p53, cyclins D1, E and A, cdk4 and cdk2, E2F-1, the CKIs, $p16^{INK4a}$ and $p21^{CIP-1}$, and the apoptosis regulators, bax, bcl-2 and bcl-x_l. In none of these proteins were we able to identify differences between svT-ag expressing and non-expressing cells (data not shown). Possible mechanisms for the continuation of proliferation despite loss of svT-ag are the constitutive activation of a signalling pathway, as has been seen with the dominant c-Ha-*ras* allele, or other elements involved in cell cycle control not tested.

4.1.2 *Cre-lox Mediated Excision*

The recombinant Cre-*lox* system developed from the coliphage P1 (Sauer and Henderson, 1988), allows integration of foreign genes into the genome and later site-specific deletion of the exogenous DNA. The fragment to be removed is flanked by *lox*P sites, an short sequence which is the target for recognition by the Cre recombinase. Figure 7a shows the non-replicating recombinant retrovirus *lox*P-HyTK-large T (Li *et al.*, 1997) which contains the SVT-ag gene placed downstrean of the CMV promoter and a hygromycin-thymidine kinase selection cassette, all sandwiched between *lox*P sites. Primary mouse hepatocytes were infected with the retrovirus and three weeks later selection was initiated by addition of hygromycin to the medium. Three to four weeks thereafter, colonies of proliferating hepatocytes were picked and clonal lines (*lox*T-ag) were established. *lox*T-ag cells were shown to produce svT-ag protein (Fig. 7b) and in monolayer culture the cells were morphologically indistinguishable from mHCT cells expressing svT-ag.

We next transfected cultures of *lox*T-ag cells with a plasmid expressing the Cre enzyme driven by the CMV promoter and a puromycin selection cassette, pCMVCre-puro (Fig. 7a) and after 48 hr, selection with puromycin was begun. Cells which were not transfected by pCMVCre-puro and those in which the Cre enzyme has failed to excise the *lox* fragment continue to express svT-ag and these must be removed in order to observe the effects of Cre. The HyTK cassette, also retained within the *lox* fragment, can activate ganciclovir to a genotoxin and therefore ganciclovir was added to the medium two weeks after transfection. Thus we were able to excise the svT-ag gene from the genome of proliferating hepatocytes. The resulting cells, 10 days after ganciclovir treatment, were larger, had irregular cell margins with short cytoplasmic

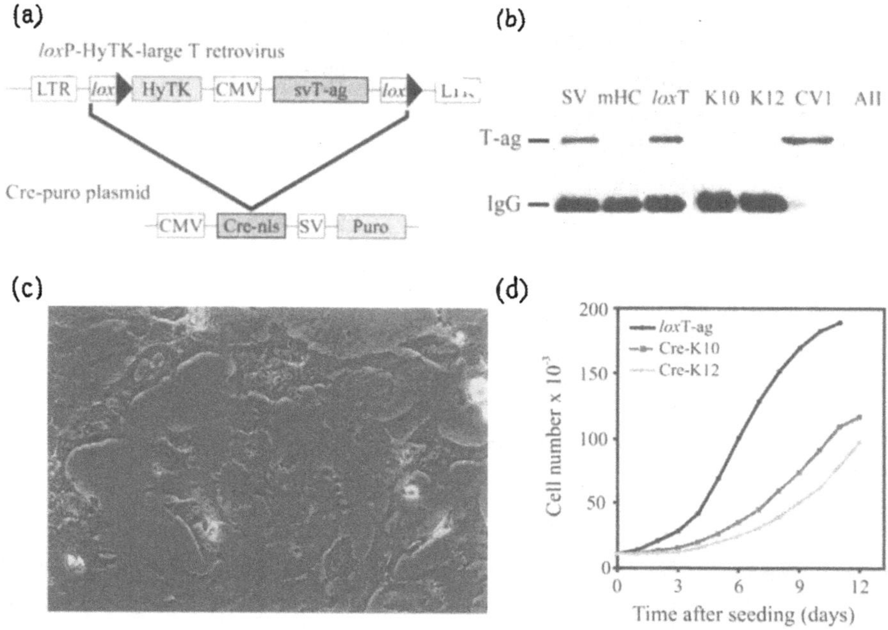

Figure 7. (a) the retrovirus *lox*P-HyTK-large T and plasmid pCMVCre-puro showing the excision of the fragment bordered by *lox* sites by Cre recombinase. (b) immunoprecipitation with antibody PAb121 (PharMingen) of svT-ag from lysates of *lox*T-ag and Cre-K10 and -K12 cells; SV, HepSV40; mHC, primary mouse hepatocytes; CV1, svT-ag transformed monkey kidney cell line; AII, svT-ag negative mouse hepatocyte cell line. (c) monolayer culture of *lox*T-ag cells transfected with pCMVCre-puro and selection with ganciclovir. Scale bar = 50 μm. (d) growth curves of *lox*T-ag and Cre-K10 and -K12 cell lines.

processes and many contained several irregularly shaped nuclei (Fig. 7c). Importantly, cell proliferation ceased very soon after treatment. This morphology is very reminiscent of primary mouse hepatocytes after one week in monolayer culture, yet these cells had already been passaged many times with an accumulated culture life of six months. After a further three weeks, the ganciclovir treated cultures had entered senescence, however, several small colonies of proliferating cells were observed. These were isolated, grown in non-selective media and could be established as cell lines (Cre). Representative cell lines, Cre-K10 and -K12, grew more slowly than *lox*T-ag lines (Fig. 7d) and were negative for svT-ag protien (Fig. 7b). Morphologically, Cre monolayers resembled those of the more organised svT-ag negative mHCT-K5 cells (Fig. 6c). Once again it is likely that Cre cells, like mHCT-K5 cells, have accumulated changes in growth regulatory genes to compensate for the loss of svT-ag.

TABLE 2. Colony growth in soft-agar.
For each cell line 5×10^3 cells were seeded in 0.35% agarose in medium. Small colonies were defined as those showing a definite morphology at 100x magnification and large colonies at 40x. Efficiency is colonies/cells plated. HepSV40 was included for comparison.

Cell line	Small	Large	Efficiency (%)
HepSV40	546	65	12.2
mHCT-K1	24	0	0.48
mHCT-K2	15	0	0.30
mHCT-K5	0	0	0.0
loxT-ag	42	5	0.94
Cre-K11	0	0	0.0
Cre-K12	0	0	0.0
HepZ, p7	0	0	0.0
HepZ, p40	0	0	0.0

In these experiments we are attempting to immortalize but not transform primary cells (Fig. 2). The transformed phenotype was assessed by measuring growth in soft-agar and the generation of tumours in nude mice. Table 2 shows that mHCT-K1, mHCT-K2 and *lox*T-ag cells, all of which express svT-ag, are capable of growth in soft-agar in contrast to svT-ag negative lines, mHCT-K5 and Cre. Similarly, svT-ag expressing cells were also tumorigenic, though the growth rate of the primary tumours was very slow, whereas svT-ag non-expressers were non-tumorigenic (Fig. 8a, b).

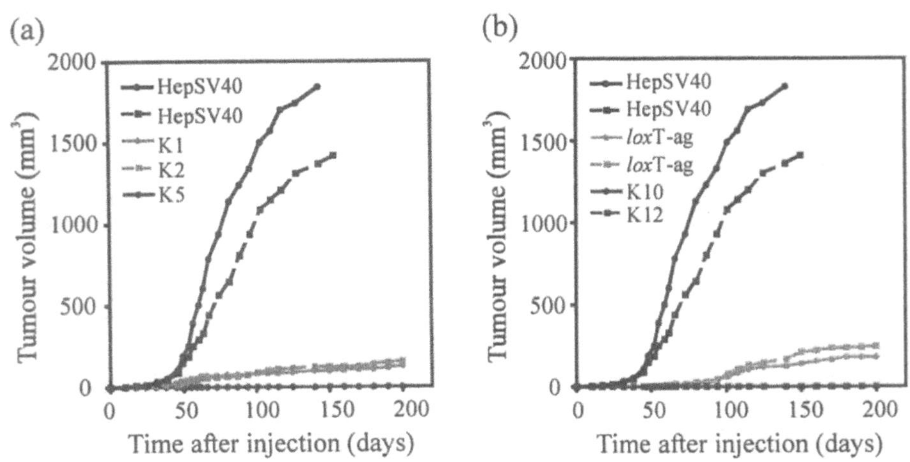

Figure 8. Tumour induction in nude mice. Mice were injected subdermally with 5×10^6 cells from (a) mHCT-K1, -K2 or -K5 and (b) *lox*T-ag, Cre-K10 or -K12 lines. HepSV40 is included for comparison.

It is thus possible to immortalize primary hepatocytes using svT-ag without accompanying transformation and tumorigenesis by allowing for removal of the oncoprotein. However, under the conditions employed for monolayer culture, changes in the expression of genes involved in growth regulation may become permanent, ie. genomic DNA alterations, upon removal of the oncoprotein. These alterations must also be considered potentially tumorigenic if the cells are to be employed in the clinic.

5. Targeting Using Antisense Sequences

5.1 IN VITRO APPLICATION OF ANTISENSE-RB1 OLIGONUCLEOTIDE

As a more direct method of reducing the expression of pRB and p53 we propose the use of phosphorothioate-based antisense DNAs; sequences identical to the non-coding strand of DNA which can form a duplex with the corresponding mRNA (Strauss *et al.*, 1992). Inhibition of translation follows as a result of duplex formation and eventual degradation by RNase H (Walder and Walder, 1988). The selection of a target sequence must take into account the formation of mRNA secondary structures which may limit oligonucleotide accessibility. However, the sequence surrounding the initiation codon has been shown to be an effective target in several instances (Fig. 9a). Introduction of antisense-*RB1* oligonucleotide into human embryonic lung (HEL) fibroblasts promotes proliferation of cells in parallel with the specific reduction of cellular pRB protein (Fig. 9b, c). Analysis of pRB after incubation of HEL cells with different antisense oligonucleotides demonstrated that targeting of the initiation codon is more efficient at reducing protein synthesis than oligonucleotides which bind downstream sequences (Strauss *et al.*, 1992). After several days the oligonucleotide became inactivated and growth rate returned to normal. The use of antisense oligonucleotide to target the pRB checkpoint appears to be justified in cycling cells.

5.2 IN VIVO APPLICATION OF ANTISENSE-RB1 OLIGONUCLEOTIDE

We next attempted to stimulate proliferation in a more ambitious setting, that of the hyperplastic liver after PH in the mouse. The kinetics of hepatocyte proliferation after PH are well documented and are known to respond to changes in diet, age and health. While these parameters all act via their respective signalling pathways on cell cycle checkpoints, it has not been shown what effect removal of the pRB protein has on compensatory hyperplasia. Mice nullizygous for *RB1* die *in utero* (Clarke *et al.*, 1992) but we can knock out pRB synthesis in the liver of normal adult mice by administration of antisense-*RB1* oligonucleotide. Complete cessation of pRB synthesis was achieved by intravenous injection of antisense oligonucleotide shortly after PH whereas an oligonucleotide with the control complementary sense sequence had no effect on pRB levels (Fig. 10a). Expression of cell cycle proteins, as determined by Western blotting and immunoprecipitation, in the livers of control and antisense oligonucleotide treated mice is depicted in Fig. 10b. A long lag phase precedes the first appearance of

(a)

sense-*RB1* oligonucleotide   5′-GTCATGCCGCCCAAAACC-3′

-10   +1

RB1 mRNA   ---GAAAGGCGUCAUGCCGCCCAAAACCC---

antisense-*RB1* oligonucleotide   3′-CAGTACGGCGGGTTTTGG-5′

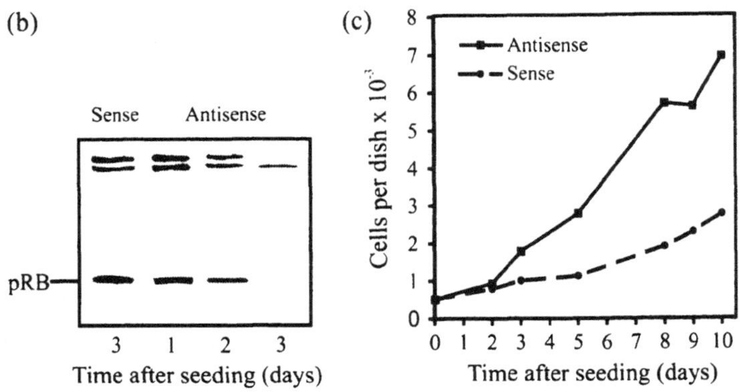

Figure 9. (a) sequences of oligonucleotides and position of base pairing with the *RB1* mRNA. (b) stimulation of proliferation in HEL after treatment with 10 μM antisense-*RB1* oligonucleotide. (c) decrease in pRB immunoprecipitated from HEL after treatment with antisense-*RB1* oligonucleotide.

cyclin D1 at 18 h after PH, marking late G1 phase. PCNA, an auxilliary subunit of DNA polymerase δ (Prelich and Stillman, 1988), is induced fractionally later at the G1 to S phase transition. S phase itself is measured by incorperation of 5′-bromo-2′-deoxyuridine (BrdU) into DNA during semi-conservative replication (Fig. 10c) and late S to G2 phase is marked by the inducton of cyclin A. Expression of cdk1 heralds the transition from G2 to M phase. In control mice the consecutive inducton of these proteins is evident and defines the kinetics of entry of quiescent hepatocytes into the cell cycle. One cell cycle is completed by around 42 h when a second wave of cyclin D1 induction begins, signalling a second G1.

In mice given antisense oligonucleotide, a disturbance of the induction patterns of the cell cycle proteins is immediately visible. The lag phase, the transition from G0 to G1, is much shorter with cyclin D1 induction now beginning at 6 h after PH. An equally premature though weaker peak of PCNA follows along with a moderate level of BrdU incorperation into DNA (Fig. 10c). However, cyclin A and cdk1 are not induced at this time. A second G1 phase begins at 30-36 h with cyclin D 1 and PCNA induction preceding that of cyclin A and cdk1 as expected. Accompanying the second cell cycle is a large peak of BrdU incorperation equivalent to the peak observed in control mice and representing a labelling index of around 70% (Fig. 10c). It appears that with the removal of the growth repressor pRB, hepatocytes have attempted to enter S phase early.

There is evidence for the participation of E2F in the induction of PCNA (Rohde *et al.*, 1996) and the precocious induction of PCNA might be attributed to excess free E2F in the absence of pRB. The full and rapid induction of cyclin D1 on the other hand points to a more direct effect, perhaps through loss of transcriptional repression by pRB or pRB-E2F complex. What is certain is that approximately 30% of hepatocytes enter S phase prematurely but may not proceed to M phase as cyclin A and cdk1 is not induced. The cell cycle begins again at 30 h and this time is successfully completed. Importantly, mice receiving oligonucleotide were not inhibited in their compensatory growth in the long term, nor was there any damage to the liver or any functional deficit detectable.

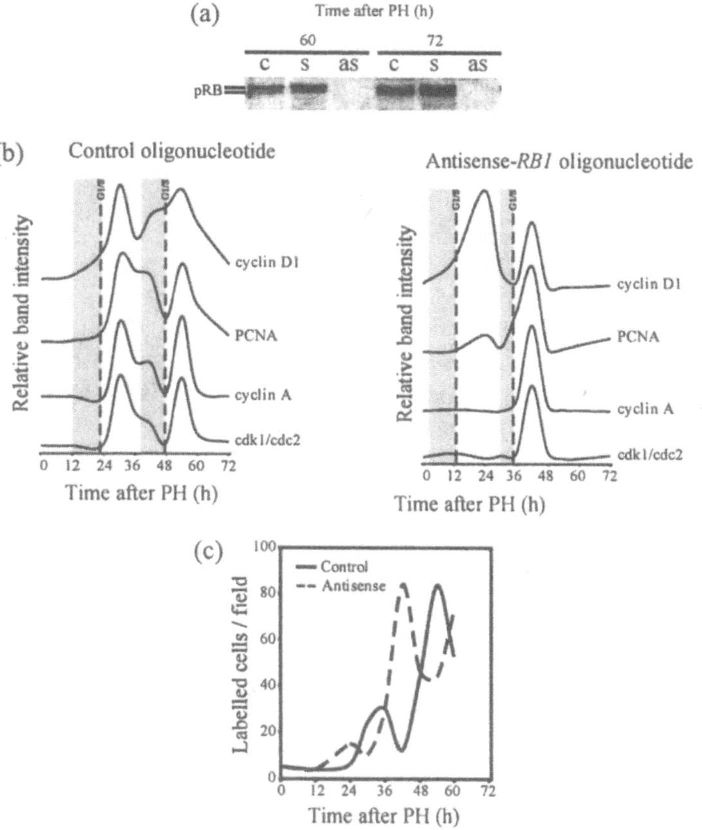

Figure 10. (a) inhibition of pRB induction after partial hepatectomy (PH) in the mouse by antisense-*RB1* oligonucleotide. Mice were injected intravenously with 250 μg oligonucleotide dissolved in 100 μl saline solution immediately after PH. (b) effect of deregulation of the pRB checkpoint on expression of cell cycle proteins during the G0 to S transtion. Immunoblots were scanned and band intensities determined and plotted against time. (c) BrdU-labelling in hepatocytes of mice given sense- or antisense-*RB1* oligonucleatide after PH.

In summary, we demonstrate that antisense-*RB1* oligonucleotide is able to block pRB synthesis *in vitro* and *in vivo* in the liver and in both cases the G1 to S phase transition was accelerated. In culture, primary cells stimulated to proliferate yet lacking pRB arrest in G1 or enter PCD under control of p53. During compensatory hyperplasia however, cells traverse S but do not appear to enter G2 or M phase. Whatever the mechanism, hepatocytes entering the cell cycle too early are clearly missing signals essential to their completing the process. In attempting to generate immortalized yet non-transformed cells, we must consider which additional stimuli to employ in addition to the suppression of the pRB function. The supression of p53 synthesis is one possibility and one which might also be achieved with antisense-*TSP53* oligonucleotide.

6. Extention of the Lifespan of Primary Hepatocytes

6.1 GROWTH STIMULATING GENES

The proliferative capacity of hepatocytes in primary culture is very short ususally comprising between 2 to 3 population doublings yet the number of cell divisions a hepatocyte may undergo *in vivo* is certainly greater. For example in murine models of liver degeneration, repopulation of the diseased organ by allogenic hepatocytes demonstrates that the transplanted cells had divided up to fifteen times (Rhim *et al.*, 1994; Overturf *et al.*, 1996). During serial transplantation of hepatocytes which had repopulated the mouse liver, cells were estimated to have undergone 69 cell doublings (Overturf *et al.*, 1997). There is then a great proliferative potential to be exploited in hepatocytes if the correct culture formulation can be achieved.

An important distinction should be made concerning the effects of cell cycle modulation strategies on the lifetime of the hepatocyte. Transient stimuli can be employed to extend the proliferative life of cultured primary cells and these might include antisense oligonucleotides, transient transfection of plasmids, non-integrating virus mediated gene delivery, as with recombinant adenovirus, in combination with growth factors. One of the plasma borne growth factors made available to the liver during compensatory hyperplasia is hepatocyte growth factor (HGF) (Fausto *et al.*, 1995; Michalopoulos and DeFrances, 1997) and this was included in the medium used for immortalization of primary hepatocytes in our experiments. A permanent mitogenic stimulus can be achieved by integration of growth promoting genes carried by recombinant retrovirus, adeno-associated virus (AAV) or on plasmids.

The first genes that we introduced into hepatocytes were those for cyclin D1 together with an antisense-*RB1* fragment containing construct (Fig. 11) at a 1:1 ratio. The protein encoding construct was diven by a strong viral promoter whilst antisense-*RB1* was under the control of a recombinant albumin promoter. The rationale behind the use of the albumin promoter is based on the fact that hepatocytes specifically express albumin and growth promoting sequences driven by this promoter should provide the cells with a growth advantage over liver non-parenchymal cells in the culture. Furthermore, should integration of the plasmids take place resulting in a permanent growth stimulus, the cells must be able to utilize the albumin promoter in order to stimulate cell division and thus proliferating cells should also be able to use their endogenous promoter to express the albumin protein.

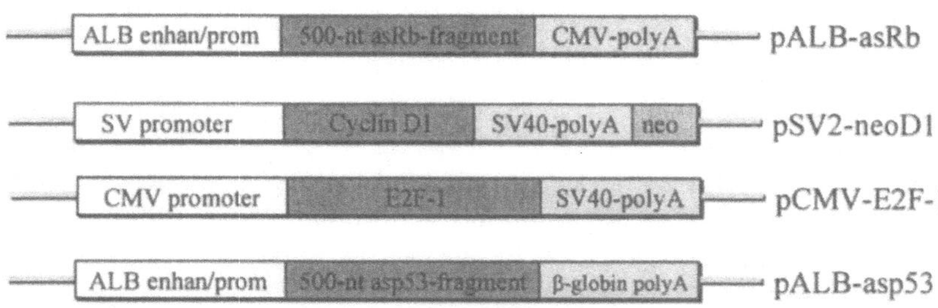

Figure 11. Plasmids transfected into human hepatocytes. SV and CMV, SV40 and CMV promoter sequences; ALB enhan/prom, recombinant enhancer-promoter consrtuct from the human albumin gene; pA or polyA, polyadenylation sequence; neo, neomycin phosphotransferase gene.

In our first experiments primary human hepatocytes were isolated from healthy tissue surrounding tumour explant material taking care to ensure that no tumour cells remained. Cells were seeded into collagen I coated plastic dishes in William's medium E containing 10% foetal bovine serum (FBS), insulin and hydrocortisone with HGF (10 ng/ml). Primary hepatocytes have a densely packed 'cobblestone' like morphology with cells containing dense organelle-rich cytoplasm and one or two and occasionally four nuclei. Over 3-4 days in culture the hepatocytes are able to proliferate and organise themselves into simple rows or rings with intercellular spaces running between rows of several adjacent cells. These 'canaliculi' recapitulate the bile ductular strutures seen *in vivo*. After 48-72 h primary human hepatocytes are in late G1 phase and preparing for DNA-replication (Gomez-Lechon *et al.*, 1995), a point at which an optimal window for transfection of exogenous DNA exists. One or more recombinant plasmids were introduced into cells by lipofection and medium replaced every two days without antibiotic selection. In this way the only selection placed on the cells was the ablility to proliferate in a monolayer configuration.

The combination of cyclin D1 with antisense-*RB1* (Fig. 11) was chosen in order to reduce the mitogenic threshold of the pRB checkpoint, firstly, by suppressing pRB synthesis and secondly, by increasing phosphorylation of residual pRB through increased cyclin D1-cdk4 activity. However, no effect was observed on the proliferative capacity of the transfected hepatocytes. Inactivation of pRB frees E2F (see section 3.1.1), yet the transcription factor may form complexes with other pRB related 'pocket' proteins such as p107 and p130 in which the function of E2F is also inactivated (Schwarz *et al.*, 1993). In support of this, it is known that the viral oncoproteins are also capable of sequestering p107 as well as pRB thereby ensuring the activation of E2F (Chellappan, 1991). For this reason the ectopic expression of E2F-1 (Fig. 11) was attempted in addition to cyclin D1 and antisense-*RB1* and these plasmids were transfected at a 1:1:1 ratio.

After one to two weeks non-transfected cells began to die off leaving small colonies of proliferating cells. Colony types were classified in five groups according to morphology:
(i) Polygonal epitheloid cells with dense granular cytoplasm and one or two regular circular nuclei with few nucleoli. Cells at the edge of colonies often displayed ruffled margins and pseudopodia. Within the colonies short intercellular canaliculi were observed. (Fig. 12a). These cells are the most reminiscent of primary hepatocytes.
(ii) Similar to (i) except flatter in appearance and with fewer extracellular spaces. These cells may be derived from the bile duct epithelia (Fig. 12b).
(iii) Irregularly shaped, mononuclear cells with perinuclear granular material. Several nucleoli were present. Colony edges are smooth and there are no large intercellular spaces (Fig. 12c). The cells show features of endothelial cell types.
(iv) Very large irregularly shaped cells with one or two nuclei and a large cytoplamic : nuclear ratio. Cells contained complex net-like structure in a somewhat denser cytoplasmic area which also included the nuclei. Whorls of filamentous material were also present in some cells. Cells were in contact either along opposite facing membranes or through cytomplasmic extensions forming short extracellular spaces (Fig. 12d). These cells may be derived from sinusoidal endothelial cells.
(v) Large polygonal cells with a large cytoplamic : nuclear ratio. Cytoplasm was clear except for defined granular areas often perinuclear. Associated with these areas were few or many translucent droplets, probably containing lipid. Singular or double nuclei were seen with many nucleoli. Intercellular junctions are observed but no large intercellular spaces (Fig. 12e). These colonies may be derived from Ito or fat-storing cells or from some other non-parenchymal support cell.

Other colony types derived from fibroblasts or proliferation-competent non-parenchymal cells were also observed but these were also present in non-transfected control cultures and cultures transfected with empty vector DNA. Colony size increased over the following three to four weeks to 1-5 x 10^3 cells at which point colony types (i) and (ii) ceased growing abruptly. Proliferation of colony types (iii) - (v) had already lessened by this time and ceased soon afterwards. We focused our interest on colonies of type (i) and (ii) which made up the majority of all colonies and were more likely to have stemmed from liver epithelia. Over the subsequent two weeks (weeks 7 to 8 after transfection) all observed colonies underwent a spontaneous and total degeneration lasting two to three days in each case. We observed the typical features of apoptosis in cells during this period including decreased cell-cell contact, membrane blebbing and nuclear deposition and fragmentation (Fig. 12f). The induction of PCD is in keeping with the unscheduled release of E2F after suppression of the pRB checkpoint (see section 3.1.3).

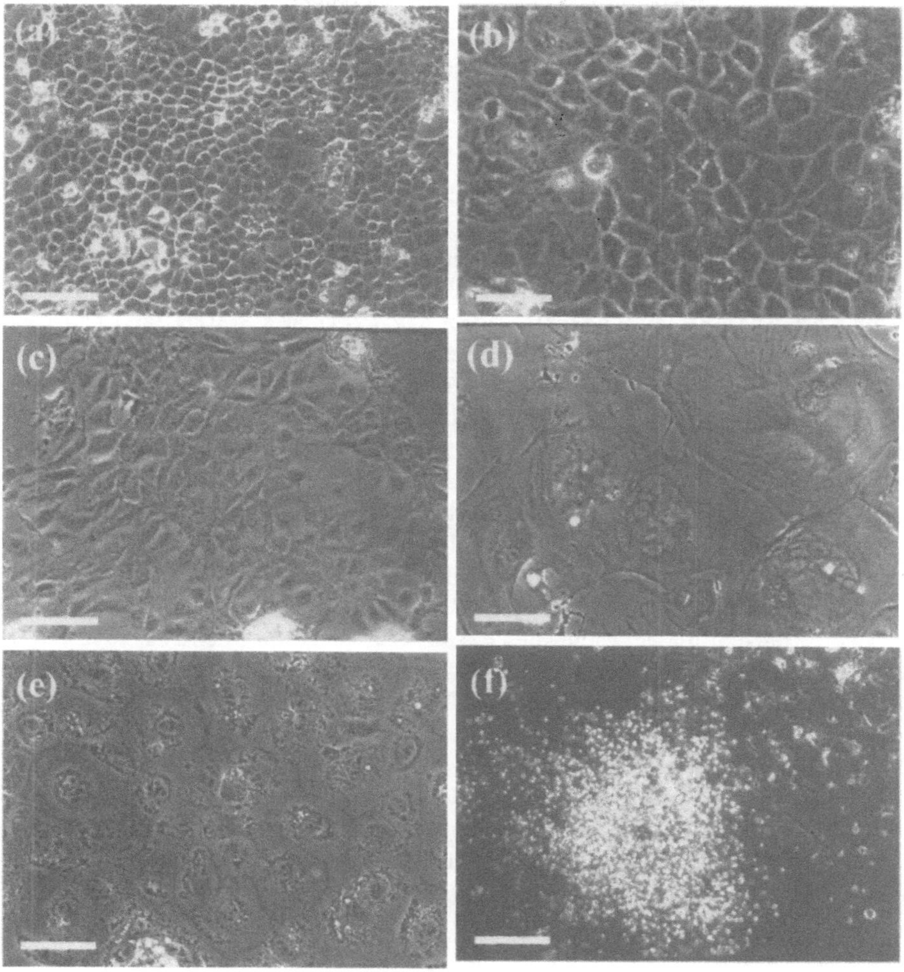

Figure 12. (a) - (e) colony types produced by the transfection of plasmids pALB-asRB, pSV2-neoD1 and pCMV-E2F-1 into primary human hepatocytes. (f) apoptosis in a hepatocyte colony seven weeks after transfection. Scale bar = 125 μm (a, f), 50 μm (b, c), 25 μm (d, e).

6.2. ANTI-APOPTOSIS GENES

The problem of apoptosis in cultured cells has been addressed in many systems and the solution depends upon the nature of the stimulus inducing PCD. In productive bioreactors, for example, the stresses resulting from high density culture and rapidly changing media conditions often take a toll on cell survival. The introduction of 'survival' genes has extended the life of cells and ultimately increased productivity (Al-Rubeai and Singh, 1998). However, these genes are usually members of the

BAX-BCL-2 family of apoptosis regulators which lie downstream of initiators of PCD and upstream of the effectors of apoptosis, the caspases (Cheng *et al.*, 1997). This is sufficient for cells which are already immortalized and have subdued the PCD response to deregulation of the pRB checkpoint during their establishment (Fig. 5). Primary cells have yet to overcome this control function and to do this we must prevent the PCD cascade from being activated. In conjunction with pRB checkpoint suppression, the p53 protein is the most obvious target to allow unhindered proliferation.

We introduced an antisense-*TSP53* fragment under the control of a recombinant albumin promoter (Fig. 11), as for the antisense-*RB1* fragment described (section 6.1). The construct pALB-asp53 was cotransfected into primary human hepatocytes with the plasmids pALB-aspRB, pSV2-neoD1 and pCMV-E2F-1 at a 1:1:1:1 ratio and cells cultured as before without antibiotic selection. Three weeks after transfection we observed the outgrowth of similar colonies to those described in section 6.1 (ie. types (i) - (iv)) and by week 8, a large number of these had undergone apoptosis. Nevertheless, a small number of colonies (a frequency of 1.6×10^{-6}) continued to grow and were eventually passaged. These colonies were of type (i) and resembled primary hepatocytes, except that they were smaller and had a somewhat organelle sparse cytoplasm. In one of these experiments a cell line was established from such a colony and named HepZ. Although we could not demonstrate this at the molecular level in individual colonies, it would appear that the inclusion of the antisense-p53 fragment was able to prevent the activation of PCD in some colonies. This lead, firstly, to a further extension of the proliferative life of the colony and secondly, to a passageable state which we equate with immortalization.

6.3 THE HEPZ CELL LINE

During the first passages, population doubling time of HepZ shortened rapidly with the result that cells had to be passaged twice a week. Cell division was mitogen and serum-dependent up until the third passage and by passage seven growth became independent of insulin. HepZ shows a strong requirement for cell-cell contact for survival. This characteristic is still seen in cultures as old as passage 70. Single cells and those at the margins of colonies display an extended morphology with cytoplasmic extensions yet within the colonies cells are polygonal and form intercellular canaliculi (Fig. 13a). Monolayer cultures of HepZ grow to a high density but do not form three-dimensional foci as would be expected from transformed cells. Confluent regions of cells are very difficult to seperate by standard trypsin/EDTA treatment indicative of very strong cell-cell junctions. In contrast, confluent cells may be freed from plastic dishes by simple agitation due to weak adherence to plastic. Survival in sparse cultures is enhanced if the cells are plated on a collagen substratum and furthermore, cells grow more evenly and with a homogenous morphology (Fig. 13b).

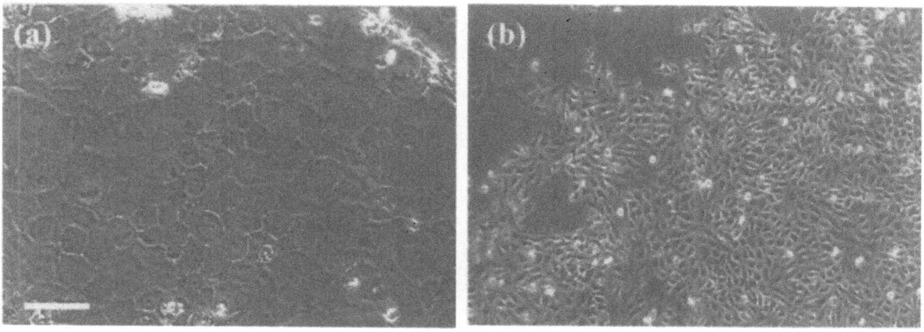

Figure 13. (a) HepZ cells in monolayer culture. (b) HepZ cells on collagen type I substratum. Scale bar = 50 μm (a), 125 μm (b).

6.3.1. *The Cell Cycle Profile of HepZ Cells*

While the extension of hepatocyte primary culture lifespan appeared to correlate with the transient transfection of growth stimulating genes and suppression of the pRB checkpoint, the events resulting in the permanent proliferation of HepZ cannot be assumed to result from genomic integration of the plasmids carrying those genes. We therefore undertook an intensive investigation of the expression of proteins involved in cell cycle control and compared the results to those obtained from liver, primary hepatocytes and various liver cell lines. In passage 7 HepZ cells we found that pRB and cyclin D1 protein was expressed at normal levels compared to other cultured cells of hepatocyte or non-hepatocyte origin (Fig. 14a). The inference from this is that HepZ cells have no repression of pRB synthesis, as would be expected if the plasmid pALB-aspRB had become integrated into the genome, nor is cyclin D1 overexpressed, as would result from the acquisition of a second, exogenous gene. Immunoblotting of E2F-1 revealed significantly elevated levels in HepZ cells indcative of plasmid integration (Fig. 14a). Such high levels would explain how HepZ escapes pRB growth control assuming that the levels of E2F-1 protein exceed the capacity of endogenous 'pocket' proteins to sequester and inactivate it. This raises the question of how these cells have dealt with the subsequent threat of PCD (see section 3.1.3) and this warrants analysis of the status of *TSP53*. We could detect p53 protein on immunoblots using some antibodies yet not with others, specifically clone DO-1 (Vojtesek *et al.*, 1992) and PAb1801 (Tuck and Crawford, 1989). Both of these antibodies recognise epitopes at the *N*-terminus of the p53 protein and we reasoned that the epitopes were structurally altered or absent ie. that *TSP53* had been mutated. To confirm this we immunoprecipitated p53 from cell lysates using wild-type and mutant specific antibodies (PAb240) (Gannon *et al.*, 1990) and could show that only the mutant specific antibody was able to bind the protein (Fig. 14a). Levels of the protein were equivalent to those seen in Huh-7 cells which also carry a mutant p53.

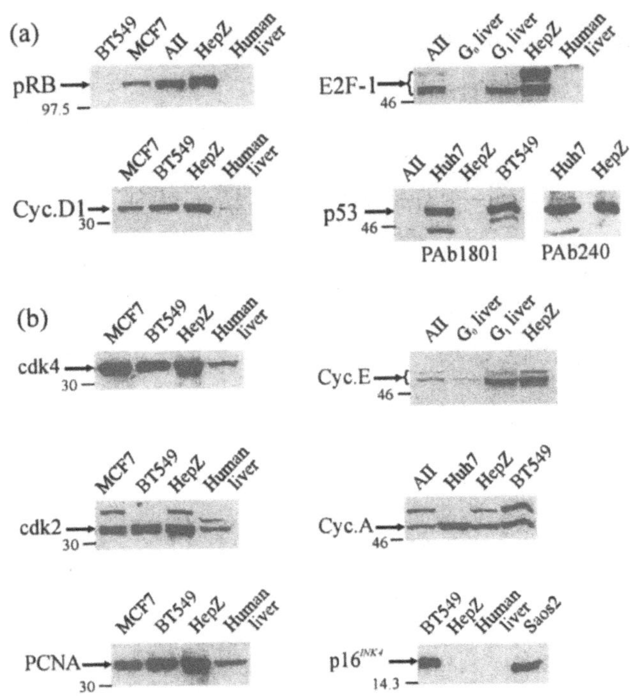

Figure 14. Immunoanalysis of cell cycle proteins in HepZ. (a) the target proteins for immortalization. Retinoblastoma protein (pRB) was detected with antibody G3-245 (PharMingen); cyclin D1 (Cyc. D1) with DCS-6 (Lukas *et al.*, 1994); E2F-1 with C-20 (Santa Cruz Biotechnology); p53 was immunoprecipitated with PAb240 and detected on Western blots with either PAb1801 or with PC35 (Oncogene Research Products). The human mammory tumour cell lines BT549 and MCF7 served as negative and positive controls respectively for pRB. AII is an hepatocyte-derived cell line from a p53 nullizygous mouse and Huh-7 is derived from a human hepatoma and contains a mutant p53 protein (Puisieux *et al.*, 1993) and these serve as negative and positive controls respectively for p53 expression. G0 liver and G1 liver are lysates of mouse liver prior to and after partial hepatectomy respectively; E2F-1 is induced in G1 liver but is not detectable in G0. Human liver is lysate of normal human liver. (b) cell cycle marker proteins. cdk4 was detected with antibody C-22; cdk2 with M2; cyclin E with M-20 (Santa Cruz Biotechnology); PCNA with PC-10 (Oncogene Research Products); cyclin A (Upstate Biotechnology); p16^{INK4} with DCS-50 (Lukas *et al.*, 1995). Proliferating cells all express cdk4, cdk2, PCNA and cyclinA. Cyclin E is induced in G1 liver. BT549 and the osteosarcome cell line, Saos2, are positive for p16^{INK4}. Molecular weight markers are indicated left.

Further analyses showed that at the level of expression all other cell cycle proteins were unaffected. Cyclin E and cdk4, whose overexpression has also been reported in immortalized cells, were present at normal levels as were those of cdk2, PCNA and cyclin A, p21$^{CIP\text{-}1}$ and p16^{INK4a} (Fig. 14b). By selecting for an inactivated p53, HepZ cells endure the high levels of E2F-1 expression and this is concurrent with the hypothesis that immortalization requires both G1 checkpoint repression and neutralization of cell cycle linked PCD.

We determined whether HepZ cells showed any characteristics of transformation by examining their ability to grow in soft-agar and to produce tumours in nude mice. No growth in soft-agar was observed (Table 2) either with early (passage 7) or late (passage 40) passage cells. Early passage cells did not generate tumours in nude mice but small, slow growing tumours were detected after a long lag phase in mice injected with late passage cells (Fig. 15). This is an important finding and warns us that cells held for long periods in monolayer culture may acquire a transformed phenotype while showing no morphological changes or further aberrations in cell cycle protein expression (see section 3). However, the growth of tumours in nude mice might be the result, not of injection of transformed cells but that the immortalized cells injected underwent changes effecting transformation *in situ*. This would also explain the long lag phase preceding the macroscopic appearance of the tumours. Clarification must await the determination of telomerase activities in early and high passage cells as well as from isolated tumours. In favour of this hypothesis is the fact that late passage cells do not grow in soft-agar.

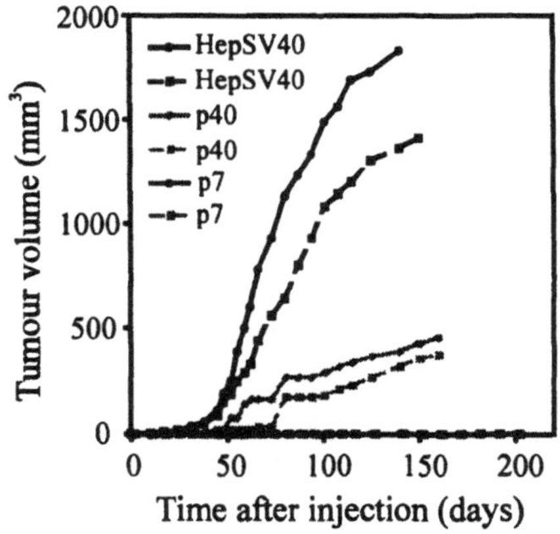

Figure 15. Tumour induction by HepZ in nude mice. Mice were injected subdermally with 5 x 10^6 cells from passage 7 (p7) or passage 40 (p40). HepSV40 was included for comparison.

6.3.2. *Functional Characterization of the HepZ Cell Line*

Determination of the synthesis of the hepatocyte-specific proteins albumin and α_1-antitrypsin, and of cytokeratins 8 an 18 was performed by immunofluorescence. The foetal liver synthesises α-fetoprotein (AFP) whose role is then taken over by albumin after birth. Reinduction of AFP synthesis in adult somatic cells is associated with the phenotypic dedifferentiation accompanying transformation and tumorigenesis, hence AFP production was also included in the analysis. As shown in Table 3, primary human hepatocytes were positive for all of the proteins tested except for AFP and first passage HepZ cells displayed an identical spectrum. By passage 3, however, α_1-antitrypsin expression in HepZ became weaker and in later passages (passage 40 and above) secreted α_1-antitrypsin levels were extremely weak, as determined by ELISA measurements of culture supernatants. The down-regulation of cell specific activities correlates with the loss or repression of hepatotropic signals in culture (see section 2.2 and 2.3). The two tumour derived cell lines HepG2 and Huh-7 are widely used for studies involving hepatocyte function and these were also assessed for specific protein synthesis. The cytokeratins and albumin were present in both cell lines but α_1-antitrypsin could not be detected. Furthermore, the foetal specific protein AFP was produced by these cell lines reflecting their origins in neoplasias.

TABLE 3. Expression of hepatocyte specific proteins in primary human hepatocytes and hepatocyte derived cell lines.

Protein	Hepatocytes	HepZ, p7	HepG2	Huh-7
Cytokeratin 8	+	+	+	+
Cytokeratin 18	+	+	+	+
Albumin	++	++	++	++
AFP	-	-	++	++
α1-antitrypsin	++	+	-	-

HepZ cells possess characteristics of differentiated hepatocytes and are non-tumorigenic, at least at early passage. HepZ expresses high levels of E2F-1 which disrupts the G1 checkpoint for progression into S phase. Unlike neoplastic transformation or expression of viral oncoproteins, this targeted deregulation does not lead to loss of tissue-specific gene expression nor to transformation and the selection of a mutant *TSP53* has negated the E2F-dependent induction of PCD (Fig. 16). While this combination may not be ideal for cells of different ontogeny, for hepatocytes it provides a starting point for the generation of safe, differentiated cell lines for use in the clinic.

10. The Outlook for Immortalized Hepatocytes

It should be possible to select the characteristics required of a hepatocyte line by choosing a suitable immortalizing strategy and culture configuration. Whereas earlier, only few tools were available to achieve the changes required for continuous proliferation and these non-specific, such as chemical mutagenesis or viral oncogene transfection, now we can target precisely the part of the cell cycle that we wish to modulate. Be it relief of growth suppression or regulatable cell cycle arrest in a specific phase, we can choose from a range of genes encoding the motors and brakes of mitosis. Maintaining liver-specific gene expression will require further study of the essential

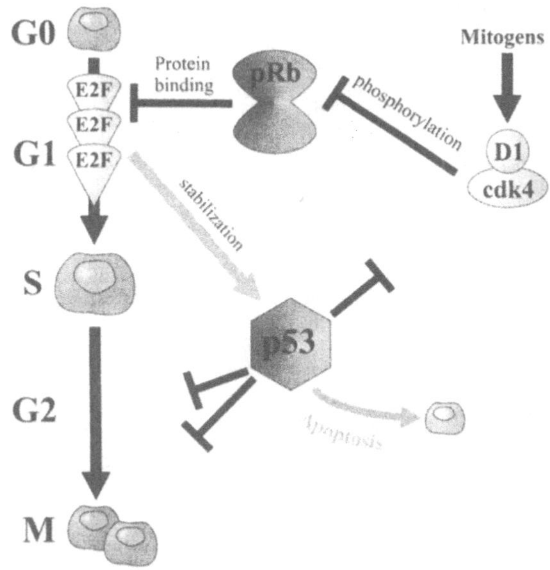

Figure 16. Deregulation of the G1 checkpoint in HepZ cells. Overexpression of E2F-1 combined with mutation and inactivation of the p53 protein leads to immortalization. Compare with Fig. 4.

factors in organ architecture and systemic signals which regulate the phenotype and this will to lead to toward the development of more efficient three-dimensional matrix culture systems. Cooperaton between cell biologists and engineers has already helped to build the foundations for this future.

8. Acknowledgements

The authors would like to thank Dr. P. Loeser, HepaVec AG, Berlin, Germany, for the pCMV-Tagori plasmid and Dr. L.P. Li, Max-Delbrueck Center for Molecular Medicine, Berlin, Germany, for the *lox*P-HyTK-large T retrovirus.

9. References

Aden, D.P., Fogel, A., Plotkin, S., Damjanov, I., and Knowles, B.B. (1979) Controlled synthesis of HBsAg in a differentiated human liver carcinoma-derived cell line. *Nature (London)* **282**, 615 - 616.
Alexander, J.J. (1984) In vitro studies of human hepatocellular carcinoma cell lines. *Adv. Hepatitis Res.*, 190 - 195.
Al-Rubeai, M., and Singh, R.P. (1998) Apoptosis in cell culture. *Curr. Opin. Biotechnol.* **9**, 152 - 156.

Arias, I.M., Boyer, J.L., Fausto, N., Jakoby, W.B., Schachter, D.A., and Shafritz, D.A., eds. *The Liver - Biology and Pathobiology*,Raven Press Ltd., New York, 1994.

Beaune, P.H., and Lecoeur, S. (1997) Immunotoxicology of the liver: adverse reactions to drugs. *J. Hepatol.* **26 Suppl. 2**, 37 - 42

Bodnar, A.G.M., Ouellette, M., Frolkis, M., Holt, S.E., Chiu, C.P., Morin, G.B., Harley, C.B., Shay, J.W., Lichtsteiner, S., and Wright, W.E. (1998) Extension of life-span by introduction of telomerase into normal human cells. *Science*, **279**, 349 - 352.

Brehm, A., Miska, E.A., McCance, D.J., Reid, J.L., Bannister, A.J., and Kouzarides, T. (1998) Retinoblastoma protein recruits histone deacetylase to repress transcription. *Nature* **391**, 597 - 601.

Bulera, S.J., Haas, M.J., Sattler, C.A., Li, Y., Pitot, H.C. (1997) Cell lines with heterogeneous phenotypes result from a single isolation of albumin-sv40 T-antigen transgenic rat hepatocytes. *Hepatology* **25**, 1192 -1203.

Bull. P.C., Thomas, G.R., Rommens, J.M., Forbes, J.R. and Cox, D.W. (1993) The Wilson disease gene is a putative copper transporting P-type ATPase similar to the Menkes gene. *Nature Genet.*, **5**, 327 - 337.

Chellappan, S.P., Hiebert, S., Mudryj, M., Horowitz, J.M., and Nevins, J.R. (1991) The E2F transcription factor is a cellular target for the RB protein. *Cell* **65**, 1053 - 1061.

Chen, S.C., Mullon, C., Kahaku, E., Watanabe, F., Hewitt, W., Eguchi, S., Middleton, Y., Arkadopoulos, N., Rozga, J., Solomon, B., and Demetriou, A.A. (1997) Treatment of severe liver failure with a bioartificial liver. *Ann. N. Y. Acad. Sci.* **831**, 350 - 360.

Cheng, E.H., Kirsch, D.G., Clem, R.J., Ravi, R., Kastan, M.B., Bedi, A., Ueno, K. and Hardwick, J.M. (1997) Conversion of Bcl-2 to a Bax-like death effector by caspases. *Science* **278**, 1966 - 1968.

Clarke, A.R., Maandag, E.R., van Roon, M., van-der-Lugt, N.M., van-der-Valk, M., Hooper, M.L., Berns, A. te Riele, H. (1992) Requirement for a functional Rb-1 gene in murine development. *Nature* **359**, 328 - 330.

Dambrough, C., Slater, S., Vass, M., and MacDonald, C. (1992) Immortalization of murine primary spleen cells by v-myc, v-ras, and v-raf. *Exp. Cell. Res.* **201**, 273 - 283.

DeCaprio, J.A., Ludlow, J.W., Lynch, D., Furukawa, Y., Griffin, J., Piwnica-Worms, H., Huang, C.M., and Livingstone, D.M. (1988) SV40 large T antigen forms a specific complex with the product of the retinoblastoma susceptibility gene. *Cell* **54**, 275 - 283.

De Stanchina, E., McCurrach, M.E., Zindy, F., Shieh, S-Y., Gerardo, F., Samuelson, A.V., Prives, C., Roussel, M.F., Sherr, C.J., and Lowe, S.W. (1998) E1A signaling to p53 involves the p19 ARF tumor suppressor. *Genes Dev.* **12**, 2434 - 2442.

Draetta, (1994) G. Mammalian G_1 cyclins. *Curr. Opin. Cell Biol.* **6**, 842 - 846.

Dyson, N., Howley, P.M., Munger, K., and Harlow, E. (1989) The human papilloma virus-16 E7 oncoprotein is able to bind to the retinoblastoma gene product. *Science* **243**, 934 - 937.

Enosawa, S., Suzuki, S., Kakefuda, T., and Amemiya, H. (1996) Examination of 7-ethoxycoumarin deethylation and ammonia removal activities in 31 hepatocyte cell lines. *Cell. Transplant.* **5 (Suppl. 1)**, S39 - 40

Fausto, N., and Webber, E.M. (1994) Liver regeneration. In: *The Liver: Biology and Pathobiology*. 3rd ed., Arias, I.M., Boyer, J.L., Fausto, N., Jakoby, W.B., Schachter, D.A., and Shafritz, D.A., eds., Raven Press Ltd., New York, 1059 - 1084.

Fausto, N., Laird, A.D., and Webber, E.M. (1995) Role of growth factors and cytokines in hepatic regeneration. *FASEB J.* **9**, 1527 - 1536.

Farrell, G.C. (1997) Drug-induced hepatic injury. *J. Gastroenterol. Hepatol.* **12**, S242 - 250.

Friedman, J.M., Chung, E.Y., and Darnell Jr., J.E. (1984) Gene expression during liver regeneration. *J. Mol. Biol.* **179**, 37 - 53

Friend, S.H., Horowitz, J.M., Gerber, M.R., Wang, X.F., Bogenmann, E., Li, F.P., and Weinberg, R.A. (1987) Deletions of a DNA sequence in retinoblastomas and mesenchymal tumors: Organization of the sequence and its encoded protein. *Proc. Natl. Acad. Sci. USA* **84**, 9059 - 9063.

Gannon, J.V., Greaves, R., Iggo, R., Lane, D,.P. (1990) Activating mutations in p53 produce a common conformational effect. A monoclonal antibody specific for the mutant form. *EMBO J.* **9**, 1595 - 1602.

Gerlach, J.C. (1996) Development of a hybrid liver support system: a review. *Int. J. Artif. Organs.* **19**, 645 - 654.

Goldstein, J.L., and Brown, M.S. (1973) Familial hypercholesterolemia: pathogenesis of a receptor disease. *Johns Hopkins Med. J.* **143**, 8 - 16.

Gomez-Lechon, M.J., Castelli, J., Guillen, I., O'Connor, E., Nakamura, T., Fabra, R., and Trullenque, R. (1995) Effects of hepatocyte growth factor on the growth and metabolism of human hepatocytes in primary culture. *Hepatology* **21**, 1248 - 1254.

Greenbaum, L.E., Cressman, D.E., Haber, B.A., and Taub, R. (1995) Coexistence of C/EBPα, β, growth-induced proteins and DNA synthesis in hepatocytes during liver regeneration. *J. Clin. Invest.* **96**, 1351 - 1365.

Grossman, M., Raper, S.E., Kozarsky, K., Stein, E.A., Engelhardt, J.F., Muller, D., Lupien, P.J., and Wilson, J.M. (1994) Successful ex vivo gene therapy directed to liver in a patient with familial hypercholesterolaemia. *Nat. Genet.* **6**, 335 - 341.

Guillouzo, A. (1998) Liver cell models in in vitro toxicology. *Environ. Health Perspect.* **106 Suppl 2** 9, 511 - 532.

Guguen-Guillouzo, C., Clement, B., Baffet, G., Beaumont, C., Morl-Chaney, E., Glaise, D., and Guillouzo, A. (1983) Maintenance and reversibility of active albumin secretion by adult rat hepatocytes co-cultured with another liver epithelial cell type. *Exp. Cell. Res.* **143**, 47 - 55.
Gunsalus, J.R., Brady, D.A., Coulter, S.M., Gray, B.M., and Edge, A.S.B. (1997) Reduction of serum cholesterol in Watanabe rabbits by xenogenic hepatocellular transplantation. *Nature Med.* **3**, 48 - 53.
Hainaut, P.; Soussi, T. Shomer, B., Hollstein, M., Greenblatt, M., Hovig, E., Harris, C.C., and Montesano, R. (1997) Database of p53 gene somatic mutations in human tumors and cell lines: updated compilation and future prospects. *Nucleic Acids Res.* **25**, 151-7
Hannon, G.J., and Beach, D. (1994) $P15^{INK4B}$ is a potential effector of TGF-β-induced cell cycle arrest. *Nature* **371**, 257 - 261.
Harland, R.C., and Platt, J.L. (1996) Prospects for xenotransplantation of the liver. *J. Hepatol.* **25**, 248 - 258.
He, L., Isselbacher, K.J., Wands, J.R., Goodman, H.M., Shih, C., and Quaroni, A. (1984) Establishment and characterization of a new human hepatocellular carcinoma cell line. *In Vitro* **20**, 493 - 504.
Helin, K., Lees, J.A., Vidal, M., Dyson, N., Harlow, E., and Fattaey, A. (1992) A cDNA encoding a pRB-binding protein with properties of the transcription factor E2F. *Cell* **70**, 337 - 350.
Hering, S., Griffin, B.E., and Strauss, M. (1991) Immortalization of human fetal sinusoidal liver cells by polyoma virus large T antigen. *Exp. Cell Res.* **195**, 1 - 7.
Herwig, S., and Strauss, M. (1997) The retinoblastoma protein: a master regulator of cell cycle, differentiation and apoptosis. *Eur. J. Biochem.* **246**, 581 - 601
Higgins, G.M., and Anderson, R.M. (1931) Experimental pathology of the liver. I. Restoration of the liver of the white rat following partial surgical removal. *Arch. Pathol.* **12**, 186 - 202.
Hinds, P.W., Mittnacht, S., Dulic, V., Arnold, A., Reed, S.I., and Weinberg, R.A. (1992) Regulation of retinoblastoma protein functions by ectopic expression of human cyclins. *Cell* **70**, 993 - 1006.
Höhne, M., Piasecki, A., Ummelmann, E., and Paul, D. (1987) Transformation of differentiated neonatal rat hepatocytes in primary culture by polyoma virus early region sequences. *Oncogene* **1**, 337 -345.
Hoffmann, B., Piasecki, A., and Paul, D. (1989) Proliferation of fetal rat hepatocytes in response to growth factors and hormones in primary culture. *J. Cell Physiol.* **139**, 654 - 662.
Horowitz, J.M., Park, S.H, Bogenmann, E., Cheng, J-C., Yandell, D.W., Kaye, F.J., Minna, J.D., Dryja, P., and Weinberg, R.A. (1990) Frequent inactivation of the retinoblastoma antioncogene is restricted to a subset of human tumor cells. *Proc. Natl. Acad. Sci. USA* **87**, 2775 - 2779.
Horwich, A.L. Inherited hepatic enzyme defects as candidates for liver-directed gene therapy. *Curr. Topics Microbiol. Immun.* **168** (1991), 185 - 200.
Huang, L.C., Clarkin, K.C., and Wahl, G.M. (1996) Sensitivity and selectivity of the DNA damage sensor responsible for activating p53-dependent G1 arrest. *Proc. Natl. Acad. Sci. USA.* **93** ,4827 - 4832.
Hunter, T., and Pines, J. Cyclins and cancer II: cyclin D and cdk inhibitors come of age. *Cell* **79** (1994), 573 - 582.
Kaelin, W.G., Krek, W., Sellers, W.R., DeCaprio, J.A., Ajchenbaum, F., Fuchs, C.S., Chittenden, T., Li, Y., Farnham, P.J., Blanar, M.A., Livingston, D.M., and Flemington, E.K. (1992) Expression cloning of a cDNA encoding a retinoblastoma-binding protein with E2F-like properties. *Cell* **70**, 351 - 364.
Kamb, A., Gruis, N.A., Weaver-Feldhaus, J., Lui, Q., Harshman, K., Tavtigan, S.V., Stockert, E., Day, R.S., Johnson, B.E., and Skolnick, M.H. (1994) A cell cycle regulator potentially involved in genesis of many tumor types. *Science* **264**, 436 - 440.
Kato, J., Matsushime, H., Hiebert, S.W., Ewen, M.E., and Sherr, C.J. (1993) Direct binding of cyclin D to the retinoblastoma gene product (pRb) and pRb phosphorylation by the cyclin D-dependent kinase cdk4. *Genes Dev.* **7**, 331 - 342.
Kreuzburg-Duffy, U.C. and MacDonald, C. (1994) Establishment and characterization of murine macrophage-like cell lines following transformation with simian virus 40 DNA deleted at the origin of replication. *J. Immunol. Methods* **174**, 33 - 51.
Kiyono, T., Foster, S.A., Koop, J.I., McDougall, J.K., Galloway, D.A., and Klingelhutz, A.J. (1998) Both $Rb/p16^{INK4a}$ inactivation and telomerase activity are required to immortalize human epithelial cells. *Nature* **396**, 84 - 88.
Lane, D.P., and Benchimol, S. (1990) p53: oncogene or antioncogene? *Genes Dev.* **4**, 1 - 8.
Leffert, H.L., Koch, K.S., Lad, P.J., Shapiro, I.P., Skelly, H., and de Hemptinne, B. (1988) Hepatocyte regeneration, replication and differentiation, in: *The Liver: Biology and Pathobiology*, 2nd edition, Arias, I.M., Jakoby, W.B., Popper, H., Schachter, D., and Shafritz, D.A., eds., Raven Press, Ltd., New York.
Li, L.P., Schlag, P.M., and Blankenstein, T. (1997) Transient expression of SV 40 large T antigen by Cre/LoxP-mediated site-specific deletion in primary human tumor cells. *Hum. Gene Ther.* **8**, 1695 - 1700.
Li, Q., Kay, M.A., Finegold, M., Stratford-Perricaudet, L.D., and Woo, S.L. (1993) Assessment of recombinant adenoviral vectors for hepatic gene therapy. *Hum. Gene Ther.* **4**, 403 - 409.
Liebermann, D.A., Hoffman, B., and Steinman, R.A. Molecular controls of growth arrest and apoptosis. (1995) p53-dependent and independent pathways. *Oncogene* **11**, 199 - 210.
Lindblad, B., Lindstedt, S, and Stehen, G. (1977) On the enzymatic defects in hereditary tyrosinemia. *Proc. Natl. Acad. Sci. USA* **74**, 4641 - 4645.
Lukas, J., Pagano, M., Staskova, Z., Draetta, G., and Bartek, J. (1994) Cyclin D1 protein oscillates and is essential for cell cycle progression in human tumour cell lines. *Oncogene*, **9**, 707 - 718.

Lukas, J., Parry, D., Aagaard, L., Mann, D.J., Bartkova, J., Strauss, M., Peters, G., and Bartek, J. (1995) Retinoblastoma-protein-dependent cell-cycle inhibition by the tumour suppressor p16. *Nature*, **375**, 503 - 506.

Lukas, J., Bartkova, J., and Bartek, J. (1996) Convergence of mitogenic signalling cascades from diverse classes of receptors at the cyclin D-cyclin-dependent kinase-pRb-controlled G1 checkpoint. *Mol. Cell Biol.* **16**, 6917 - 6925.

Magnaghi-Jaulin, L., Groisman, R., Naguibneva, I., Robin, P., Lorain, S., LeVillain, J.P., Troalen, F., Trouche, D., and Harel-Bellan, A. (1998) Retinoblastoma protein represses transcription by recruiting a histone deacetylase. *Nature* **391**, 601 - 604.

McNab, G.M., Alexander, J.J., Lecatas, G., Bey, E.M., and Urbanowicz, J.M. (1976) Hepatitis B surface antigen produced by a human hepatoma cell line. *Br. J. Cancer* **34**, 509 - 515.

Meyer, U.A. (1996) Overview of enzymes of drug metabolism. *J. Pharmacokinet. Biopharm.* **24**, 449 - 459.

Michalopoulos, G.K., and DeFrances, M.C. (1997) Liver regeneration. *Science*, **276**, 60 -66.

Morgenbesser, S.D., Williams, B.O., Jacks, T., and DePinho, R.A. (1994) p53-dependent apoptosis produced by Rb-deficiency in the developing mouse lens. *Nature* **371**, 72 - 74.

Nakabayashi, H., Taketa, K., Miyano, K., Yamane, T., and Sato, J. (1982) Growth of human hepatoma cell lines with differentiated functions in chemically defined medium. *Cancer Res.* **42**, 3858 - 3863.

Nigro, J.M., Baker, S.J., Preisinger, A.C., Jessup, J.M., Hostetter, R., Cleary, K., Binger, S.H., Davidson, N., Baylin, S., Devilee, P., Glover, T., Collins, F.S., Weston, A., Modali, R., Harris, C.C., and Vogelstein, B. (1989) Mutations in the p53 gene occur in diverse human tumour types. *Nature* **342**, 705 - 708.

Nobori, T., Miura, K., Wu, D., Lois, A., Takabayshi, K., and Carson, D. (1994) Deletions of the cyclin-dependent kinase-4 inhibitor gene in multiple human cancers. *Nature* **368**, 753 - 756.

Overturf, K., Al-Dhalimy, M., Tanguay, R., Brantly, M., Ou, C.N., Finegold, M., and Grompe, M. (1996) Hepatocytes corrected by gene therapy are selected *in vivo* in a murine model of hereditary tyrosinaemia type I. *Nature Gen.* **12**, 266 - 273.

Overturf, K., Al-Dhalimy, M., Ou, C.N., Finegold, M., and Grompe, M. (1997) Serial transplantation reveals the stem-cell-like regenerative potential of adult mouse hepatocytes. *Am. J. Pathol.*, **151**, 1273 - 1280.

Paul, D., Hohne, M., Pinkert, C., Piasecki, A., Ummelmann, E., and Brinster, R.L. (1988) Immortalized differentiated hepatocyte lines derived from transgenic mice harboring SV40 T-antigen genes. *Exp.Cell Res.* **175**, 354 - 62.

Pfeifer, A.M.A., Cole, K.E., Smoot, D.T., Weston, A., Groopman, J.D., Shields, P.G., Vignaud, J-M., Juillerat, M., Lipsky, M.M., Trump, B.F., Lechner, J.F., and Harris, C.C. (1993) Simian virus 40 large tumor antigen-immortalized normal human liver epithelial cells express hepatocyte characteristics and metabolize chemical carcinogens. *Proc. Natl. Acad. Sci. USA* **90**, 5123 - 5127.

Prelich, G., and Stillman, B. (1988) Coordinated leading and lagging strand synthesis during SV40 DNA replication in vitro requires PCNA. *Cell* **53**, 117 - 126.

Puisieux, A., Galvin, K., Troalen, F., Bressac, B., Marcais, C., Galun, E., Ponchel, F., Yakicier, C., Ji, J., and Ozturk, M. (1993) Retinoblastoma and p53 tumor suppressor genes in human hepatoma cell lines. *FASEB J.* **7**, 1407 - 1413.

Rhim, J.A., Sandgren, E.P., Palmiter, R.D., and Brinster, R.L. (1994) Replacement of disease mouse liver by hepatic cell transplantation. *Science* **263**, 1149 - 1152.

Rohde, M., Warthoe, P., Gjetting, T., Lukas, J., Bartek, J., and Strauss, M. (1996) The retinoblastoma protein modulates expression of genes coding for diverse classes of proteins including components of the extracellular matrix. *Oncogene* **12**, 2393 - 2401.

Sauer, B., and Henderson, N. (1988) Site-specific DNA recombination in mammalian cells by the Cre recombinase of bacteriophage P1. *Proc. Natl. Acad. Sci. USA.* **85**, 5166 - 5170.

Schneider, M., Marison, I.W., and von Stocker, U. (1996) The importance of ammonia in mammalian cell culture. *J. Biotech.* **46**, 161 - 185.

Schwarz, J.K., Devoto, S.H., Smith, E.J., Chellappan, S.P., Jakoi, L., and Nevins, J.R. (1993) Interactions of the p107 and Rb proteins with E2F during the cell proliferation response. *EMBO J.* **12**, 1013 - 1020.

Serrano, M., Hannon, G., and Beach, D. (1993) A new regulatory motif in cell cycle control causing specific inhibition of cyclin D/CDK4. *Nature* **366**, 704 - 707.

Sherr, C.J., and Roberts, J.M. (1995) Inhibitors of mammalian G_1 cyclin-dependent kinases. *Genes Dev* **9**, 1149 - 1163.

Sirica, A.E., and Pitot, H.C. (1980) Drug metabolism and effects of carcinogens in cultured hepatic cells. *Pharmacol. Rev.* **31**, 205 - 228.

Slansky, J.E., and Farnham, P.J. (1996) Introduction to the E2F family: protein structure and gene regulation. *Curr. Top. Microbiol. Immunol.* **208**, 1 - 30.

Strauss, M., Hering, S., Lubbe, L., and Griffin, B.E. (1990) Immortalization and transformation of human fibroblasts by regulated expression of polyoma virus T antigens. *Oncogene* **5**, 1223 - 1229.

Strauss, M., Hering, S., Lieber, A., Herrmann, G., Griffin, B.E., and W., Arnold. (1992). Stimulation of cell division and fibroblast focus formation by antisense repression of retinoblastoma protein synthesis. *Oncogene* **7**, 769 - 773.

Strauss, M., Lukas, J., and Bartek, J. (1995) Unrestricted cell cycling and cancer. *Nature Med.* **1**, 1254 - 1246.

Sussman,.N.L., and Kelly, J.H. Extracorporeal liver support: cell-based therapy for the failing liver. *Am. J. Kidney Dis.* **30 Suppl. 4** (1997), S66 - 71.

Tuck, S.P., and Crawford, L. (1989) Overexpression of normal human p53 in established fibroblasts leads to their tumorigenic conversion. *Oncogene Res.* **4**, 81 - 96.

Vojtesek, B., Bartek, J., Midgley, C.A., and Lane, D.P. (1992) An immunochemical analysis of the human nuclear phosphoprotein p53. New monoclonal antibodies and epitope mapping using recombinant p53. *J. Immunol. Methods* **151**, 237 - 244.

Walder, R.Y. and Walder, J.A. (1988). Role of RNase H in hybrid-arrested translation by antisense oligonucleotides. *Proc. Natl. Acad. Sci. USA*, **85**, 5011 - 5015.

Waxman, D.J., Morrissey, J.J., Naik, S., and Jauregui, H.O. (1990) Phenobarbital induction of cytochromes P-450. High-level long-term responsiveness of primary rat hepatocyte cultures to drug induction, and glucocorticoid dependence of the phenobarbital response. *Biochem. J.* **271**, 113 - 119

Wernass, B.A., Levine, A.J., and Howley, P.M. (1990) Association of human papillomavirus types 16 and 18 E6 proteins with p53. *Science* **248**, 76 - 79.

Weinberg, R.A. (1995) The retinoblastoma protein and cell cycle control. *Cell* **81**, 323 - 330.

Whyte, P., Buchkovich, J.K., Horowitz, J.M., Friend, S.H., Raybuck, M., Weinberg, R.A., and Harlow, E. (1988) Association between an oncogene and an anti-oncogene: The adenovirus E1A protein proteins bind to the retinoblastoma gene product. *Nature* **334**, 124 - 129.

Wollenberg, G.K., Semple,.E., Quinn, B.A., and Hayes, M.A. (1987) Inhibition of proliferation of normal, preneoplastic, and neoplastic rat hepatocytes by transforming growth factor-beta. *Cancer Res.* **47**, 6595 - 6599.

Woodworth, C.D., and Isom, H.C. (1987) Transformation of differentiated rat hepatocytes with adenovirus and adenovirus DNA. *J. Virol.* **61**, 3570 - 3579.

Xiong, Y., Hannon, G.J., Zhang, H., Casso, D., Kobayashi, R., and Beach, D. (1993) p21 is a universal inhibitor of cyclin dependent kinases. *Nature* **366**, 701 - 704.

Zindy, F., Eischen, C.M., Randle, D.H., Kamijo, T., Cleveland,, J.L., Sherr, C.J., and Roussel, M.F. (1998) Myc signaling via the ARF tumor suppressor regulates p53-dependent apoptosis and immortalization. *Genes Dev.* **12**, 2424 - 2433

INDEX

GPSR Compliance

The European Union's (EU) General Product Safety Regulation (GPSR) is a set of rules that requires consumer products to be safe and our obligations to ensure this.

If you have any concerns about our products, you can contact us on ProductSafety@springernature.com

In case Publisher is established outside the EU, the EU authorized representative is:

Springer Nature Customer Service Center GmbH
Europaplatz 3
69115 Heidelberg, Germany

FSC
www.fsc.org
MIX
Papier | Fördert gute Waldnutzung
FSC® C083411

Kolibri 360 GmbH
Ferdinand-Jühlke-Straße 7
99095 Erfurt, Deutschland
produktsicherheit@kolibri360.de